AF754765

Methods in Molecular Biology

Series Editor
John M. Walker
School of Life and Medical Sciences
University of Hertfordshire
Hatfield, Hertfordshire, UK

For further volumes:
http://www.springer.com/series/7651

For over 35 years, biological scientists have come to rely on the research protocols and methodologies in the critically acclaimed *Methods in Molecular Biology* series. The series was the first to introduce the step-by-step protocols approach that has become the standard in all biomedical protocol publishing. Each protocol is provided in readily-reproducible step-by-step fashion, opening with an introductory overview, a list of the materials and reagents needed to complete the experiment, and followed by a detailed procedure that is supported with a helpful notes section offering tips and tricks of the trade as well as troubleshooting advice. These hallmark features were introduced by series editor Dr. John Walker and constitute the key ingredient in each and every volume of the *Methods in Molecular Biology* series. Tested and trusted, comprehensive and reliable, all protocols from the series are indexed in PubMed.

Molecular Dermatology

Methods and Protocols

Edited by

Natalia V. Botchkareva

Monasterium Laboratory Skin and Hair Research Solutions GmbH, Münster, Germany

Gillian E. Westgate

Centre for Skin Sciences, University of Bradford, Bradford, West Yorkshire, UK

Humana Press

Editors
Natalia V. Botchkareva
Monasterium Laboratory Skin and Hair
Research Solutions GmbH
Münster, Germany

Gillian E. Westgate
Centre for Skin Sciences
University of Bradford
Bradford, West Yorkshire, UK

ISSN 1064-3745 ISSN 1940-6029 (electronic)
Methods in Molecular Biology
ISBN 978-1-0716-0647-6 ISBN 978-1-0716-0648-3 (eBook)
https://doi.org/10.1007/978-1-0716-0648-3

© Springer Science+Business Media, LLC, part of Springer Nature 2020
This work is subject to copyright. All rights are reserved by the Publisher, whether the whole or part of the material is concerned, specifically the rights of translation, reprinting, reuse of illustrations, recitation, broadcasting, reproduction on microfilms or in any other physical way, and transmission or information storage and retrieval, electronic adaptation, computer software, or by similar or dissimilar methodology now known or hereafter developed.
The use of general descriptive names, registered names, trademarks, service marks, etc. in this publication does not imply, even in the absence of a specific statement, that such names are exempt from the relevant protective laws and regulations and therefore free for general use.
The publisher, the authors, and the editors are safe to assume that the advice and information in this book are believed to be true and accurate at the date of publication. Neither the publisher nor the authors or the editors give a warranty, expressed or implied, with respect to the material contained herein or for any errors or omissions that may have been made. The publisher remains neutral with regard to jurisdictional claims in published maps and institutional affiliations.

Cover illustration: Taken from Chapter 5, Figure 1. See chapter for further details.

This Humana imprint is published by the registered company Springer Science+Business Media, LLC part of Springer Nature.
The registered company address is: 1 New York Plaza, New York, NY 10004, U.S.A.

Preface

The skin is the largest organ in the body and plays important roles in protection from a variety of environmental insults as well as in the control of temperature and water balance, defense from external insults, sensation, and physical communication. The rapid development of novel methodologies has allowed scientists to achieve significant progress in our understanding of the molecular and cellular mechanisms that control skin physiology and its contribution to cutaneous pathologies. This volume offers readers an opportunity to learn about and use the most important laboratory techniques that underlie dermatological investigation and includes in situ, in vitro, ex vivo, and in vivo models of skin and its appendages. Moreover, in recent years, investigative dermatology has been revolutionized by the establishment of unique techniques that unravel epigenetic mechanisms in skin cells, such as gene organization in the nucleus and gene expression regulation by the non-coding transcriptome, and such techniques are included in this volume. Lastly, we include several methodologies for mimicking or modulating skin pathologies such as wound healing, skin cancer, hair follicle disorders, and effects of visible light (photobiomodulation). This volume will be an essential resource for researchers working in the areas of fundamental as well as applied dermatological science.

Münster, Germany — *Natalia V. Botchkareva*
Bradford, UK — *Gillian E. Westgate*

Contents

Contributors

DANIEL ABERDAM • *INSERM U976, Hôpital Saint-Louis, Université de Paris, Paris, France*
MOHAMMED I. AHMED • *School of Science and Technology, Nottingham Trent University, Nottingham, UK*
RICHARD BAKER • *Faculty of Life Sciences, The Centre for Skin Sciences, University of Bradford, Bradford, UK*
CARMIT BAR • *Department of Cell, Developmental, and Regenerative Biology, Black Family Stem Cell Institute, Icahn School of Medicine at Mount Sinai, New York, NY, USA*
MARTA BERTOLINI • *Monasterium Laboratory Skin & Hair Research Solutions GmbH, Muenster, Germany*
ELEONORA CANDI • *IDI-IRCCS, Rome, Italy; Department of Experimental Medicine, University of Rome "Tor Vergata", Rome, Italy*
IRENE CASTELLANO-PELLICENA • *The Charles Institute of Dermatology, University College Dublin, Dublin, Ireland*
ANNA CELLI • *Department of Dermatology, University of California, San Francisco, CA, USA*
JANIN EDELKAMP • *Monasterium Laboratory Skin & Hair Research Solutions GmbH, Muenster, Germany*
ELENA EZHKOVA • *Department of Cell, Developmental, and Regenerative Biology, Black Family Stem Cell Institute, Icahn School of Medicine at Mount Sinai, New York, NY, USA*
MARKUS FEHRHOLZ • *Monasterium Laboratory, Skin and Hair Research Solutions GmbH, Muenster, Germany*
MICHAEL Y. FESSING • *Faculty of Life Sciences, Centre for Skin Sciences, School of Chemistry and Biosciences, University of Bradford, Bradford, UK*
JENNIFER GHERARDINI • *Monasterium Laboratory Skin & Hair Research Solutions GmbH, Muenster, Germany*
CLAIRE A. HIGGINS • *Department of Bioengineering, Imperial College London, London, UK*
FRANCISCO JIMÉNEZ • *Mediteknia Hair Transplant Clinic and Hair Lab, Gran Canaria, Canary Islands, Spain; Universidad Fernando Pessoa Canarias, Gran Canaria, Spain*
EKATERINA P. KALABUSHEVA • *Koltzov Institute of Developmental Biology of Russian Academy of Sciences, Moscow, Russia; Pirogov Russian National Research Medical University, Moscow, Russia*
NING XU LANDÉN • *Dermatology and Venereology Division, Department of Medicine (Solna), Karolinska Institute, Stockholm, Sweden; Ming Wai Lau Centre for Reparative Medicine, Stockholm Node, Karolinska Institute, Stockholm, Sweden*
ANNA MARIA LENA • *Department of Experimental Medicine, University of Rome "Tor Vergata", Rome, Italy*
AYELET LEVY • *INSERM U976, Hôpital Saint-Louis, Université de Paris, Paris, France*
SUMMIK LIMBU • *Department of Bioengineering, Imperial College London, London, UK*
MARA MANCINI • *IDI-IRCCS, Rome, Italy*
ANDREI N. MARDARYEV • *Faculty of Life Sciences, Centre for Skin Sciences, School of Chemistry and Biosciences, University of Bradford, Bradford, UK*

KEVIN J. MCELWEE • *Centre for Skin Sciences, University of Bradford, Bradford, UK; Department of Dermatology and Skin Science, University of British Columbia, Vancouver, BC, Canada*

CHARLES MIGNON • *Philips Research, Eindhoven, The Netherlands*

HANNA NIEHUES • *Department of Dermatology, Radboud Institute for Molecular Life Sciences, Radboud University Medical Center, Nijmegen, The Netherlands*

ISABELLE PETIT • *INSERM U938, CRSA, Hôpital Saint-Antoine, Sorbonne Université, Paris, France; INSERM U976, Hôpital Saint-Louis, Université de Paris, Paris, France*

ILARIA PICCINI • *Monasterium Laboratory Skin & Hair Research Solutions GmbH, Muenster, Germany*

MAXIMILIAN E. PICKUP • *School of Science and Technology, Nottingham Trent University, Nottingham, UK*

GIJS RIKKEN • *Department of Dermatology, Radboud Institute for Molecular Life Sciences, Radboud University Medical Center, Nijmegen, The Netherlands*

JÖRG SCHEFFEL • *Department of Dermatology and Allergy, Allergie-Centrum-Charité, Charité—Universitätsmedizin Berlin, Berlin, Germany*

HANNA SIISKONEN • *Department of Dermatology and Allergy, Allergie-Centrum-Charité, Charité—Universitätsmedizin Berlin, Berlin, Germany; Department of Dermatology, Kuopio University Hospital and University of Eastern Finland, Kuopio, Finland*

M. JULIE THORNTON • *Faculty of Life Sciences, The Centre for Skin Sciences, University of Bradford, Bradford, UK*

YOSHIKAZU UCHIDA • *Northern California Institute for Research and Education, San Francisco, CA, USA; Department of Dermatology, University of California, San Francisco, CA, USA*

NATALLIA E. UZUNBAJAKAVA • *Philips Research, Eindhoven, The Netherlands*

V. JULIAN VALDES • *Department of Cell Biology, Instituto de Fisiología Celular, Universidad Nacional Autónoma de México (UNAM), Mexico City, Mexico*

ELLEN H. VAN DEN BOGAARD • *Department of Dermatology, Radboud Institute for Molecular Life Sciences, Radboud University Medical Center, Nijmegen, The Netherlands*

MAX VAN LESSEN • *Monasterium Laboratories, Skin and Hair Research Solutions, Münster, Germany*

EKATERINA A. VOROTELYAK • *Koltzov Institute of Developmental Biology of Russian Academy of Sciences, Moscow, Russia; Pirogov Russian National Research Medical University, Moscow, Russia; Lomonosov Moscow State University, Moscow, Russia*

EDDY H. C. WANG • *Department of Dermatology, Columbia University Medical Center, New York, NY, USA*

RACHAEL WILLIAMS • *Faculty of Life Sciences, The Centre for Skin Sciences, University of Bradford, Bradford, UK*

JIANMIN WU • *Institute of Genomic Medicine, Wenzhou Medical University, Wenzhou, China*

Chapter 1

Isolation of Epidermal Keratinocytes from Human Skin: The Scratch-Wound Assay for Assessment of Epidermal Keratinocyte Migration

Irene Castellano-Pellicena and M. Julie Thornton

Abstract

The migration of epidermal keratinocytes is the basis for skin reepithelialization during wound healing. The in vitro scratch-wound assay using monolayers of primary human epidermal keratinocytes is a straightforward and effective method to assess their migratory capacity. The mechanical scratch of a confluent monolayer directly disrupts the adhesion of the keratinocytes to one another and to the underlying matrix, resembling the physical trauma of a wound in an in vitro assay. The keratinocytes will undergo an epithelial-to-mesenchymal transition, which will confer an ability to migrate toward each other to cover the gap by restructuring cell–cell and cell–extracellular matrix connections. However, a good scratch-wound method and protocol to ensure scratch reproducibility is essential, particularly when using primary cell cultures where donor variability may also impact on results.

Key words Primary human epidermal keratinocytes, Scratch-wound assay, Migration, Epithelial–mesenchymal transition (EMT)

1 Introduction

The benefits of cell culture for setting up systematic studies has been widely accepted [1]. Cell culture assays are mainly carried out with cell lines rather than primary cells despite their limitations for translation. The human epidermal keratinocyte (EK) is a very particular cell type whose fate is rapid differentiation, cornification and finally desquamation [2]. Only basal epidermal keratinocytes or bulge stem cell progenies within the hair follicle retain high proliferative potential [2]. Katayama et al. [3] reported significant differences between primary human EKs and the immortalized human epidermal cell line HaCaT in terms of their RNA expression profile; these differences were mainly limited to genes regulating the cell cycle. Compared to primary human EKs, HaCaT cells showed upregulation of genes expressed during the G1/S- and

Natalia V. Botchkareva and Gillian E. Westgate (eds.), *Molecular Dermatology: Methods and Protocols*, Methods in Molecular Biology, vol. 2154, https://doi.org/10.1007/978-1-0716-0648-3_1,
© Springer Science+Business Media, LLC, part of Springer Nature 2020

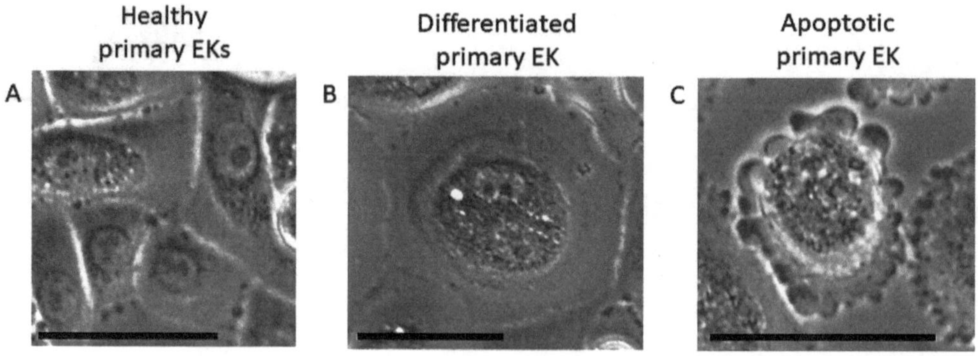

Fig. 1 Example of primary EKs in culture. (**a**) Healthy, proliferating primary EKs. (**b**) Differentiated primary EK, more abundant at high passages (>3). (**c**) Apoptotic primary EK. Scale bar = 50 μm

S-phase, while negative regulators of the S-phase entry were downregulated. There were also differences in the expression of transcription factors related to regulation of the cell cycle, such as Amyloid Beta Precursor Protein Binding Family B Member 1 (APBB1), Extra Spindle Pole Bodies Like 1, Separase (ESPL1), Transcription Factor 19 (TCF19), and Zinc Finger Protein 300 (ZNF300) [3]. This study also reported smaller but still significant differences in the transcriptome when comparing early and late passage human primary EKs due to their differentiation and senescence at higher passage numbers [3]. Indeed, primary EK differentiation increases with each passage in culture. Cells become bigger and slow down their rate of proliferation (Fig. 1b), up to a point where apoptosis is inevitable (Fig. 1c). This is the main reason why primary human EKs should be used at an early passage (Fig. 1a), as they resemble the original tissue to a much greater extent than late passage primary human EKs. Therefore, while bearing in mind the limitations of cell culture itself, the use of low passage primary cells over cell lines will yield more representative and meaningful data.

The isolation of EKs is a key step to ensure a good supply of primary human EKs for cell culture. In order to get an EK culture that is fibroblast-free, the separation of the epidermis from the dermis is essential and should be performed by the highly efficient enzyme, dispase. This enzyme is able to gently cleave collagen IV, the main protein of the *lamina densa* of the basement membrane that separates the epidermis from the underlying dermis [4, 5]. Once the epidermis has been separated from the dermis, the use of a second enzyme, trypsin, will break the cell–cell interactions of the epidermal sheet resulting in a cell suspension of individual epidermal cells. The use of keratinocyte growth media, which contains keratinocyte growth factors will induce proliferation in keratinocytes only, and other epidermal cell types such as melanocytes [6, 7], immune cells (e.g., Langerhans cells) [8], and the neuroendocrine Merkel cells will be lost [9].

Restoration of the epidermal barrier is required when the skin has been damaged, (i.e., after wounding). In particular, reepithelialization is the first step for restoration of the skin barrier, and it is the process of reestablishing a first layer of keratinocytes and preventing infection [10]. In the basal layer, this process involves mobilization, mitosis and migration of keratinocytes from the surrounding epidermis, followed by cellular differentiation and full epidermal barrier restoration. After wounding, cell contact inhibition is lost, which induces the keratinocytes that are immediately adjacent to the wound margin to migrate. This process of epithelial–mesenchymal transition (EMT), allows epithelial cells that are in contact with a basement membrane to acquire a mesenchymal cell phenotype by activating biochemical changes related to an enhanced migratory capacity, invasiveness, resistance to apoptosis, and production of extracellular matrix (ECM) proteins [11]. There are three types of EMT in human biology; the first type occurs during development, type 2 refers to the EMT which involves epidermal keratinocytes during wound healing as part of the repair process, and finally type 3 is the EMT which takes place during cancer and metastasis [12].

During type 2 EMT, downregulation of desmosomes (E-cadherin downregulation) occurs to enable keratinocyte migration, and simultaneously new fibronectin-recognizing integrins (e.g., $\alpha5\beta1$ and $\alpha v\beta6$) are expressed [13]. Expression of particular cytokeratins becomes altered; specific keratins (K16 and K17) are upregulated [14, 15] together with the mesenchymal cell marker, vimentin [16, 17]. Specific transcription factors, SNAI1, SLUG, ZEB1, ZEB2, and TWIST that promote EMT transition are all upregulated in EK upon EMT induction [18]. EMT in wound healing is associated with inflammation; once inflammation has been resolved and the wound site is covered by keratinocytes again, the reverse mesenchymal–epithelial transition (MET) occurs [19] by activation of the GRHL2, OVOL1, and OVOL2 transcriptions factors [18].

Epidermal migration can be studied in vitro using primary human EKs. The scratch-wound assay is the most effective technique which mimics reepithelialization in vitro. The primary keratinocytes experience the trauma (i.e., mechanical scratching) which leaves a cell-free area for the keratinocytes to migrate across to restore the monolayer. Indeed, primary EKs are able to mimic in vivo EK migration during wound healing particularly well in terms of cell–matrix and cell–cell interactions by epithelial cells [20]. Standardization by using a scratch-wound tool is highly recommended. In conclusion, the scratch-wound assay is an economical and straight forward method for the study of cell migration.

2 Materials

1. 1× phosphate-buffered saline (PBS) without Ca^{2+} and Mg^{2+}.
2. 1× PBS with Ca^{2+} and Mg^{2+}.
3. Transporting media: 500 ml Minimum Essential Medium (MEM) supplemented with 5 ml of 100% GlutaMAX (5× final concentration), 50 ml of fetal bovine serum (FBS) (10% final concentration), 5 ml of penicillin/streptomycin (Pen/-Strep) (500 U/ml and 500 μg/ml final concentration, respectively), and 5 ml of amphotericin B/sodium deoxycholate (12.5 μg/ml and 10.25 μg/ml final concentration, respectively), both contained in Fungizone antimycotic. Store at 4 °C for no longer than 1 month.
4. Tissue wash solution: 500 ml sterile 1× PBS. Add 500 U/ml penicillin and 500 μg/ml streptomycin (Pen/Strep) and 12.5 μg/ml of amphotericin B and 10.25 μg/ml of sodium deoxycholate, both contained in Fungizone antimycotic. Store at 4 °C for no longer than 1 month.
5. Complete primary keratinocyte growth medium (PromoCell) supplemented with 0.004 ml/ml of bovine pituitary extract (BPE), 0.125 ng/ml of epidermal growth factor (EGF) (recombinant human), 5 μg/ml of insulin (recombinant human), 0.33 μg/ml of hydrocortisone, 0.39 μg/ml of epinephrine, 10 μg/ml of transferrin (recombinant human), 0.06 mM of $CaCl_2$. Store at 4 °C for no longer than 1 month.
6. Complete Dulbecco's Modified Eagle's Medium (DMEM) supplemented with 5× GlutaMAX, 10% fetal bovine serum (FBS), 100 U/ml penicillin and 100 μg/ml streptomycin (Pen/Strep) and 2.5 μg/ml of amphotericin B and 2.05 μg/ml of sodium deoxycholate, both contained in Fungizone antimycotic. Store at 4 °C for no longer than 1 month.
7. 50 and 15 ml conical tubes.
8. Petri dishes (35- and 100-mm diameter).
9. Sterile dissection kit (forceps, curved microdissection forceps, scissors, and scalpel handle). Autoclave instruments before starting the protocol.
10. Disposable scalpel blades (size 24).
11. Dispase II (2.4 U/ml). Dissolve dispase in sterile 1× PBS without Ca^{2+} and Mg^{2+}supplemented with 1× antibiotics (100 U/ml penicillin and 100 μg/ml streptomycin (Pen/-Strep) (GIBCO), 2.5 μg/ml of amphotericin B and 2.05 μg/ml of sodium deoxycholate, both contained in Fungizone antimycotic) and aliquot in volumes of 4 ml. Store at −20 °C.

12. Trypsin-EDTA (0.25%), phenol red.
13. Trypsin-EDTA (0.05%), phenol red.
14. Solution of 0.5 mM ethylenediaminetetraacetic acid (EDTA). Make a 1:1000 dilution of 0.5 M EDTA in 1× PBS without Ca^{2+} and Mg^{2+}. Filter the final solution using a 0.22 μm pore size filter unit.
15. Trypsin neutralizing solution (TNS).
16. Routine benchtop centrifuge.
17. Vortexer.
18. Tissue culture flasks—size T25 cm^2 and T75 cm^2.
19. Cell culture incubator set up at 37 °C and 5% CO_2 in air.
20. Laminar flow cabinet class II (for cell culture).
21. Cell counter (hemocytometer or automatic counter).
22. 12-Well cell culture plates.
23. Fine-tipped marker pen.
24. Scratch device or yellow plastic tip (*see* **Note 1**).
25. Inverted microscope phase contrast coupled to a brightfield camera.

3 Methods

Seek ethical approval at your institution and obtain all human skin samples with informed patient consent.

Work under sterile conditions. Ideally, use one laminar flow hood for human tissue processing and primary cell extraction only.

3.1 Human Skin Tissue Processing and Primary EK Extraction

Work with the tissue kept as much as possible at 4 °C .

1. Place the skin tissue in transporting media immediately after surgery and transfer to the laboratory. Skin tissue can be kept in transport media at 4 °C for up to 24 h before processing.
2. Wash the tissue in cold Tissue Wash Solution in a 50 ml tube, before transferring into a 100 mm tissue culture dish with Tissue Wash Solution (Fig. 2—1, 1). Remove any fat using microdissection forceps and scissors and dispose of it via the appropriate channels (*see* **Note 2**).
3. Tissue with a minimum size of 10 cm^2 is required to yield sufficient number of keratinocytes for cell culture. Cut the skin into small squares (≈0.5 cm^2) using the microdissection forceps and scalpel.
4. Thaw an aliquot (4 ml) of 2.4 U/ml dispase and add the small skin tissue pieces into the 15 ml tube. Make sure there is enough dispase to cover all the skin pieces. If the starting tissue

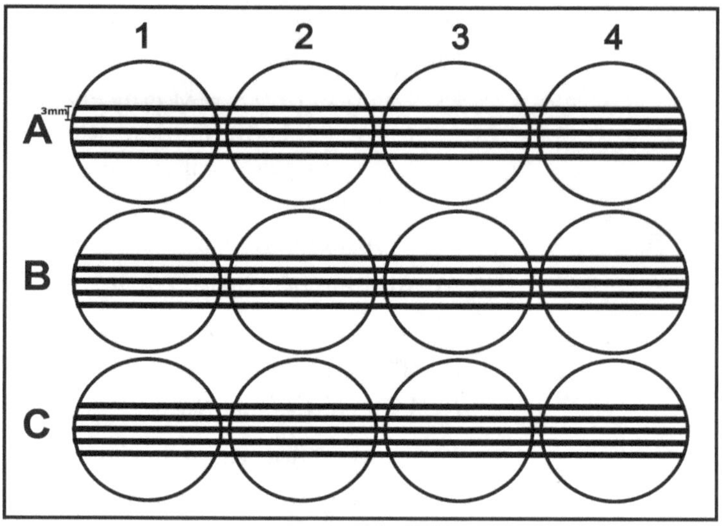

Fig. 2 Layout of a 12-well plate for a scratch assay. Lines drawn on the back of the plate are essential to ensure the pictures are always taken at the same coordinates on the plate

is large (>50 cm^2), then two aliquots of dispase will be required. Incubate overnight at 4 °C.

5. Transfer the skin to a 100 mm petri dish and cover with cold PBS. Peel the epidermis from the dermis using a pair of micro-dissection normal forceps and a pair of curved forceps. Transfer the epidermis into a clean 35 mm dish with 2 ml of cold PBS. Discard the dermis or retain it to use for dermal fibroblast isolation.
6. When the epidermis has been removed, transfer it to a 15 ml falcon tube containing 3 ml of 0.25% trypsin. Incubate for 5–10 min at 37 °C and vortex immediately to get a single cell suspension.
7. Add 3 ml of TNS (equal volume to the trypsin) to neutralize the trypsin.
8. Remove any epidermis that has not still trypsinized and place it in a fresh 15 ml tube with 3 ml of 0.25% trypsin in order to improve the cell yield. Keep the first cell suspension on ice.
9. Repeat the trypsinization step one more time (three in total).
10. Centrifuge all three 15 ml falcon tubes containing the epidermal cell suspension (≈6 ml) for 5 min at 300 × *g*.
11. Mix (2:1) primary keratinocyte growth media (14 ml) with complete DMEM (7 ml). Warm the media in the water bath at 37 °C.
12. Resuspend each of the three cell pellets in 1 ml of 2:1 keratinocyte–DMEM media prepared in **step 11**. Add the cell

suspension from each of the three pellets together and count the number of cells present using a hemocytometer.

13. Add the 3 ml of cell suspension into 12 ml of 2:1 keratinocyte–DMEM media (minimum density of 13,000 cells/cm^2) and transfer into a T75 cm^2 cell culture flask (if there are not enough cells, use a T25 cm^2 cell culture flask instead). Incubate overnight at 37 °C/5% CO_2.

3.2 Culture and Passage of Primary Human EKs

1. Check that cells have adhered to the flask using an inverted phase contrast microscope.
2. Remove the media and replace with keratinocyte growth media; keep changing the media every second day until the EKs reach 80% confluency. This might take from 2 days to 1 week, depending on the initial seeding density (*see* **Note 3**).
3. Warm up 1× PBS to 37 °C, 0.5 mM of EDTA, 0.05% trypsin–EDTA, TNS, and primary keratinocyte growth media to passage EKs from P0 to P1 with a split ratio of 1:3.
4. Remove media and add 0.5 mM EDTA (*see* Table 1 for volumes). Incubate for 4 min at 37 °C to disrupt cell–cell and cell–matrix interactions.
5. Remove EDTA, add 0.05% trypsin–EDTA and incubate for 2 min at 37 °C (Table 1) (*see* **Note 4**).
6. Check cell detachment under the phase contrast microscope.
7. Add TNS to neutralize the trypsin (Table 1). Add 6 ml of PBS to make ensure all the cells are in suspension and then centrifuge the cell suspension at 200 × *g* for 5 min.
8. Dissolve the cell pellet in 1 ml of primary keratinocyte growth media and pipette thoroughly up and down to resuspend the EKs. Dilute the cells in 36 ml of keratinocyte growth media and split between three new T75 cm^2 flasks (26,000 cells/cm^2).

3.3 The Scratch-Wound Healing Assay

After one passage (P1) of the isolated EK, there should be enough cells to perform a scratch-wound assay depending upon how many plates are required. If not, then continue to passage the EKs taking care to prevent cell differentiation by never allowing them to reach a confluency above 80%.

1. Prepare the 12-well plates for a scratch assay (*see* **Note 5**). Using a fine marker pen, draw a line in the back of the plate going through the center part of the wells. Draw two extra lines on top of the center line, 3 mm apart. Draw another two extra lines on bottom of the center line, again 3 mm apart (*see* Fig. 1).
2. Trypsinize the primary EKs as previously described in Subheading 3.2 and count the cells using an automatic cell counter or a manual hemocytometer [21].

Table 1
Volumes (ml) of EDTA, 0.05% trypsin–EDTA, TNS added during keratinocyte passaging in the two different flask formats used, T25 cm^2 and T75 cm^2

Volumes added	T25 flask (ml)	T75 flask (ml)
0.05 mM EDTA	1.5	2.5
0.05% Trypsin/EDTA	1	2
TNS	2	4

3. Seed the cells at a density of 100,000 cells/well in a 12-well plate. Add 1 ml of keratinocyte growth media per well.
4. Monitor cell growth until the keratinocytes have reached a confluency of approximately 80%; this might take from 24 to 48 h. Remove the media and replace with EGF and BPE free keratinocyte growth media for 24 h to reduce cell proliferation (*see* **Note 6**).
5. Scratch the cell monolayers. The simplest method for performing a scratch is by using a plastic pipette tip applying enough force to disrupt the cells without damaging the plastic (*see* **Note 7**). To enhance reproducibility, the use of a scratch device is advisable (*see* **Note 1**).
6. Wash the cell monolayer with 1 ml of 1×PBS with Ca^{2+} and Mg^{2+} and then add 1 ml of fresh keratinocyte growth media without EGF and BPE.
7. Immediately after the scratch (time zero) take five images of the scratch in each well using an inverted phase contrast microscope with an objective of 10× magnification and a camera attachment. The images should be taken just above each of the marked lines (Fig. 2) avoiding photographing the lines as it will interfere with the image analysis. Return the plate to the incubator at 37 °C and 5% CO_2 in air.
8. Repeat at 4, 12, and 21 h after the initial scratch was produced (*see* Fig. 3).
9. In order to further understand the migratory behavior of EKs on molecular level, RNA extraction can be done for the analysis of the changes in the expression of different markers of cell migration (*see* **Note 8**). Use nonscratched wells as the control.

3.4 Analysis and Interpretation of the Data

1. Analyze the area of the gap that has not been covered by the keratinocytes using the appropriate software (*see* **Note 9**).
2. Calculate the gap area that has been closed (*see* **Note 10**):

   Migrated area = Initial area (Ai) − Area at a specific time point (At).

   To express this as a percentage:

   Percentage of gap closed (%): 100 − (At/Ai)∗100.

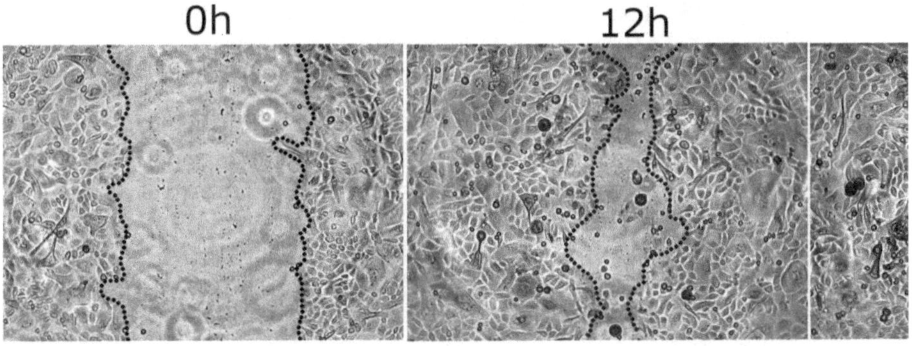

Fig. 3 Primary human EK migration over 21 h period after scratch (at 0, 12, and 21 h) [14]. Human EKs were established from female facial skin (49-year-old donor, passage 3)

3. First, calculate the mean closure in each well using the five pictures taken per well. Secondly, calculate the mean of the replicate wells belonging for each culture conditions tested (e.g., control vs. test compound).
4. Perform statistical analysis (if several groups are compared at different time points; a two-way ANOVA with repeated measures is recommended when the data is normally distributed).
5. Experiments must be repeated using EKs obtained from different donors to confirm significant results (*see* **Note 11**) to account for donor variability.

4 Notes

1. There are some commercial scratch devices available, but most of them are only compatible with 96- or 24-well plates. Otherwise, a custom-made scratch tool can be made in house. This could be a metallic tool that should be autoclaved between every use or a tool which uses reusable plastic tips [22].
2. The fat tissue has a yellow color while the dermis is white.
3. Human primary EKs will readily differentiate in vitro. In order to maintain their proliferative potential, it is important to be very meticulous while maintaining them. It is crucial to never let them grow above 80% confluency by changing the media every 2 days, or even more frequently if the culture medium changes color from pink to yellow (an indication of higher cell numbers).
4. EDTA will break the cell–cell attachment but not cell–flask attachments. The use of EDTA will reduce the amount of time EKs will have to be exposed to trypsin in the following step.

5. The scratch-wound assay can also be performed in 6-well plates, but when using primary cells, the number of cells may be a limitation.
6. Removal of EGF and BPE from the cell culture media will ensure a reduction in the proliferation capacity of primary human EKs; therefore, only migration will occur under these conditions. However, if a stricter method is required, mitomycin C, a DNA cross-linker, can be used to inhibit proliferation [23]. For primary EKs, mitomycin C at a concentration of 10 μg/ml will stop proliferation without inducing cellular toxicity [24].
7. Either a small/medium (yellow) or large (blue) pipette tip can be used to make the scratch. It is important to change the tip between every scratch/well to prevent carryover of scratched/damaged cells between the wells, which at the same time will interfere with the new scratch, reducing reproducibility between the wells.
8. Unscratched monolayers can be cultured in parallel for the gene expression study by qRT-PCR (perform three RNA extractions per condition/well) to quantify changes in the expression of markers EMT and migration, such as vimentin, keratin 16 or 17 (KRT16 or KRT17) and ZEBs, which are upregulated during primary human EK migration [15, 16]. If protein distribution analysis is required, cells can be cultured in chamber slides or coverslips and molecular markers localized by immunocytochemistry (ICC).
9. Use any software that allows area measurements. As an example, the freely available software, Image J can be used [25]. The pixels should be converted to mm by clicking "Analyse" followed by "Set scale." The area can be selected manually and measured by clicking "Analyse" followed by "Measure" or alternatively a scratch macro can be used.
10. Sometimes the migration is shown as distance migrated rather than area migrated, however cells do not move in 1D on the plastic surface but rather in 2D; therefore, the calculation of migration area is recommended for more accurate and meaningful results.
11. Variability when using human primary cells from different donors can be high. When analyzing migration of primary human EKs, 45% variability can be seen in primary EKs derived from female facial skin from three donors (passages between 2 and 4) (Fig. 4). Donor variability on the speed of closure can significantly impact the analysis when data from multiple donors is combined. Variability is also introduced with differences in donor age, sex, and different anatomical regions. As a solution, individual data for each donor can be shown or

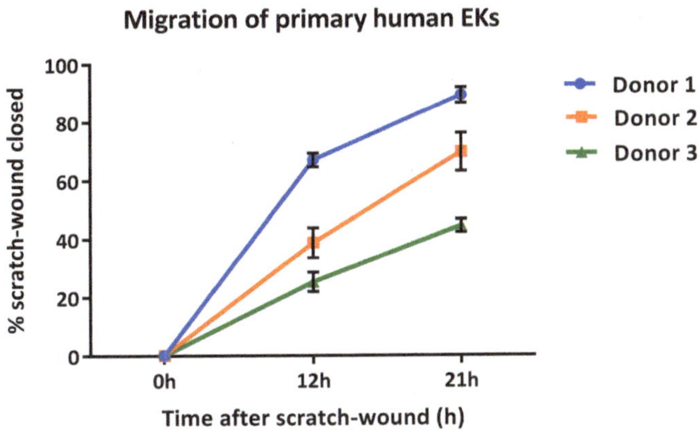

Fig. 4 Migration of primary human EKs from different donors [14]. Human EKs were established from female facial skin (49–53 years of age, passage 3–5)

normalization has to be performed. For normalization, the migration of all the treatment groups is divided by the migration of the nontreated group within each donor cells; afterward, mean and statistical analysis of different donor cells can be done.

References

1. Freshney RI (2015) Culture of animal cells: a manual of basic technique and specialized applications. John Wiley & Sons, New York
2. Fuchs E, Raghavan S (2002) Getting under the skin of epidermal morphogenesis. Nat Rev Genet 3(3):199–209. https://doi.org/10.1038/nrg758
3. Katayama S, Skoog T, Jouhilahti EM, Siitonen HA, Nuutila K, Tervaniemi MH, Vuola J, Johnsson A, Lonnerberg P, Linnarsson S, Elomaa O, Kankuri E, Kere J (2015) Gene expression analysis of skin grafts and cultured keratinocytes using synthetic RNA normalization reveals insights into differentiation and growth control. BMC Genomics 16:476. https://doi.org/10.1186/s12864-015-1671-5
4. Parkin JD, San Antonio JD, Pedchenko V, Hudson B, Jensen ST, Savige J (2011) Mapping structural landmarks, ligand binding sites, and missense mutations to the collagen IV heterotrimers predicts major functional domains, novel interactions, and variation in phenotypes in inherited diseases affecting basement membranes. Hum Mutat 32(2):127–143. https://doi.org/10.1002/humu.21401
5. Abreu-Velez AM, Howard MS (2012) Collagen IV in normal skin and in pathological processes. N Am J Med Sci 4(1):1–8. https://doi.org/10.4103/1947-2714.92892
6. Singh SK, Kurfurst R, Nizard C, Schnebert S, Perrier E, Tobin DJ (2010) Melanin transfer in human skin cells is mediated by filopodia—a model for homotypic and heterotypic lysosome-related organelle transfer. FASEB J 24(10):3756–3769. https://doi.org/10.1096/fj.10-159046
7. Tarafder AK, Bolasco G, Correia MS, Pereira FJ, Iannone L, Hume AN, Kirkpatrick N, Picardo M, Torrisi MR, Rodrigues IP, Ramalho JS, Futter CE, Barral DC, Seabra MC (2014) Rab11b mediates melanin transfer between donor melanocytes and acceptor keratinocytes via coupled exo/endocytosis. J Invest Dermatol 134(4):1056–1066. https://doi.org/10.1038/jid.2013.432
8. Romani N, Clausen BE, Stoitzner P (2010) Langerhans cells and more: langerin-expressing dendritic cell subsets in the skin. Immunol Rev 234(1):120–141. https://doi.org/10.1111/j.0105-2896.2009.00886.x
9. Boulais N, Misery L (2008) The epidermis: a sensory tissue. Eur J Dermatol 18(2):119–127
10. Pastar I, Stojadinovic O, Yin NC, Ramirez H, Nusbaum AG, Sawaya A, Patel SB, Khalid L, Isseroff RR, Tomic-Canic M (2014)

Epithelialization in wound healing: a comprehensive review. Adv Wound Care (New Rochelle) 3(7):445–464. https://doi.org/10.1089/wound.2013.0473
11. Kalluri R, Neilson EG (2003) Epithelial-mesenchymal transition and its implications for fibrosis. J Clin Investig 112 (12):1776–1784. https://doi.org/10.1172/JCI200320530
12. Stone RC, Pastar I, Ojeh N, Chen V, Liu S, Garzon KI, Tomic-Canic M (2016) Epithelial-mesenchymal transition in tissue repair and fibrosis. Cell Tissue Res 365(3):495–506. https://doi.org/10.1007/s00441-016-2464-0
13. Koivisto L, Heino J, Häkkinen L, Larjava H (2014) Integrins in wound healing. Adv Wound Care 3(12):762–783. https://doi.org/10.1089/wound.2013.0436
14. Castellano-Pellicena I, Uzunbajakava NE, Mignon C, Raafs B, Botchkarev VA, Thornton MJ (2018) Does blue light restore human epidermal barrier function via activation of Opsin during cutaneous wound healing? Lasers Surg Med 51(4):370–382. https://doi.org/10.1002/lsm.23015
15. Fitsialos G, Chassot AA, Turchi L, Dayem MA, LeBrigand K, Moreilhon C, Meneguzzi G, Busca R, Mari B, Barbry P, Ponzio G (2007) Transcriptional signature of epidermal keratinocytes subjected to in vitro scratch wounding reveals selective roles for ERK1/2, p38, and phosphatidylinositol 3-kinase signaling pathways. J Biol Chem 282(20):15090–15102. https://doi.org/10.1074/jbc.M606094200
16. Velez-delValle C, Marsch-Moreno M, Castro-Muñozledo F, Galván-Mendoza IJ, Kuri-Harcuch W (2016) Epithelial cell migration requires the interaction between the vimentin and keratin intermediate filaments. Sci Rep 6:24389. https://doi.org/10.1038/srep24389
17. Haensel D, Dai X (2018) Epithelial-to-mesenchymal transition in cutaneous wound healing: where we are and where we are heading. Dev Dyn 247(3):473–480. https://doi.org/10.1002/dvdy.24561
18. Nieto MA, Huang RY, Jackson RA, Thiery JP (2016) EMT: 2016. Cell 166(1):21–45. https://doi.org/10.1016/j.cell.2016.06.028
19. Kalluri R, Weinberg RA (2009) The basics of epithelial-mesenchymal transition. J Clin Invest 119(6):1420–1428. https://doi.org/10.1172/JCI39104
20. Liang CC, Park AY, Guan JL (2007) In vitro scratch assay: a convenient and inexpensive method for analysis of cell migration in vitro. Nat Protoc 2(2):329–333. https://doi.org/10.1038/nprot.2007.30
21. Phelan MC (2007) Basic techniques in mammalian cell tissue culture. Curr Protoc Cell Biol. Chapter 1:Unit 1.1. https://doi.org/10.1002/0471143030.cb0101s36
22. Malinda KM, Kusakabe AO, Wysocki AB, Sullivan JV (2001) Scratch wound assay device. US6309818B1
23. Xue M, Campbell D, Jackson CJ (2007) Protein C is an autocrine growth factor for human skin keratinocytes. J Biol Chem 282 (18):13610–13616. https://doi.org/10.1074/jbc.M610740200
24. Pomari E, Dalla Valle L, Pertile P, Colombo L, Thornton MJ (2015) Intracrine sex steroid synthesis and signaling in human epidermal keratinocytes and dermal fibroblasts. FASEB J 29(2):508–524. https://doi.org/10.1096/fj.14-251363
25. Schneider CA, Rasband WS, Eliceiri KW (2012) NIH image to ImageJ: 25 years of image analysis. Nat Methods 9:671. https://doi.org/10.1038/nmeth.2089

Chapter 2

Isolation of Different Dermal Fibroblast Populations from the Skin and the Hair Follicle

Rachael Williams and M. Julie Thornton

Abstract

The establishment of primary cells from fresh tissue is a widely used method for investigating human tissue in vitro. The skin harbors different cell populations in the dermis and the hair follicle, which can be isolated for downstream analysis. Here we describe the isolation of four dermal fibroblast populations from human haired skin and their maintenance in culture. The four cell populations for which isolation is described are papillary dermal fibroblast cells, reticular dermal fibroblast cells, hair follicle dermal sheath cells, and hair follicle dermal papilla cells.

Key words Cell culture, Dermal fibroblast, Dermal papilla, Dermal Sheath, Hair follicle, Papillary dermis, Reticular dermis

1 Introduction

At the end of the nineteenth century both Virchow and then Duvall first described fibroblasts based on their morphology [1, 2]. Fibroblasts are mesenchyme-derived cells that have an elongated morphology [1–3]. They are the most common cell type of human connective tissue and play multiple roles including the synthesis of extracellular matrix (ECM), providing scaffolds for other cells, release of growth factors, regulating inflammation and mediate scar formation, and tissue fibrosis [4].

The dermis is the connective tissue layer of the skin and comprises two major subtypes, the papillary dermis and the reticular dermis [5]. The papillary dermis is a thin layer 300–400 μm deep that sits below the epidermis with the papillae rising into the rete ridges of the epidermis, creating peg-like structures. The reticular dermis sits below the papillary and above the adipose tissue and constitutes the major portion of the dermis [6]. The two parts of the dermis are composed of fibroblasts with different phenotypes, with papillary dermal fibroblasts producing thin, unorganized

Natalia V. Botchkareva and Gillian E. Westgate (eds.), *Molecular Dermatology: Methods and Protocols*, Methods in Molecular Biology, vol. 2154, https://doi.org/10.1007/978-1-0716-0648-3_2,
© Springer Science+Business Media, LLC, part of Springer Nature 2020

collagen fibers while the reticular fibroblasts produce organized thick collagen fiber bundles [5]. Papillary dermal fibroblasts also divide faster and have less contact inhibition so grow to a higher density, in contrast reticular fibroblasts have higher contractibility [6].

The hair follicle is a fascinating organ with regard to how it will regenerate multiple times within the human life span [7]. The hair follicle (HF) sits within the dermis, at sites of terminal hair growth, and reaches down into the dermal adipose tissue. HFs have specialized fibroblast populations in the dermal papilla (DP) at the base of the bulb, and in the connective tissue dermal sheath (DS) which encapsulates the hair follicle. The DP and DS are derived from the dermal progenitors that also give rise to the papillary fibroblasts [8]. Each hair follicle regenerates in the form of a cycle which consists of a growth phase (anagen), a regression phase (catagen,), a resting phase (telogen), and then hair shedding (exogen), with this cycle continuing to remodel the hair follicle and produce a new hair fiber [9]. The DP regulates the hair follicle cycle and secretes growth factors that cause proliferation and differentiation in the surrounding bulb matrix keratinocytes, initiating and regulating fiber growth [7].

The DS comprises three collagen layers and embedded in the middle of the three collagen layers, composed predominantly of collagen VI, reside the DS fibroblasts [10]. The DS is believed to have the ability to replenish the adjacent DP cell pool, as implicated by DS cells migrating and undergoing a phenotypic transition to become DP cells, as demonstrated in experimentally amputated rat whisker follicles [11]. Although they are of the same embryological origin as DP cells, and have this capability to conserve the hair follicle, it is thought it is a reserve mechanism, since recruitment into undamaged papilla is yet to be observed, and they also lack some of the inductive capabilities of the DP cells [12].

The isolation, growth, and maintenance of dermal fibroblast cells in cell culture is relatively simple. However, the isolation of hair follicle fibroblast cells is more challenging as it does require follicle microdissection, which may require some practice; once established their maintenance is comparable to that of their dermis-derived counterparts. The DP and dermal DS lose their hair follicle inductivity after several passages in 2D cultures [13], so it has been suggested that for downstream studies that the cells are grown in a 3D configuration, such as spheroids, in order to mimic the in vivo hair follicle inductivity profile [14].

Here we describe the widely accepted methods for the isolation of dermal papillary and reticular fibroblasts, as well as hair follicle dermal papilla and sheath cells and their maintenance in culture.

2 Materials

2.1 Tissue Preparation

1. *Transport media*: Dulbecco's Modified Eagle's Medium (DMEM) low glucose with 1000 mg/L glucose, L-glutamine, and sodium bicarbonate, liquid, sterile-filtered, suitable for cell culture, supplemented with 10% fetal calf serum (FCS), 5% penicillin/streptomycin (solution contains 5000 U/mL of penicillin and 5000 μg/mL of streptomycin) 1% Fungizone (2.50 μg of amphotericin B and 2.05 μg of sodium deoxycholate per mL) and 1% GlutaMAX (2 mM final concentration).
2. *Phosphate buffered saline (PBS)*: phosphate buffered saline tablets, PH 7.4 (five tablets per 100 mL, or as directed by manufacturer, in sterile water and then autoclaved for cell culture use).
3. *Tissue wash*: PBS containing 5% penicillin/streptomycin (solution contains 5000 U/mL of penicillin and 5000 μg/mL of streptomycin) and 1% Fungizone (2.50 μg of amphotericin B and 2.05 μg of sodium deoxycholate per mL of distilled water).
4. Trypsin (0.25%).
5. Cell culture flask (size dependent on the amount of sample available; e.g., 2 cm^2 of tissue, once dissected into 3 mm cubes can be placed in T75. 1 cm^2 = T25).
6. Small petri dishes (35 mm).

2.2 Cell Culture Maintenance

1. Cell culture media: DMEM containing 10% FCS, 0.2% primocin (100 μg/mL) and 1% GlutaMAX (2 mM final concentration).
2. 1× Hank's Solution (10× Hank's Solutions diluted 1:10 in sterile water).
3. Trypsin–EDTA (0.05%).

2.3 Equipment

1. Lamina flow hood.
2. Dissecting microscope with cool light (Magnification: 10–800×).
3. Scalpel and 15 mm blade.
4. Fine watchmaker forceps (size no. 4).
5. Flat forceps (100 mm).
6. Fine gauge needles (26 G).
7. Centrifuge (bench top).

3 Methods

3.1 Tissue Preparation

1. Fresh tissue samples should ideally be received approximately 3 h postsurgery and stored in transport media at 4 °C. Samples should be processed within 24 h of surgery.
2. Under a cell culture lamina flow hood, wash the samples in 15 mL tissue wash in a 100 mm petri dish, changing the wash three times before placing in 15 mL PBS.
3. On the lid of a culture dish place the sample with forceps and shave any hair using a scalpel and dispose of the hair (*see* **Note 1**).
4. Cut the tissue into 5 mm cubes using a scalpel (*see* **Note 2**).
5. Remove the adipose layer cutting directly underneath the dermis by placing the sample in between flat forceps and cut away the adipose tissue using the scalpel (Fig. 1). For dermal papilla and dermal sheath isolation put this tissue to one side in PBS or dispose of in the capped tube.

3.2 Papillary and Reticular Dermal Fibroblast Isolation

1. To isolate papillary fibroblasts the skin is cut underneath the epidermis (as close as possible) and placed in a dish 35 mm with the epidermis facing up. Add 2 mL DMEM supplemented with either 10% or 20% FCS, and GlutaMAX. Alternatively the epidermis can be removed by floating in 3 mL trypsin (0.25%) at 6 °C overnight, and then in 0.05% trypsin in the incubator at 37 °C for 1 h before peeling away the epidermis with watchmaker forceps (*see* **Note 3**) (Fig. 2).
2. To isolate reticular fibroblasts, use the skin left over from where the papillary dermis was removed and place in a 35 mm dish or T75 flask (depending on the amount of material harvested). Add 2 mL (35 mm) or 7 mL (T75) DMEM supplemented with either 10% or 20% serum and 1% GlutaMAX (Fig. 3) (*see* **Note 4**).
3. Place in the incubator at 37 °C, humidified with 5% CO_2 and do not disturb for 10 days.
4. After 10 days cells should have started to explant out from the tissue (*see* **Note 5**). Carefully remove the skin with forceps and replace the culture media every 2–3 days until the cells reach confluency.

3.3 Hair Follicle Dermal Sheath and Dermal Papilla Isolation

1. Take the adipose tissue containing the anagen hair follicles stored in PBS.
2. Set up the dissecting microscope with the optimum magnification for example so that the bottom of the follicle is in focus.

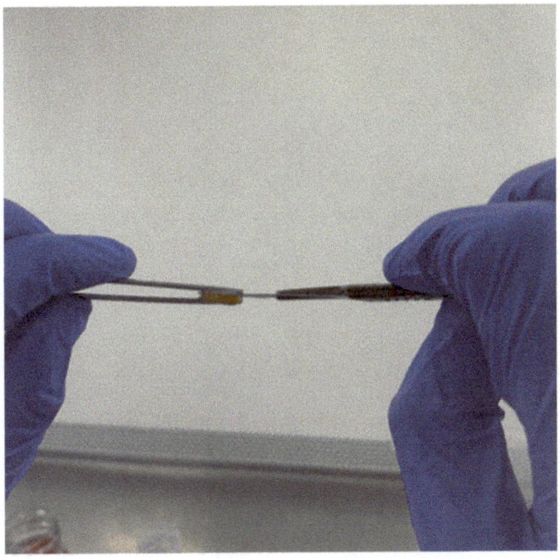

Fig. 1 Technique for the cutting of the adipose tissue beneath dermis by placing between flat forceps and using a scalpel blade

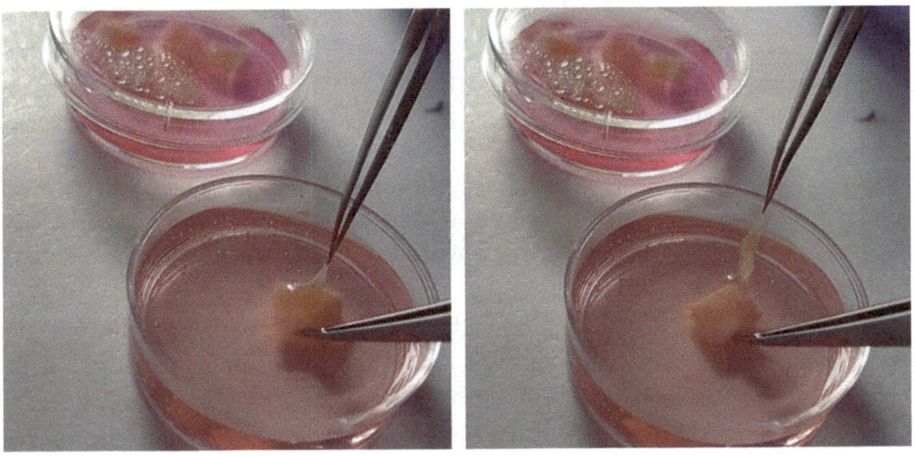

Fig. 2 Peeling off the epidermis post 24 h in trypsin 0.25% at 6 °C followed by 1-h incubation at 37 °C in 0.05% trypsin

3. Using a scalpel and holding the sample with flat forceps cut the 5 mm adipose tissue into smaller pieces, cutting in between the follicles, to avoid damaging them (Fig. 4a, b).
4. Take a cluster of follicles and using fine gauge (26 G) needles gently brush the adipose tissue from around the follicles (Fig. 4c).
5. Once clean, transfer the hair follicles to a separate dish containing 1 mL of culture media (*see* **Note 6**).
6. Take an individual follicle and dissect it just above the bulb using the needles (Figs. 4a and 5d).

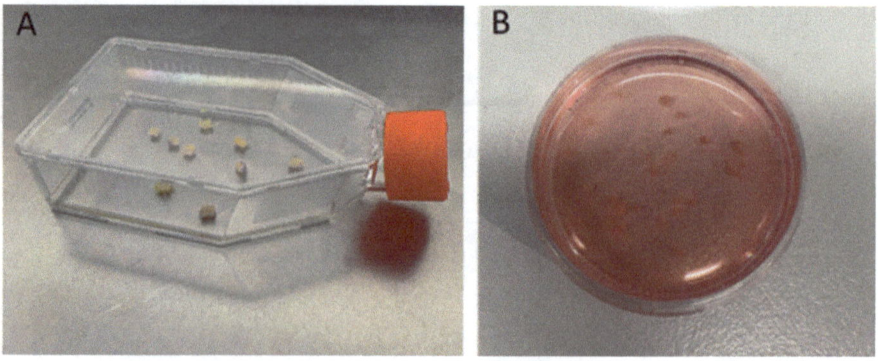

Fig. 3 (**a**) Dermis tissue in a flask with growth media. (**b**) Dermis taken from a smaller piece of starting material in culture media in a small 35 mm culture dish

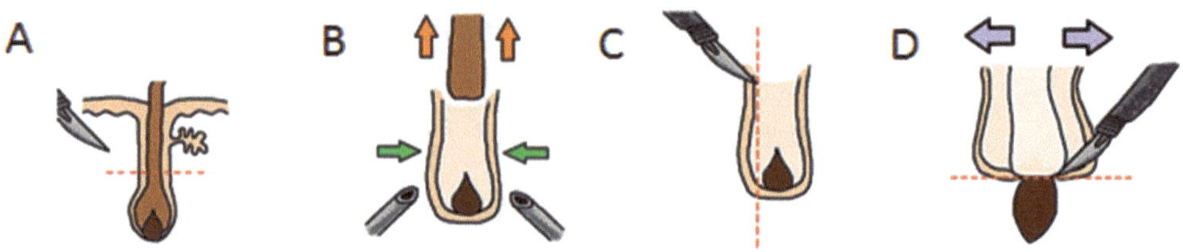

Fig. 4 Summary schematic of DP isolation using micro dissection. (**a**) Indicates cut line above the hair follicle. (**b**) Squeezing of the bulb to remove matrix. (**c**) Cutting of the DS. (**d**) Opening of the DS and removal of the DP

7. Squeeze the follicle bulb gently either side to remove the matrix (the pigmented part), which will leave the lower part of the dermal sheath with the dermal papilla intact (Figs. 4b and 5e).
8. Cut down one edge of the sheath with a needle and then open-up the cup like structure until the DP (this is a tear drop structure) hangs out (Figs. 4c and 5f).
9. Cut the DP away from the sheath where the two join and transfer the DP to another dish containing media (*see* **Note 7**) (Figs. 4d and 5g).
10. Press the DP into the plastic culture dish so it is anchored down (*see* **Note 8**) (Fig. 5h).
11. Move the remaining dermal sheath to another dish containing media and scratch it into the plastic dish using needles to anchor it (*see* **Note 9**).
12. Repeat the process for both cell types until two dishes, one containing 6–10 DPs and the other 5 or more DS are generated.
13. Leave them undisturbed in the incubator at 37 °C humidified under 5% CO_2, for 7 days, checking the media and gently topping up if necessary.

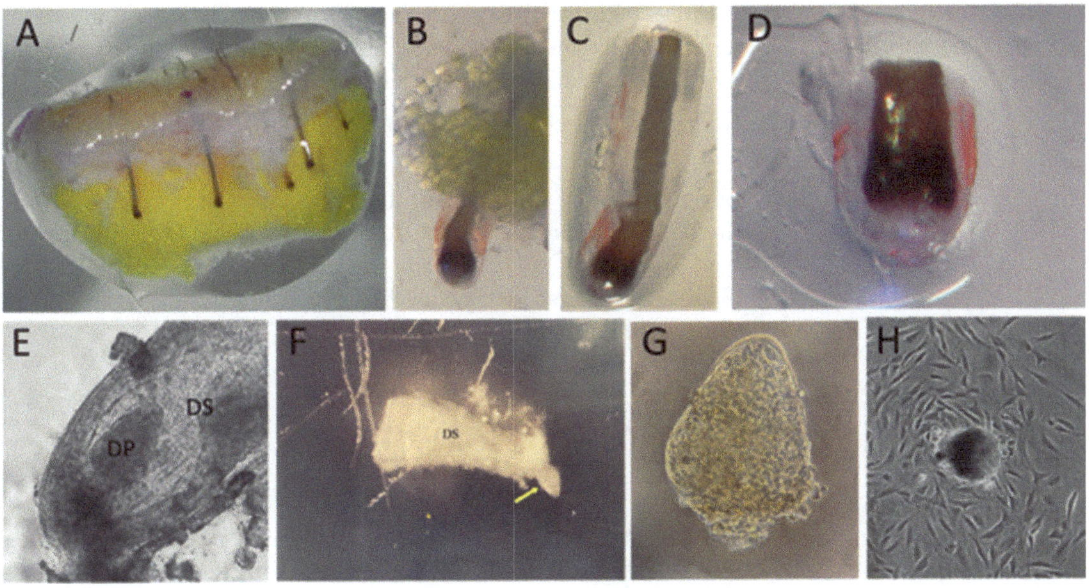

Fig. 5 Hair follicle dissection images—from tissue to explant. (**a**) Tissue cut into 3 mm pieces, showing anagen hair follicle bulbs in the adipose (mag. ×10). (**b**) Isolated follicle prior to fat removal (mag. ×50). (**c**) A clean isolated follicle (mag. ×50). (**d**) Bottom of the bulb containing the matrix (mag. ×50). (**e**) Bottom of the bulb with the matrix removed revealing the DP (mag. ×800). (**f**) DS after been cut and opened to reveal the DP (mag. ×80). (**g**) Isolated DP (mag. ×800). (**h**) Cells explanting from the DP after maintained culture (mag. ×128)

14. After 7 days change the media every 2 days by gently tilting the dishes to remove and replenishing the media, taking care not to dislodge the anchored DS or DPs.
15. Cell explants can generally be seen between 14 and 45 days later (*see* **Note 10**).

3.4 Cell Culture Maintenance and Passage

1. Once confluent wash the cells twice for 5 min in 10 mL PBS before adding 10 mL 1× Hank's Solution for 10 min at 37 °C (*see* **Note 12**).
2. Remove the Hank's solution and add 3 mL of trypsin–EDTA (0.05%) to the T75 flask (*see* **Note 11**) to cover the cell monolayer and incubate at 37 °C for 5 min to dissociate the cells from the culture vessel.
3. Tap the flask on a surface to fully dissociate the cells and check under the microscope to confirm that all cells are detached (*see* **Note 12**).
4. Add a minimum of 6 mL of cell culture media (double that of the trypsin) to neutralize and stop the activity of the trypsin.
5. Remove the cell suspension and add to a sterile capped 20 mL tube.

6. Add an additional 5 mL of media to the flask to collect any residual cells and transfer the cell suspension to the same tube to ensure all cells are removed from the flask.
7. Centrifuge the cells at 180 RCF for 3 min at room temperature.
8. Either passage (split 1:2, or 1:3, depending on requirements) the flasks by resuspending the cells in the required amount of media and add to whatever size required culture vessels (*see* **Note 13**) or use the pellet for any downstream assays (e.g., Western blot or RNA extraction).

4 Notes

1. Human tissue should be handled and disposed of according to the Human Tissue Act (HTA) (United Kingdom) or as per local regulations. Disposing of tissue in a capped tube allows for contained disposal.
2. At this point if any tissue is required for immunohistochemistry it can be snap frozen in liquid nitrogen and stored at −80 °C or processed for paraffin embedding.
3. The epidermis can be used to harvest epidermal keratinocytes.
4. Take care not to add too much media as the samples will float and will be unable to firmly attach to the base of the dish. If using 35 mm culture dishes, place the culture dishes in a square dish or larger petri dish and then sandwich this between two dishes filled with 15 mL PBS to stop the samples from drying out in the incubator.
5. Dermal fibroblast explants may be under the tissue itself, so the tissue should be gently removed from the dish after 10 days to check for explants and reduce the risk of infection. Dispose of the tissue according to HTA policy.
6. Use different dishes for ease. Put clean follicles in one dish, follicles with the matrix removed in another, all in media. This means each stage can be done in batches rather than processing each follicle individually, that is, cutting the DP out one after the other.
7. The DP will float and eventually sink if the grip is lost. Wait a minute and move the focus on the microscope up and down until the tear-drop structure comes into focus.
8. Try to avoid damaging the DP too much as it will be less likely to explant.
9. Unlike the DP the DS can be firmly scratched into the plate.

10. DP cells will grow in aggregates for the first 1–4 passages, and dermal sheath cells will grow in webbed formations between passages 1 and 5. Cells should be used before reaching passage 5, due to loss of cell phenotype. Reticular fibroblasts can be used to higher passage numbers (passage 15); however, with multiple passages, papillary fibroblasts can differentiate to a reticular phenotype (beyond passage 6) [15].
11. Scale the amount of media/wash and Hank's solution according to culture vessel size.
12. Due to fibroblasts secreting extracellular matrix, some colonies may be still bound to the flask, if this is the case, remove the trypsin from the flask containing the cells and place in a capped tube and add media to the tube (2× the volume of trypsin) to neutralize the action of trypsin. Then repeat the trypsin step on the flask to remove the remaining cells.
13. Leave the cells to attach to the vessel overnight and check the cells the following day before changing to fresh media.

Acknowledgments

This work was supported by the Centre for Skin Sciences, University of Bradford, UK.

References

1. Virchow R (1859) Die Cellularpathologie in ihrer Begründung auf physiologische und pathologische Gewebelehre. Verlag von August Hirschwald, Berlin
2. Duval MM (1889) Atlas d'embryologie. G. Masson, Paris
3. Croft CB, Tarin D (1970) Ultrastructural studies of wound healing in mouse skin. I. Epithelial behaviour. J Anat 106:63–77
4. Fernandes IR, Russo FB, Pignatari GC, Evangelinellis MM, Tavolari S, Muotri AR, Beltrão-Braga PCB (2016) Fibroblast sources: where can we get them? Cytotechnology 68:223–228
5. Cormack DH (1987) The integumentary system. In: Ham's histology, 9th edn. J.B. Lippincott Company, Philadelphia, pp 450–474
6. Sorrell JM, Caplan AI (2004) Fibroblast heterogeneity: more than skin deep. J Cell Sci 117:667–675
7. Stenn KS, Paus R (2001) Controls of hair follicle cycling. Physiol Rev 81:449–494
8. Driskell RR, Lichtenberger BM, Hoste E, Kretzschmar K, Simons BD, Charalambous M, Ferron SR, Herault Y, Pavlovic G, Ferguson-Smith AC, Watt FM (2013) Distinct fibroblast lineages determine dermal architecture in skin development and repair. Nature 504:277–281
9. Alonso L, Fuchs E (2006) The hair cycle. J Cell Sci 119:391–393
10. Ito M, Sato Y (1990) Dynamic ultrastructural changes of the connective tissue sheath of human hair follicles during hair cycle. Arch Dermatol Res 282:434–441
11. Oliver RF (1966) Regeneration of dermal papillae in rat vibrissae. J Invest Dermatol 47:496–497
12. Jahoda CA, Reynolds AJ (2001) Hair follicle dermal sheath cells: unsung participants in wound healing. Lancet 358:1445–1448
13. Higgins CA, Chen JC, Cerise JE, Jahoda CAB, Christiano AM (2013) Microenvironmental reprogramming by three-dimensional culture enables dermal papilla cells to induce de novo human hair-follicle growth. Proc Natl Acad Sci U S A 110:19679–19688

14. Higgins CA, Richardson GD, Ferdinando D, Westgate GE, Jahoda CAB (2010) Modelling the hair follicle dermal papilla using spheroid cell cultures. Exp Dermatol 19:546–548
15. Janson D, Saintigny G, Mahé C, El Ghalbzouri A (2013) Papillary fibroblasts differentiate into reticular fibroblasts after prolonged in vitro culture. Exp Dermatol 22:48–53

Chapter 3

Isolation of Epidermal and Hair Follicle Melanocytes

Richard Baker and M. Julie Thornton

Abstract

Here we describe the isolation of epidermal melanocytes and hair follicle melanocytes from human skin tissue. Epidermal and hair follicle melanocytes are two distinct populations of melanocytes which are contained within the skin and the hair follicle, with differing yet overlapping roles. Epidermal melanocytes are normally isolated from the epidermis of vellus-haired skin tissue (e.g., face, breast, abdomen), while hair follicle melanocytes are derived from the outer root sheath (ORS) of the middle/lower terminal anagen hair follicles of the scalp. These methods utilize ethically sourced human skin tissue obtained from donors undergoing plastic surgery.

Key words Cell culture, Melanocyte, Hair follicle melanocyte, Pigment, Hair follicle, Outer root sheath, Melanin, Epidermis

1 Introduction

Epidermal melanocytes are pigment producing cells responsible for the production and transfer of melanin to keratinocytes in the epidermis. They will rarely proliferate in vivo and reside in the basal layer, forming epidermal melanin units; a functional unit of cells composed of one melanocyte and 36 keratinocytes [1]. The melanin produced by the melanocytes is transferred to the keratinocytes in the epidermal melanin units, and serves to protect these cells from DNA damage caused by exposure to UV irradiation.

Hair follicle melanocytes consist of two distinct populations; pigmented, dendritic, nonproliferative melanocytes derived from the bulb region of the follicle and amelanotic, nondendritic, highly proliferative cells which are located in the ORS of the hair follicle [2]. The pigmented cells located in the bulb are more differentiated and supply melanin to the cortical keratinocytes in the bulb matrix. They are more densely populated in the hair follicle than in the epidermis, with a ratio of one melanocyte to five precortical matrix cells in the hair bulb, and 1:1 in the basal layer of the hair bulb matrix [3, 4].

Natalia V. Botchkareva and Gillian E. Westgate (eds.), *Molecular Dermatology: Methods and Protocols*, Methods in Molecular Biology, vol. 2154, https://doi.org/10.1007/978-1-0716-0648-3_3,

© Springer Science+Business Media, LLC, part of Springer Nature 2020

However, it is the amelanotic cells that are responsible for maintaining pigment in the hair follicle, because it is these cells which are required for sustaining the population of melanocytes in the bulb region of the hair follicle as when the hair follicle goes through each cycle, the lower part, including the bulb, is remodeled each time [5, 6]. These amelanotic melanocytes are immature and only begin producing pigment once they differentiate upon reaching the bulb region. The ORS niche provides a pool of stem cells to the hair follicle and epidermis and it is from this reservoir of cells that the epidermal melanocytes are also repopulated in the event of wound healing, or in conditions such as vitiligo [7].

The isolation and maintenance of primary human epidermal melanocytes in culture is a relatively easy and reliable procedure. A good yield of cells can easily be obtained from relatively small pieces of tissue (e.g., 2 cm^2). This method outlines the removal of the adipose layer from below the reticular dermis and subsequently the removal of the epidermis from the dermis by treating with trypsin. The epidermal melanocytes can then be isolated from the epidermal sheet by incubating with trypsin–ethylenediaminetetraacetic acid (EDTA) solution.

The isolation of hair follicle melanocytes is a more difficult technique, due in part to the relatively low numbers of ORS melanocytes present in a human hair follicle [8, 9]. To successfully establish hair follicle melanocytes in culture, it is the immature, amelanotic melanocytes that are isolated since the bulb melanocytes are already differentiated and nonproliferative [10, 11]. This technique describes the isolation of these cells from human scalp tissue and the establishment of these cells in culture. Following removal of the dermis and epidermis, the adipose layer containing the lower portion of the hair follicle is digested using collagenase, which after washing leaves just the individual hair follicles. The ORS of these follicles is broken down following incubation with 0.05% trypsin/EDTA before seeding into 3.5 cm petri dishes and purifying using Geneticin treatment [12, 13].

2 Materials

2.1 Equipment

1. Class II Laminar flow cabinet.
2. 3.5 cm petri dishes.
3. 10 cm^2 square culture plates.
4. 30 mL plastic universals.
5. T75 cell culture flask.
6. Scalpel with No. 11 scalpel blades.
7. Fine watchmaker forceps (No. 4).
8. 10 cm forceps.

9. Humidified CO_2 incubator.
10. Benchtop cell culture centrifuge.
11. Cell strainer (40 μm).

2.2 Reagents

1. Phosphate buffered saline, pH 7.4 (PBS).
2. Transport Medium: 500 mL Dulbecco's Modified Eagle's Medium (DMEM) low glucose with 1000 mg/L glucose, 2 mM GlutaMAX, 10% fetal calf serum (FCS), penicillin/streptomycin Solution (5000 U/mL penicillin and 5000 μg/mL streptomycin), and 1% Fungizone (2.5 μg/mL of amphotericin B and 2 μg/mL of sodium deoxycholate).
3. Tissue Wash Solution: PBS pH 7.4, supplemented with penicillin/streptomycin solution (5000 U/mL penicillin and 5000 μg/mL streptomycin) and 1% Fungizone (2.5 μg/mL of amphotericin B and 2 μg/mL of sodium deoxycholate).
4. 0.25% Trypsin.
5. 0.05% trypsin–EDTA.
6. Complete Keratinocyte Serum Free Medium (KSFM): KSFM supplemented with 50 μg/mL bovine pituitary extract, 5 ng/mL EGF (Human Recombinant), 2 mM GlutaMAX, 100 μg/mL Primocin.
7. Complete Minimum Essential Medium (MEM): MEM supplemented with 2% FCS, 2 mM GlutaMAX, 100 μg/mL Primocin, 1% nonessential amino acid Mix, 5 ng/mL endothelin-1, 5 ng/mL bFGF.
8. Complete Melanocyte Culture Medium: two parts Complete MEM: one part Complete KSFM.
9. Geneticin (G418) Solution (50 mg/mL).
10. Collagenase (Type V).

3 Methods

Fresh tissue samples should be placed in transport medium immediately postsurgery, and stored at 4 °C. Samples should be processed within 24 h of surgery.

All work to be carried out in a Class II laminar flow cabinet, unless otherwise stated. Any excess tissue should be handled and disposed of appropriately according to the Human Tissue Act or as per the local regulations.

3.1 Isolation of Epidermal Melanocytes

1. Transfer skin samples into a 10 cm^2 square culture plate. Wash the tissue twice with 20 mL tissue wash.

2. Using a scalpel, remove any haired tissue retaining the non-haired tissue (*see* **Note 1**). Cut the sample into approximate 0.5 cm^2 cubes.
3. Pick up the tissue cubes using a flat pair of forceps to orientate it so the longitudinal section through the tissue is visible (Fig. 1). With a scalpel cut through the tissue to remove the adipose tissue (Fig. 2).
4. Wash the tissue again with 20 mL of Tissue Wash before placing into a 3.5 cm Petri Dish containing 3 mL of 0.25% trypsin. The skin should be placed into the trypsin with the epidermis uppermost and incubated for 4 h at 37 °C (*see* **Note 2**).
5. Using fine forceps gently peel the epidermis from the dermis and float it into 5 mL PBS in a 10 cm^2 square culture plate (Fig. 3).
6. Once the entire epidermal layer has been removed, place all the epidermal samples into a plastic 30 mL universal containing 3 mL 0.05% trypsin–EDTA and incubate for 5 min at 37 °C.
7. Vortex the samples for 1 min before removing the supernatant and transferring it into 10 mL of melanocyte culture medium, taking care not to aspirate any solid pieces of epidermis. Add a further 3 mL of 0.05% Trypsin/EDTA to the universal tube containing the epidermis and repeat the trypsin treatments until the trypsin solution is transparent after vortexing. This typically takes three trypsinization steps.
8. Centrifuge the cell suspension at 180 RCF for 5 min, then discard the supernatant and resuspend the cell pellet in 10 mL melanocyte medium. Transfer the cell suspension to a T75 flask containing a total of 10 mL melanocyte culture medium and incubate at 37 °C in 5% CO_2.
9. Replace the culture medium after 2 days. The culture should be monitored microscopically for keratinocyte and fibroblast contamination. If any contamination is observed add G418 to the culture medium (100 μg/mL) which will selectively kill the more rapidly dividing fibroblasts (*see* **Note 4**) [13]. G418 treatment should be applied for 48 h. Upon completion of the G418 treatment the culture medium should be replaced with fresh, untreated medium.
10. The cells should be routinely cultured with the replacement of fresh melanocyte culture medium every 3 days. A pure melanocyte culture can generally be obtained within 2 weeks. The cells should be passaged at a ratio of 1:3 and will proliferate for approximately ten passages.

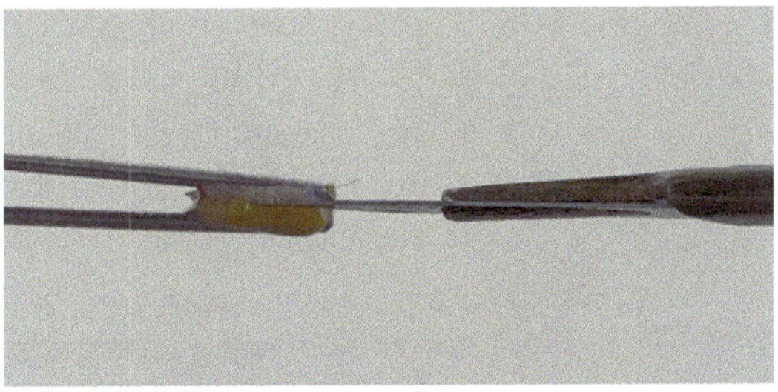

Fig. 1 Technique for the cutting of the adipose tissue in the lower dermis by placing between flat forceps and using a scalpel blade

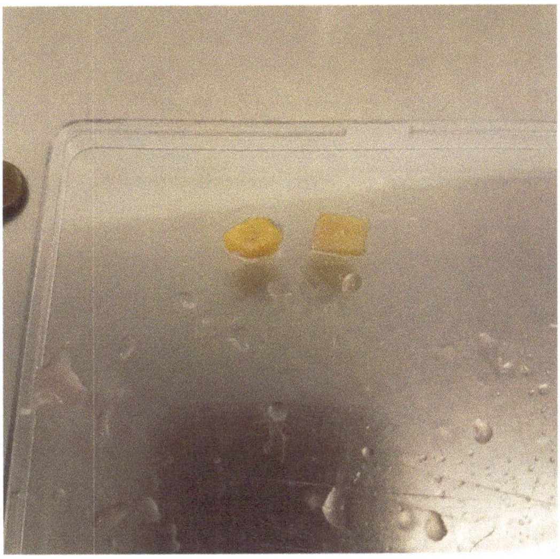

Fig. 2 Tissue cube with the epidermis and dermis separated from the adipose tissue

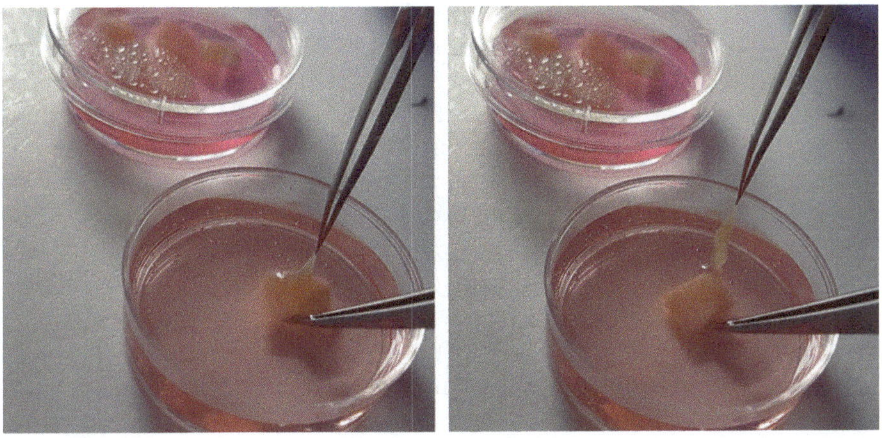

Fig. 3 Peeling off the epidermis after 16 h in trypsin 0.25% at 4 °C followed by 1 h incubation at 37 °C in 0.05% trypsin–EDTA

3.2 Isolation of Hair Follicle Melanocytes

1. Transfer hairy scalp skin samples into a 10 cm^2 square culture plate. Wash the tissue twice with 20 mL tissue wash.
2. Using a scalpel, remove any nonhaired tissue and shave off hair close to the epidermis. Cut the sample into approximate 0.5 cm^2 cubes (*see* **Note 1**).
3. Pick up the tissue cubes using a flat pair of forceps, orientating so you can see a longitudinal section through the tissue. Using a scalpel cut through the tissue at the very bottom of the dermis at the dermal–adipose junction (Fig. 1), retaining the adipose tissue containing the hair follicles (Fig. 2) (*see* **Note 3**).
4. Place the adipose tissue cubes into 10 mL transport medium in a plastic universal tube. Add collagenase to the universal tube to a final concentration of 0.5%, and vortex until the collagenase has completely dissolved. Incubate at 37 °C for 1 h, vortexing for 1 min every 20 min.
5. Pour the contents of the universal tube into a 10 cm^2 square culture dish. Remove any fatty connective tissue using forceps and put it back into the universal tube.
6. Using a serological pipette aspirate the collagenase solution from the culture plate trying not to disturb the hair follicles which will sink to the bottom of the plate. Put the collagenase solution back into universal tube with the remaining fatty/connective tissue.
7. Resuspend the follicles in the culture plate in melanocyte culture medium and transfer to a clean plastic universal tube.
8. Repeat this process of incubating the tissue with collagenase solution and removing follicles for a total of three times. The incubation time can be reduced to 40 min for all subsequent incubations.
9. Once all the follicles have been collected carefully remove the culture medium taking care not to disturb the hair follicles which have sunk to the bottom of the dish. Wash with 20 mL PBS and allow the follicles to sink to the bottom of the dish once more. Carefully remove the PBS and repeat this wash step ten times, or until all the fat has been removed from the follicles (Fig. 4). Resuspend the hair follicles in 5 mL 0.05% trypsin–EDTA and incubate at 37 °C for 5 min.
10. Vortex the hair follicle solution and carefully remove the trypsin/hair follicle cell suspension and add it to 10 mL melanocyte culture medium in a clean universal tube. Add 3 mL of fresh 0.05% trypsin–EDTA and repeat this trypsinization step a total of three times, until only the hair shafts remain postvortexing (Fig. 5).

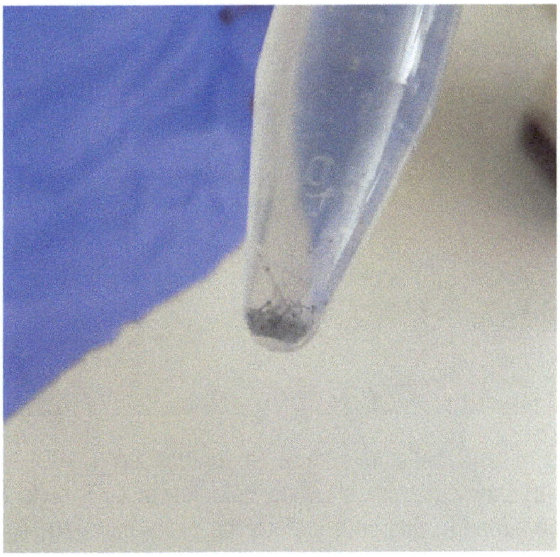

Fig. 4 Hair follicles collected from the tissue after collagenase treatment and ten washes with PBS

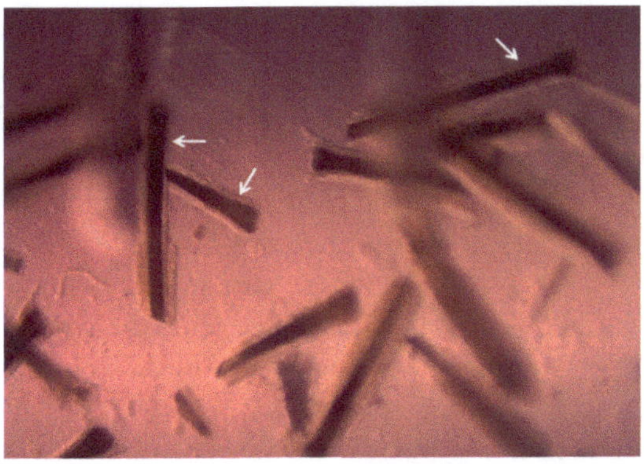

Fig. 5 Hair follicles undergoing 0.05% trypsin–EDTA treatment. The ORS is beginning to break down. Further trypsinization will remove additional ORS material and leave just the hair shafts remaining

11. Take the cell suspension and pass it through a 40 μm cell strainer to remove any large pieces of hair follicle tissue, which may contaminate the culture.
12. Centrifuge the cell suspension at 180 RCF for 5 min. Discard the supernatant and resuspend the cell pellet in 3 mL melanocyte medium. Add the cell suspension to a 3.5 cm petri dish and incubate at 37 °C in 5% CO_2 in air.
13. Replace the culture medium after 2 days. The culture should be monitored microscopically for keratinocyte and fibroblast contamination (Fig. 6). If any contamination is observed add G418 to the culture medium (100 μg/mL) which will selectively kill the more rapidly dividing fibroblasts (*see* **Note 4**) [13].

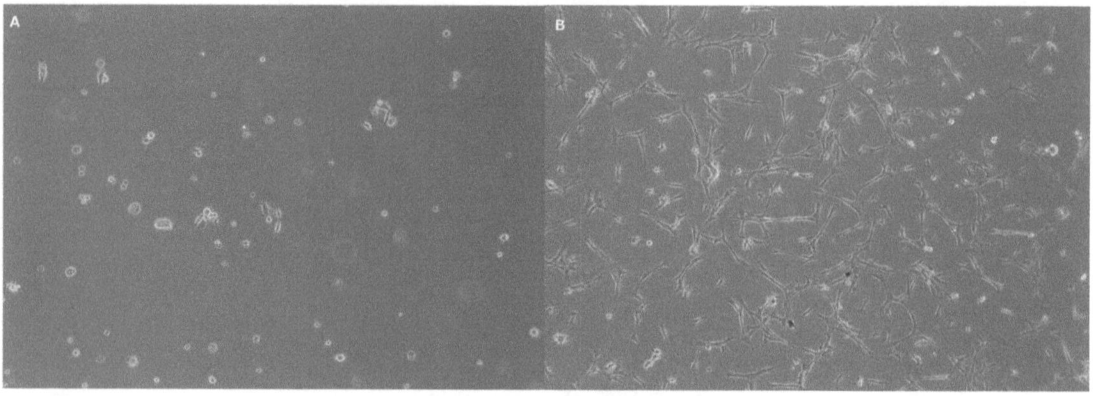

Fig. 6 Images of ORS cells in culture (magnification ×100). (**a**) Day 1 culture showing cells attaching postisolation. (**b**) Day 7 culture showing colonies of ORS cells forming. If several cell types are observed G418 should be added at this point before the cells reach 100% confluence

G418 treatment should be applied for 48 h. Upon completion of the G418 treatment the culture medium should be replaced with fresh, untreated medium.

14. The cells should be routinely differentially trypsinized at 70% confluence (*see* **Note 5**) with further G418 treatments if any contamination of the culture by fibroblasts persists.
15. Once a pure melanocyte culture is established, the cells should be routinely cultured with the replacement of fresh melanocyte culture medium every 3 days (*see* **Note 6**). A pure melanocyte culture can generally be obtained within 3–4 weeks with an established culture within 8 weeks. The cells should be passaged at a ratio of 1:3 and will proliferate for approximately ten passages.

4 Notes

1. Only nonhaired (vellus) tissue should be used for the isolation of epidermal melanocytes since terminal hair follicles can prevent the easy removal of the epidermal sheet.
2. The epidermal sheet can take longer than 4 h to become completely detached. If it does not peel off easily using forceps, return the tissue to the incubator for an additional 30 min before attempting again. An alternate method is to leave the tissue in the fridge at 4 °C overnight (instead of incubating for 4 h at 37 °C).
3. It is important that enough of the dermis is removed so the follicles are not retained in the connective tissue when they are incubated in the collagenase solution.

4. G418 treatment should be added to the culture when aggregations of fibroblast contamination begin to form. If left unchecked, it can be difficult to remove all the fibroblast contamination from the culture once it has become established. Keratinocyte contamination does not require the immediate addition of G418 because they will differentiate as the culture progresses.
5. Differential trypsinization can be used to limit fibroblast contamination. Melanocytes detach more quickly in 0.05% trypsin–EDTA solution than fibroblasts and should detach in approximately 90 s (this should be monitored microscopically). Fibroblasts remain attached for a longer period so the detached melanocyte cells can be collected for further culture.
6. ORS melanocytes can be hard to identify as melanocytes in early culture as they are amelanotic and do not have typical melanocyte morphology. Premelanosome Gp100 staining with NKI/beteb or P-mel 1 antibodies can be performed to confirm the presence of melanocytes in the culture.

Acknowledgments

This work was supported by the Centre for Skin Sciences, University of Bradford, UK.

References

1. Fitzpatrick TB, Breathnach AS (1963) The epidermal melanin unit system. Dermatol Wochenschr 147:481–489
2. Tobin DJ, Bystryn JC (1996) Different populations of melanocytes are present in hair follicles and epidermis. Pigment Cell Res 9:304–310
3. Slominski A, Wortsman J, Plonka PM, Schallreuter KU, Paus R, Tobin DJ (2005) Hair follicle pigmentation. J Invest Dermatol 124:13–21. https://doi.org/10.1111/j.0022-202X.2004.23528.x
4. Tobin DJ, Paus R (2001) Graying: gerontobiology of the hair follicle pigmentary unit. Exp Gerontol 36:29–54
5. Commo S, Gaillard O, Bernard BA (2004) Human hair greying is linked to a specific depletion of hair follicle melanocytes affecting both the bulb and the outer root sheath. Br J Dermatol 150:435–443. https://doi.org/10.1046/j.1365-2133.2004.05787.x
6. Tobin DJ, Hagen E, Botchkarev VA, Paus R (1998) Do hair bulb melanocytes undergo apotosis during hair follicle regression (catagen)? J Investig Dermatol 111:941–947. https://doi.org/10.1046/j.1523-1747.1998.00417.x
7. Nishimura EK (2011) Melanocyte stem cells: a melanocyte reservoir in hair follicles for hair and skin pigmentation. Pigment Cell Melanoma Res 24:401–410. https://doi.org/10.1111/j.1755-148X.2011.00855.x
8. Ortonne JP, Schmitt D, Thivolet J (1980) PUVA-induced repigmentation of vitiligo: scanning electron microscopy of hair follicles. J Invest Dermatol 74:40–42
9. Ortonne J-P (1989) Non-keratinocyte cells in hair follicles (with special emphasis to human hair follicles). In: Van Neste D, Lachapelle JM, Antoine JL (eds) Trends in human hair growth and alopecia research. Springer Netherlands, Dordrecht, pp 75–87. https://doi.org/10.1007/978-94-011-7873-0_9
10. Tobin DJ, Colen SR, Bystryn JC (1995) Isolation and long-term culture of human hair-follicle melanocytes. J Invest Dermatol 104:86–89

11. Takada K, Sugiyama K, Yamamoto I, Óba K, Takeuchi T (1992) Presence of amelanotic melanocytes within the outer root sheath in senile white hair. J Investig Dermatol 99:629–633. https://doi.org/10.1111/1523-1747.ep12668031
12. Savkovic V, Dieckmann C, Milkova L, Simon JC (2012) Improved method of differentiation, selection and amplification of human melanocytes from the hair follicle cell pool. Exp Dermatol 21:948–950. https://doi.org/10.1111/exd.12038
13. Halaban R, Alfano FD (1984) Selective elimination of fibroblasts from cultures of normal human melanocytes. In Vitro 20:447–450. https://doi.org/10.1007/BF02619590

Chapter 4

Isolation and Culture of Human Skin Mast Cells

Hanna Siiskonen and Jörg Scheffel

Abstract

Mast cells are intriguing immune cells monitoring the interfaces of our body with the environment. Long-term cultures of mast cells originating from hematological malignant cells have been traditionally used to investigate mast cell biology, but these cells are not optimal to study the characteristics of human skin mast cells because of their origin and also due to the changes caused by long-term culture. To overcome these limitations, freshly isolated mast cells from human skin have been recently introduced as a research tool and provide a better resource for studies investigating the properties of mast cells in the human skin. Here, the isolation procedure of human skin mast cells from skin to cell culture is explained in detail and clarified in several figures.

Key words Mast cells, Human skin, Isolation, Primary cell culture, MACS

1 Introduction

Mast cells (MCs) originate from a common progenitor cell in the bone marrow and differentiate in tissues to specific subtypes, which reside in the tissues (reviewed in [1]). Compared to other tissue resident immune cells such as dendritic cells or macrophages MCs are of comparably low abundance. Therefore, MC lines HMC-1 and LAD2 or peripheral stem cell- or cord blood-derived MCs have been intensively used to study mast cell biology (*see* the list of several publications referenced in [2]). HMC-1 cells originate from the peripheral blood of a patient with MC leukemia [3], while LAD2 cells originate from the bone marrow from a patient with MC sarcoma/leukemia [4]. As with other cultured cell, long-term culture can modify the properties and responses of the cells and they may no longer represent the original cell type. In addition, as both of these cell lines originate from hematological, malignant condition, they represent healthy skin MCs only poorly and for example, their enzyme contents are different [2]. When the special features of skin MCs are to be investigated, it is of utmost importance to use valid primary cells as close to MCs in vivo as possible. Due to these reasons, it has become

Natalia V. Botchkareva and Gillian E. Westgate (eds.), *Molecular Dermatology: Methods and Protocols*,
Methods in Molecular Biology, vol. 2154, https://doi.org/10.1007/978-1-0716-0648-3_4,
© Springer Science+Business Media, LLC, part of Springer Nature 2020

essential to find methods to isolate MCs directly from human skin. Here we describe the current method used in our laboratory to isolate and culture human skin MCs (hsMCs). However, it should be kept in mind that the work with freshly isolated MCs also has its limitations. Due to their low abundance, only a limited number of cells from a certain individual are available and comparable large amount of tissue is required to isolate a suitable number of cells. Moreover, there is a donor-to-donor variability in addition to differences based on the tissue source [1, 5]. As with all primary cell cultures, phenotypes change over time in culture and therefore cells are best for use shortly after isolation.

2 Materials

Regular isolation of hsMCs requires a constant source of skin—ideally, a close collaboration with a surgical team (*see* **Note 1**). Skin can be obtained from certain cosmetic or medical plastic surgeries such as reduction mammoplasty, abdominoplasty, circumcision, or blepharoplasty (*see* **Note 2**).

According to the Declaration of Helsinki (59th WMA General Assembly, Seoul, October 2008) the work needs an ethical approval from the ethics administration of the institution and a written informed consent with sufficient, clear and appropriate information regarding use of the skin in research purposes must be given and signed by the patients. The donation of the skin is voluntary and the patients must be informed about the process.

2.1 Buffer and Media

1. Transport Medium (make sure that all pieces are submerged in the media): Basal Iscove's medium, noncomplete with 15 mg phenol red (FG 0465, Biochrom, Berlin, Germany) and 2.5 μg/ml Amphotericin B. The skin should arrive ideally in the laboratory within 2–5 h after surgery (*see* **Note 3**).
2. Dispase medium: Dulbecco's Phosphate-Buffered Saline (DPBS) with Ca^{2+} and Mg^{2+} containing 2.4 U/ml Dispase Type II.
3. Digestion medium (use 10 ml/g of skin): DPBS with Ca^{2+} and Mg^{2+} containing collagenase (15 mg/g of skin), hyaluronidase from bovine testes (7.5 mg/g of skin), DNase (10 μg/ml final concentration), 1% penicillin–streptomycin, 2% fetal bovine serum, 2.5 μg/ml amphotericin B, and additional 5 mM $MgSO_4$.
4. MACS buffer: DPBS without Ca^{2+} and Mg^{2+} containing 5% fetal bovine serum and 1% penicillin–streptomycin.
5. MC culture medium: Basal Iscove's medium, noncomplete with 15 mg phenol red (FG 0465, Biochrom) supplemented

with 10% fetal bovine serum, 1% penicillin–streptomycin, 1% nonessential amino acids, and α-monothioglycerol (230 μM final concentration). For long term culture (≥24 h), add recombinant human IL-4 (20 ng/ml) and SCF (100 ng/ml).

2.2 Antibodies/Microbeads for Flow Cytometry and Magnetic Cell Enrichment

1. CD117 microbead kit (Miltenyi Biotec, Bergisch Gladbach, Germany).
2. Anti-CD117 (clone A3C6E2, Miltenyi Biotec) (*see* **Note 4**).
3. Anti-FcεRI (clone AER-37 [CRA-1]; BioLegend) (*see* **Note 5**).

2.3 Additional Equipment

1. Biosafety cabinet class 2 (*see* **Note 6**).
2. Cell culture incubator providing a humid atmosphere with constant temperature of 37 °C and 5% CO_2.
3. A basic light microscope and Neubauer (improved) counting chamber.
4. A centrifuge with swing-out rotor for 50 ml conical tubes.
5. A shaking water bath (*see* **Note 7**).
6. Equipment for magnetic cell enrichment [columns (LS), magnets and stand or AutoMACS (Miltenyi Biotec) with Chill Block 15 (Miltenyi Biotec) and cell strainer with 100, 40 (Corning) and 30 μm mesh size (Miltenyi Biotec)].
7. Stainless steel sieves, 300 and 40 μm (Retsch, Haan, Germany) (*see* **Note 8**).
8. Sterile forceps, scalpels, and scissors.
9. Petri dishes (140 mm), conical 50 and 15 ml centrifugal tubes and two sterile (plastic) containers with lid (for enzymatic digestion).

3 Methods

3.1 Tissue Digestion

Isolation of hsMCs has been adopted from [2]. All procedures are carried out at room temperature and in a class 2 biosafety cabinet, unless otherwise specified.

1. Weigh the skin before starting the procedure to calculate the required amount of Digestion medium, enzymes for tissue digestion, and potential cell yield (*see* above).
2. Place the skin specimens in a sterile petri dish (140 mm) without buffer and remove fat tissue using sterile scissors and a scalpel. If any fat was removed, control-weigh the skin again. Cut the skin into small 10 × 20 mm pieces with thin incisions at 1–2 mm intervals in a comb-like pattern (Fig. 1) to increase the surface for enzymatic digestion. The skin pieces are then

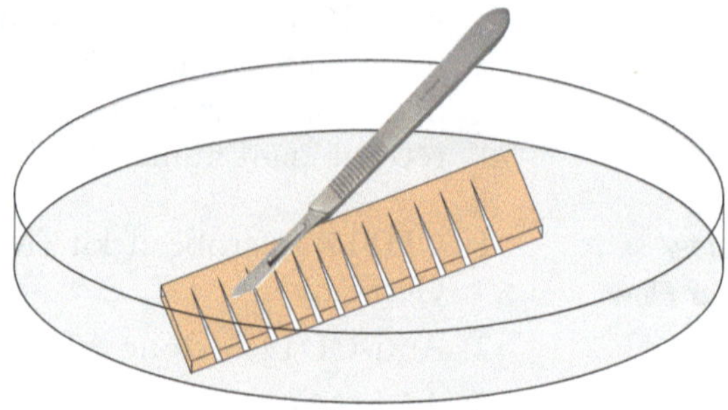

Fig. 1 Schematic illustration of skin preparation. The skin is cut with a scalpel into 5 × 20 mm strips with incisions every 1–2 mm before being immersed and incubated in cold dispase medium overnight at 4 °C

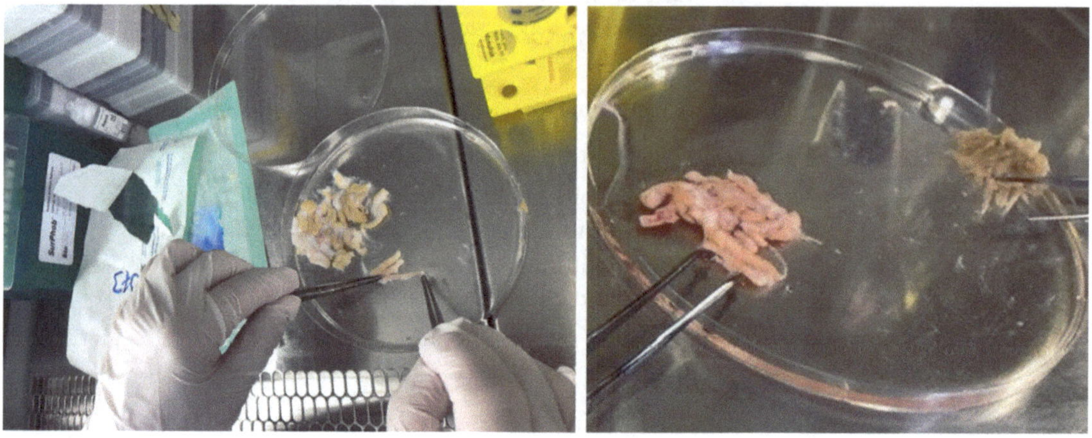

Fig. 2 After dispase treatment over night the epidermis is carefully removed with sterile forceps from the dermis. The epidermis-free dermal pieces are then collected in a new petri dish

transferred to Dispase medium and incubated overnight in a sterile container with lid at 4 °C (*see* **Note 9**).

3. On the next day, remove epidermis carefully from dermis with fine forceps and collect dermal pieces in a new petri dish (Figs. 2 and 3) (*see* **Note 10**).
4. Homogenize pieces with a sharp, curved scissor to form a uniform mass (Fig. 4) for about 25–30 min. This step is critical for a successful isolation of MCs (*see* **Note 11**). Poorly minced tissue reduces the yield of cells dramatically (*see* **Note 12**).
5. Transfer the homogenized skin into a new sterile container with digestion medium containing collagenase, hyaluronidase, and DNase, and incubate for 60 min (foreskin and eyelid) or 90 min (breast skin) with vigorous shaking in a tempered water bath at 37 °C.

Fig. 3 The dermal tissue is homogenized with scissors until it forms uniform mass. Complete homogenization of the tissue is crucial for optimal MC yield

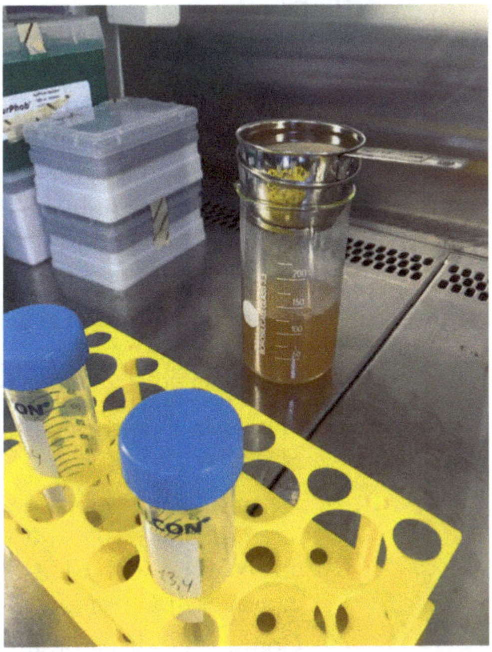

Fig. 4 After enzymatic digestion the suspension is filtered via 300 μm + 40 μm metal sieves to remove larger pieces of undigested tissue in sieve. The filtrate is centrifuged to separate the single cells and digestion medium. The remaining skin pieces are homogenized again as in **step 4** repeatedly incubated with the retained digestion medium (**steps 5** and **6**)

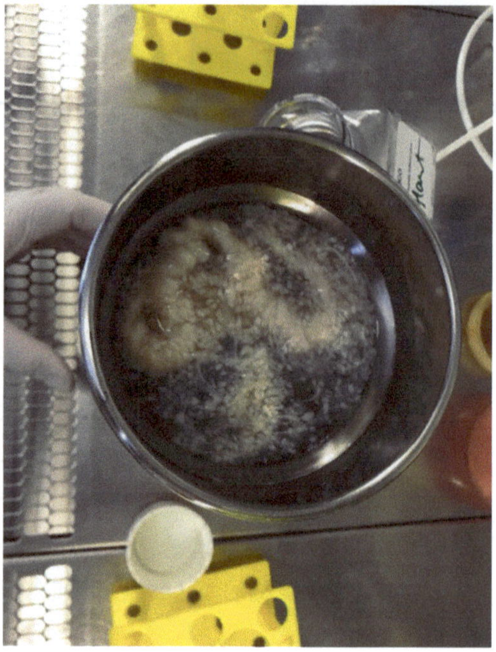

Fig. 5 Larger pieces of undigested tissue

6. Filter the suspension through 300 μm + 40 μm mesh sieves to remove larger pieces of undigested tissue (5). Centrifuge the filtrate at 400 × *g* for 15 min at +4 °C, collect the supernatant (Digestion medium) and resuspend the pellet in cold DPBS without Ca^{2+} and Mg^{2+} (*see* **Note 13**).
7. Keep the cells on ice and repeat **steps 4–6** with the retained tissue from filtration at least once using new Digestion medium.
8. Combine cells from all tissue digestion rounds in one tube, fill up to 50 ml with cold DPBS without Ca^{2+} and Mg^{2+} and wash once using centrifugation at 400 × *g* for 10 min at 4 °C (*see* **Note 14**).
9. Decant DPBS supernatant, resuspend in MACS buffer and filter through 100 and 40 μm cell strainers, sequentially. Fill up to 50 ml with MACS buffer and wash once using centrifugation at 400 × *g* for 10 min at 4 °C.
10. Count the cells, if desired.

3.2 Magnetic Cell Enrichment (MACS)

1. Decant MACS buffer supernatant, resuspend the pellet in 5 ml of MACS buffer and transfer to a 15 ml conical tube through a 30 μm cell strainer. Rinse the 50 ml tube again with 5 ml of MACS buffer transfer to the 15 ml conical tube through the 30 μm cell strainer. Rinse the cell strainer again with additional

2–3 ml MACS buffer. Centrifuge again at 400 × *g* for 10 min at 4 °C.

2. Aspirate MACS buffer supernatant and resuspend thoroughly in 1 ml final volume MACS buffer using a 1000 μl pipette (*see* **Note 15**). Add 50 μl of FcR blocking reagent (CD117 microbead kit), mix by flicking and incubate at 4 °C for 15 min. Add 50 μl of CD117 MicroBeads (CD117 microbead kit), mix by flicking and incubate at 4 °C for 10 min (occasionally mix by flicking).
3. Wash cells by adding 10 ml of MACS buffer and centrifuge at 400 × *g* for 10 min. Aspirate the supernatant completely and resuspend cells in 2 ml MACS buffer.
4. If an AutoMACS (Miltenyi Biotec) is not available, proceed with **step 6**.
5. If an AutoMACS is available, enter a maximum aspiration volume of 3000 μl and run the program "posselds" (positive selection via two columns). Proceed with **step 10**.
6. During centrifugation, insert a MACS LS column into an appropriate magnetic separator (magnet) and rinse the column with 3 ml of MACS buffer.
7. Add cell suspension from **step 3**.
8. Wash the column three times with 3 × 3 ml of MACS buffer. If desired, collect the flow through (CD117 negative fraction) and keep at 4 °C until use.
9. Remove the column from the magnetic separator and place it into a suitable collection tube (e.g., 15 ml conical tube). Add 2 ml of MACS buffer and flush out the cells by firmly pushing the plunger into the column (CD117 positive fraction). Repeat with additional 2 ml of MACS buffer. Proceed with **step 10**.
10. Take 10–50 μl aliquots to count cells and analyze by flow cytometry. To residual bulk, add 10 ml of MACS buffer and centrifuge at 400 × *g* for 10 min at 4 °C. Aspirate MACS buffer supernatant and resuspend cells by flicking or gentle pipetting.

3.2.1 Flow Cytometry

1. Take an aliquot of 20 μl from CD117 negative and positive fraction. Add 5 μl of fluorochrome labeled anti-CD117 (clone A3C6E2) and FcεRI (clone AER-37 [CRA-1]) and incubate on ice for 10 min (*see* **Note 16**).
2. Wash cells with 500 μl of MACS buffer, resuspend in 300 μl of MACS buffer containing a nuclear dye (e.g., DAPI or propidium iodide) (*see* **Note 17**) to identify cell debris and dead cells.
3. Analyze cells with an appropriate FACS analyzer (Fig. 6).

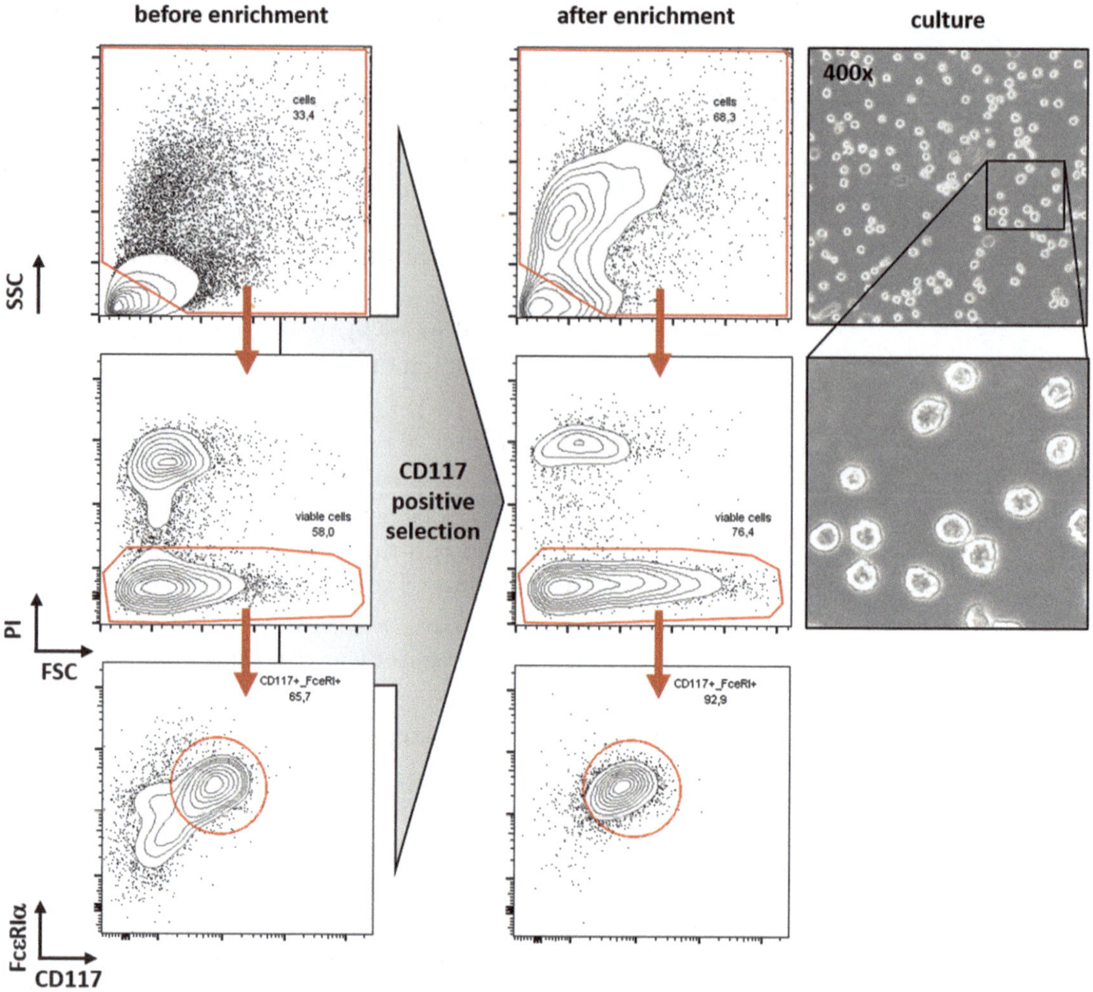

Fig. 6 Isolated skin cells were analyzed for CD117+/FcεRIa+ cells (MC) by flow cytometry before and after CD117 positive enrichment by MACS. Note that enrichment yields a purity of CD117/FcεRIa double positive MCs >90%. Microscopic pictures were taken in a 400× magnification from the enriched cells after resuspension in MC culture medium at a density of 1 × 10^6/ml cells

3.2.2 Long-Term Culture

1. For culturing, suspend cells in 1 ml of cytokine free MC culture medium per 1 × 10^6 cells and culture in a tissue culture incubator for 24 h (*see* **Note 18**).
2. After 24 h add 100 ng/ml human stem cell factor (SCF) and 20 ng/ml human interleukin-4 (IL-4).
3. Feed cells twice a week by adding SCF (100 ng/ml) and IL-4 (20 ng/ml) (*see* **Note 19**).
4. Determine cell number and expression of FcεRI and CD117 by flow cytometry and cell counting on a regular basis. Keep cells at a density of 1 × 10^6 cells per ml of medium. Cells start to proliferate between week two and three. After 4–6 weeks in

culture, the cell numbers almost double every 10 days. Proliferation of MCs starts to decline around 6 weeks postisolation and stops around week eight (*see* **Note 20**).

4 Notes

1. For transportation of the tissue, it is required to provide a waterproof primary container which is packed along with absorbent tissue in a waterproof secondary container with appropriate labels such as “Exempt Human Specimen” for specimens being tested negative for blood borne pathogens or “Biological Substance category B” including the UN3373 label for (potentially) hazardous material respectively.
2. In general, highest yields are obtained from eyelids and foreskin with approximately 0.5–2 × 10^6 cells/g of skin or breast skin with approximately 0.2–0.5 × 10^6 cells/g of skin.
3. If direct processing of the skin is not possible, smaller pieces can be immersed in tissue preservation solutions for up to 24 h (e.g., MACS® Tissue Storage Solution; Miltenyi) to limit necrosis and apoptosis.
4. Anti-CD117 antibody should be compatible with the CD117 microbead kit.
5. AER-37 is also referred to as CRA-1. The CRA-1 clone binds to the alpha subunit of the human high-affinity Fc receptor for IgE (FcεRIa) in the presence and absence of IgE. It does not prevent binding of IgE and has the capability to induce mast cell activation (degranulation).
6. Use Class II cabinets to provide both kinds of protection—sample and of the environment—especially when working with untested human material.
7. Use a water bath with shaking capability for optimal and constant temperature distribution during the dispersion step.
8. These cell strainers can be reused when autoclaved; however, there is a risk of potential contamination. Alternatively, single-use strainers can be used.
9. It is recommended to incubate up to 15 g of skin in 25 ml of Dispase medium in a 50 ml conical tube.
10. The epidermis is removed as thin, brownish sheets and the whitish dermis is left for further processing. The epidermis can be further used for generation of cultures of primary keratinocytes.
11. This step is of utmost importance for successful isolation. The homogenization of the tissue with sharp scissors should be

continued until a pink, slime-like homogenous mass is achieved.

12. Manual homogenization of the tissue with scissors is preferred over electrical tissue dissociators to obtain highest yields and best quality of cells.
13. The pellet should be resuspended gently by tapping the bottom of the tube with fingers. Harsh resuspending with a pipette should be avoided.
14. Always double-rinse all 50 ml tubes with a few milliliters of DPBS (or MACS buffer) when combining to not lose cells.
15. The amount of cells obtained will always be less than reasonable for 1 ml volume with 50 μl FcR block/beads, if you do not process more than 50 g of skin. At $\geq$50 g of skin, increase the volume to 2 ml and 100 μl of FcR block/beads.
16. Nonexperienced users should use ideally two colors that have nonoverlapping emission spectra (i.e., FITC/APC or PE/APC) to avoid errors in misinterpretation of the data due to improper compensation.
17. Non-cell permeable dyes such as DAPI or propidium iodide can be used. Alternatively, live/dead cell staining kits for flow cytometry are available from various manufacturers. Make sure that the nuclear stain is compatible with the colors used for the detection of the cellular markers.
18. If experiencing sizeable numbers of residual macrophages and fibroblasts, use TC-treated flasks for the first 24 h of culture, resuspend thoroughly by using a wide-bore pipette and transfer hsMCs to a new suspension culture flask.
19. Occasionally, add fresh medium to keep cells at 1×10^6/ml. When medium appears overused by cells, thoroughly resuspend cells using a wide-bore pipette, centrifuge at $300 \times g$ for 3 min, aspirate half of the supernatant and add the same amount of fresh medium. Thoroughly resuspend cells using a wide-bore pipette and transfer hsMCs to a new suspension culture flask.
20. Cells start to proliferate between week 2 and 3. After 4–6 weeks in culture, the cell numbers almost double every 10 days. Proliferation of MCs starts to decline around 6 weeks postisolation and stops around week 8.

References

1. Frossi B, Mion F, Sibilano R et al (2018) Is it time for a new classification of mast cells? What do we know about mast cell heterogeneity? Immunol Rev 282(1):35–46. https://doi.org/10.1111/imr.12636
2. Guhl S, Babina M, Neou A et al (2010) Mast cell lines HMC-1 and LAD2 in comparison with mature human skin mast cells—drastically reduced levels of tryptase and chymase in mast cell lines. Exp Dermatol 19(9):845–847.

https://doi.org/10.1111/j.1600-0625.2010.01103.x

3. Butterfield JH, Weiler D, Dewald G et al (1988) Establishment of an immature mast cell line from a patient with mast cell leukemia. Leuk Res 12(4):345–355
4. Kirshenbaum AS, Akin C, Wu Y et al (2003) Characterization of novel stem cell factor responsive human mast cell lines LAD 1 and 2 established from a patient with mast cell sarcoma/leukemia; activation following aggregation of FcepsilonRI or FcgammaRI. Leuk Res 27(8):677–682. S0145212602003430 [pii]
5. Huber M, Cato ACB, Ainooson GK et al (2019) Regulation of the pleiotropic effects of tissue-resident mast cells. J Allergy Clin Immunol 144 (4S):S31–S45. S0091-6749(19)30207-6 [pii]

Chapter 5

Organotypic 3D Skin Models: Human Epidermal Equivalent Cultures from Primary Keratinocytes and Immortalized Keratinocyte Cell Lines

Gijs Rikken, Hanna Niehues, and Ellen H. van den Bogaard

Abstract

The three-dimensional culturing of human keratinocytes at the air–liquid interface yields a fully stratified epidermis including a functional stratum corneum and thus enables the study on epidermal structure and function in the context of biomedical, toxicological and pharmaceutical sciences. Here we provide a step-by-step detailed protocol for the isolation of human primary keratinocytes and the development of human epidermal equivalents generated from primary keratinocytes or immortalized keratinocytes (N/TERT-1; N/TERT-2G), including widely accepted procedures for the analysis of barrier function, tissue morphology, cell proliferation, and gene expression.

Key words Skin tissue engineering, Epidermis, Keratinocyte isolation, Skin barrier, Proliferation, Laboratory protocol

1 Introduction

After the seminal discovery by Rheinwald and Green [1] that human keratinocytes can be isolated from skin and expanded in vitro, the first report on the development of skin equivalents from human adult epidermal cells and dermal substrates cultured at the air–liquid interface was published in 1981 [2]. Today, even in the absence of a dermal substrate, a fully developed stratified epidermis can be grown on a plastic inert filter. Given the tissue architecture and functional skin barrier, these human epidermal equivalents are registered alternatives for experimental animals in toxicity testing pipelines as they mimic native human epidermis [3]. For the development of human epidermal equivalents, in-house culture protocols have been optimized and refined over the last decades to facilitate the incorporation of different cell types and disease-specific phenotypes [4]. In general, human adult primary keratinocytes, isolated from surplus healthy skin or patient

Natalia V. Botchkareva and Gillian E. Westgate (eds.), *Molecular Dermatology: Methods and Protocols*, Methods in Molecular Biology, vol. 2154, https://doi.org/10.1007/978-1-0716-0648-3_5,
© Springer Science+Business Media, LLC, part of Springer Nature 2020

skin, are used for organotypic cultures, as foreskin keratinocytes render a more activated tissue state with psoriatic features [5]. However, the availability of donor skin can be limited and the short in vitro life span and interdonor variation may hamper experimental readout parameters. Immortalized keratinocyte cell lines can provide an unlimited cell source and solution to the aforementioned issues. The N/TERT-1 and N/TERT-2G keratinocyte cell line [6] has been proven to develop high quality epidermal [7] and skin equivalents [8] and faithfully mimics primary keratinocyte gene and protein expression with regard to epidermal proliferation, differentiation, skin barrier function and inflammatory responses. In contrast, the widely used HaCaT keratinocyte cell line is unsuitable for organotypic cultures as it fails to develop a fully stratified epidermis and lacks a functional stratum corneum [9].

When using organotypic epidermal cultures for biomedical research or toxicity testing, it is of utmost importance to validate the quality of the tissue architecture and function, as readout parameters may be influenced by suboptimal tissue morphology. Extensive validation and effective troubleshooting during the experimental procedures and culture period is vital for generating robust and reproducible results. In this chapter, we therefore provide a step-by-step culture protocol with detailed materials and methods, guidelines for troubleshooting, and we describe the procedures for general quality checks of tissue morphology and function to enable widespread implementation of organotypic skin cultures across laboratories.

2 Materials

2.1 Primary Adult Keratinocyte Isolation

1. Centrifuge.
2. 37 °C water bath.
3. 37 °C incubator with 5% CO_2.
4. Gamma irradiator.
5. Hemocytometer.
6. Freezing container.
7. Vented cap cell culture flasks (or cell culture 6-well plates), sterile pipettes, tubes, and general sterile cell culture consumables.
8. 1.8 mL cryovials.
9. 6 mm disposable punch biopsy.
10. Roundhead tweezers.
11. Scissors.
12. Syringe needles.
13. Styrofoam.

14. Disinfectant (Sterilium, Hartman).
15. Fresh human adult skin.
16. 3T3-J2 mouse embryonic fibroblast cell line (commercially available).
17. Phosphate buffered saline (PBS).
18. Epidermal growth factor.
19. Dimethyl sulfoxide (DMSO).
20. 0.25% trypsin–EDTA.
21. 0.4% trypan blue/PBS.
22. Antibiotics medium: add 5 mL antibiotic/antimycotic (A5955, Sigma-Aldrich) and 0.5 mL gentamycin (G1397, Sigma-Aldrich) to 500 mL High Glucose Dulbecco's Modified Eagle's Medium (D6546, Sigma-Aldrich). Medium can be stored 14 days refrigerated or aliquoted and frozen for a longer period of time at −20 °C.
23. 0.25% Trypsin/PBS: dissolve 0.25 g trypsin in 100 mL PBS and filter sterilize using a 0.2 μm pore size filter. Prepare fresh before use.
24. 3T3 cell medium: mix 500 mL Dulbecco's Modified Eagle's Medium without pyruvate with 50 mL fetal bovine serum and 5 mL penicillin–streptomycin (P4333, Sigma-Aldrich).
25. Keratinocyte growth medium: 300 mL High Glucose Dulbecco's Modified Eagle's Medium, 150 mL Ham's F12, 50 mL fetal bovine serum, 10 mL L-glutamine, 5 mL penicillin–streptomycin (pen-strep), and 5 mL adenine hydrochloride hydrate. Supplement with 0.5 mL insulin from bovine pancreas, 1 mL hydrocortisone, 0.5 mL triiodothyronine (T3), 0.5 mL cholera toxin, and 0.5 mL epidermal growth factor. *See* Table 1 for details.

    Medium can be stored 14 days refrigerated or frozen for a longer period of time.

2.2 3D Epidermal Equivalents

1. Centrifuge.
2. Hemocytometer or automated cell counter.
3. Roundhead tweezers.
4. 37 °C water bath.
5. 37 °C incubator with 5% CO_2.
6. 24-Well cell culture plates, sterile pipettes, tubes, and so on.
7. Nunc 24-well carrier plate system no inserts (141008, Thermo Scientific).
8. Nunc 24-well carrier plate system with 0.4 μm pore size cell culture inserts (141002, Thermo Scientific).
9. 6 mm disposable punch biopsy.

Table 1
Preparation of keratinocyte growth medium component stock solutions

Component	Preparation	Stock	Aliquots	Final
Adenine	Dissolve in H_2O and add concentrated HCL until solution is completely clear	4.86 mg/mL	5 mL	24.3 μg/mL
Cholera toxin	Dissolve in H_2O to 10 U/mL, serial dilute in steps of 10× in H_2O	1.0E−08 U/mL	0.5 mL	1.0E−11 U/mL
EGF	Dissolve in 0.1% BSA/PBS	10 μg/mL	0.5 mL	10 ng/mL
FBS	Ready to use	100%	50 mL	10% (v/v)
Glutamine	Ready to use	200 mM	10 mL	4 mM
Hydrocortisone	Dissolve to 5 mg/mL in 95% EtOH, further dilute in DMEM	0.2 mg/mL	1 mL	0.4 μg/mL
Insulin	Dissolve in H_2O and add concentrated HCL until solution is completely clear	5 mg/mL	0.5 mL	5 μg/mL
Pen-strep	Ready to use	10.000 U/mL	5 mL	100 U/mL
T3	Dissolve to 250 μg/mL in 0.02 M NaOH, further dilute in PBS	1.36 μg/mL	0.5 mL	1.36 ng/mL

Store aliquots at −20 °C until use

10. Phosphate buffered saline (PBS).
11. 0.4% trypan blue/PBS.
12. Rat tail collagen, type I (C3867, Sigma-Aldrich).
13. CnT-Prime medium (CnT-PR, CELLnTEC); aliquot and store at −20 °C.
14. CnT-Prime 3D Barrier medium (CnT-PR-3D, CELLnTEC): aliquot per 60 mL of medium and store at −20 °C.
15. High Glucose Dulbecco's Modified Eagle's Medium (DMEM): aliquot per 40 mL of medium and store at −20 °C.
16. 3D differentiation medium:

    Thaw an aliquot of CnT-Prime 3D Barrier medium and DMEM and combine both aliquots to obtain 100 mL of 60/40 ratio (CELLnTEC to DMEM). Store at 4 °C for a maximum of 2 weeks.

3 Methods

3.1 Primary Adult Keratinocyte Isolation (T: Troubleshooting)

1. Seed 3T3-J2 cells at a density 10,000 cells/cm^2 in 3T3 cell medium. Upon confluency passage cells (every 3 days) with 0.25% trypsin-EDTA (TE) in a 1:3 ratio.

2. One day prior to keratinocyte isolation, harvest and irradiate 3T3-J2 cells with 3295 cGy in 3T3 medium to induce nonfatal growth arrest (*see* **Note 1**). Spin down and resuspend the feeder cells in keratinocyte growth medium. Count the feeder cells and seed at 30,000 cells/cm^2 in keratinocyte growth medium using 75 mL per T225 culture flask (3.3 mL/cm^2 surface) (*see* **Note 2**).
3. For primary adult keratinocyte isolation use freshly excised tissue from plastic surgery (for single patient skin biopsies start directly at **step 4**) (*see* **Note 3**). Pin the tissue to a Styrofoam board using injection needles and clean the skin surface with disinfectant. Take 6 mm diameter full-thickness biopsy punches and place them into a generous volume of antibiotic/antimycotic medium for 4 h at 4 °C. From now aseptic handling within the flow hood is required.
4. Discard the medium and add 0.25% Trypsin/PBS to the biopsies. Add 1 mL trypsin–PBS per 2–3 biopsies; for example, collect 100 biopsies (*see* **Note 4**) in 50 mL trypsin solution. Incubate 16–20 h (overnight) at 4 °C.
5. Discard trypsin–PBS solution from the biopsies and stop the enzymatic reaction by the addition of 10% (v/v) fetal bovine serum (SH30071.03, GE Healthcare Life Sciences).
6. Place the skin biopsies into a medium size sterile petri dish and add keratinocyte growth medium for full coverage. Use two pairs of roundhead tweezers. Grab a single biopsy firmly at the dermal part, right underneath the epidermis leaving the epidermis side exposed. Hold the biopsy under medium level in de petri dish and peel off the epidermal sheet with the other tweezers and carefully scrape the surface of the biopsy (for 5–10 s) with the rounded part of the tweezers. This scraping is required to harvest the basal keratinocytes (*see* **Note 5**).
7. Repeat this procedure for every individual biopsy and put scraped biopsies into a separate petri dish.
8. Every 30 min, take the cell suspension from the petri dish into a tube and put on ice to keep the keratinocytes metabolically inactive until all biopsies are processed.
9. Add fresh keratinocyte growth medium to the petri dish before scraping more biopsies. Optional: wash the scraped biopsies once with keratinocyte growth medium to maximize cell yield before discarding them.
10. Use a 100 μm pore size cell strainer to obtain a homogeneous cell suspension. Wash the cell strainer once with keratinocyte growth medium for maximum cell yield.

11. Centrifuge the cell suspension at 400 × *g* for 7 min, discard the supernatant and resuspend the pellet in keratinocyte growth medium.
12. Count the cells using a hemocytometer. Check for cell viability by Trypan blue (1:1 cell suspension/Trypan blue) and morphology. Only count small viable circular cells (*see* **Note 6**).
13. Seed the keratinocytes onto the feeder cells at a density of 50,000 cells/cm^2 (e.g., 11.25 million cells per T225 flask) by adding the cell suspension to the conditioned keratinocyte growth medium in the flask.
14. Leave the flask for at least 48 h to enable keratinocyte adhesion. Thereafter, check the cells daily to monitor cell growth and morphology.
15. Refresh the culture medium with keratinocyte growth medium supplemented with EGF (0.5 mL EGF per 500 mL keratinocyte growth medium) when the first small colonies have formed in between the feeder cells. Thereafter, refresh with EGF supplemented medium every other day, or even daily when almost confluent, depending on the growth speed and nutrient turnover (*see* **Note 7**). Due to keratinocyte proliferation, colonies will enlarge, fuse and overgrow the feeder cells (T1).
16. Harvest keratinocytes at 95% confluency, or earlier in case of suspected differentiation (*see* **Note 8**).
17. Prepare cryovials, freezing containers, and chilled 20% DMSO in keratinocyte growth medium.
18. Wash cells once with PBS, add prewarmed 0.25% TE to just cover the flask surface. Carefully monitor under the microscope when the feeder cells in between the keratinocytes detach. Feeder cells detach quickly and should be discarded before keratinocyte harvesting.
19. Remove the detached feeders and add fresh 0.25% TE to harvest keratinocytes using 150 μL TE/cm^2.
20. Place flasks at 37 °C to increase enzymatic activity. Regularly check for detachment and tap the flask to help loosen the cells (*see* **Note 9**).
21. Collect the cell suspension, add equal volumes of keratinocyte growth medium to stop the enzymatic reaction. If required, wash the flask once to collect all cells.
22. Centrifuge the cell suspension 7 min at 400 × *g*, discard supernatant and carefully resuspend in keratinocyte growth medium.

23. Count cells with trypan blue (*see* **step 3**). Bring cell suspension to a density of two million cells per mL by the addition of keratinocyte growth medium.
24. Gently add an equal volume of 20% DMSO in keratinocyte growth medium to the cell suspension to obtain a final DMSO concentration of 10% and final cell concentration of one million per mL.
25. Keep the cell suspension on ice to prevent cytotoxicity by DMSO. Work quickly to aliquot in cryovials and place cryovials into the freezing container. Store freezing container at −80 °C for at least 4 h or overnight. For longer storage transfer cryovials to a liquid nitrogen storage facility.

3.2 3D Human Epidermal Equivalents (HEEs)

For optimal results, antibiotics should be omitted during the entire culture period. Alternatively, antibiotics may be used only during the final days of the air–liquid interface culture when epidermal differentiation is already at an advanced stage. Antibiotics may hamper keratinocyte growth and epidermal stratification [10]. Accordingly, aseptic culture methods are of utmost importance to minimize contaminations.

1. Prepare 100 μg/mL rat tail collagen solution in cold PBS. Use a roundhead tweezer to bring the inserts to the lowest position of the transwell system. Pipet 150 μL of the collagen solution into each insert. Incubate at 4 °C for at least 1 h for the collagen to adhere to the filter (*see* **Note 10**).
2. Thaw cells, quickly transfer to a 50 mL tube and slowly add at least 10 volumes of cold PBS dropwise into the 50 mL tube while shaking to dilute the DMSO to a minimum of 1%.
3. Spin cells down at 400 × *g* for 7 min, discard the supernatant, loosen the cell pellet by gentle shaking and tapping and then resuspend cells in CnT prime medium (*see* **Note 11**).
4. Count the cells using a hemocytometer. Check for cell viability by trypan blue (1:1 cell suspension/trypan blue) and morphology. Only count small viable circular keratinocytes (*see* **Note 6**).
5. Dilute the cell suspension to a concentration of one million cells per mL in CnT prime medium (*see* **Note 12**).
6. Add 600 μL of CnT Prime medium into each basolateral chamber (outside well) of a 24-well cell culture plate.
7. Carefully remove the collagen coating solution from the insert. For this, lift-up the carrier plate, hold it at an angle and use a 200 μL pipette to remove the collagen. Wash the inside of the insert with 150 μL PBS and carefully remove the PBS (*see* **Note 13**).
8. Transfer the lid and carrier plate to the 24-well plate filled with CnT Prime medium (**step 6**).

9. Homogenize the cell suspension and slowly add 150 μL (150,000 cells from a suspension of one million cells per mL) dropwise into the insert (*see* **Note 14**).
10. Add 300 μL CnT Prime medium to the basolateral chamber to yield a total of 900 μL. Incubate at 37 °C with 5% CO_2 for 48 h (*see* **Note 15**).
11. After 48 h, change the medium to 3D differentiation medium. First add 600 μL of the 3D differentiation medium into the wells of a new 24-well plate. Lift the carrier plate and hold it at an angle to carefully remove the medium inside the insert (*see* **Note 13**). Transfer the carrier plate to the 24-well cell culture plate containing fresh differentiation medium. Carefully add 150 μL 3D differentiation medium into the insert and subsequently add 300 μL medium into the basolateral chamber. Incubate at 37 °C with 5% CO_2 for 24 h (*see* **Note 15**, T2).
12. The next day, the HEEs are lifted to the air–liquid interface to induce epidermal differentiation and stratification. Pipet 1600 μL 3D differentiation medium into the wells of a new 24-well plate. Again, carefully remove the medium inside the insert as described above (*see* **Note 13**). Then the inserts must be brought to the highest position of the transwell system. Transfer the carrier plate to the cell culture plate with fresh 3D differentiation medium. Incubate at 37 °C with 5% CO_2 (*see* **Note 16**).
13. HEEs are cultured in 3D differentiation medium until the end of the culture period and are refreshed every other day (T2, T3). For regular cultures, HEEs are fully differentiated and can be harvested from day 8 of the air–liquid interface culture onwards.
14. For harvesting, take the insert from the carrier plate and place into a petri dish. Aseptic handling can be neglected from this point on. Punch out the middle of the filter using a biopsy punch. Do not rotate the biopsy punch but only gently press down to not cause any damage to the tissue (*see* **Note 17**). The filter, containing the HEE, should come lose from the plastic support and can be processed for readout analysis.

General Remarks

- The above written protocol refers to 24-well inserts. If a larger culture area is desired, 12-well inserts and well plates may be used (*see* **Note 18**).
- Besides primary human keratinocytes, immortalized N/TERT keratinocytes may be used to obtain HEEs[7] (*see* **Note 19**).

3.3 Readout Analyses

3.3.1 Proliferation Analysis

Besides the conventional staining of proliferative cells by immunohistochemistry or fluorescence (e.g., Ki67 marker expression), cell proliferation during the cell culture can be monitored and quantified by supplementing the differentiation medium with 5-ethynyl-2′-deoxyuridine (EdU) before harvesting. EdU is a dynamic fluorescent tracer which is incorporated into newly synthesized DNA during the S-phase of the cell cycle [11]. After processing for histology and sectioning, EdU incorporation in the HEEs is visualized by a Click-it EdU kit (Life Technologies) using any Alexa-Fluor label, combined with 4′,6-diamidino-2-fenylindole (DAPI) staining of all nuclei according to manufacturer's instructions (Fig. 1).

3.3.2 Dye Permeation

To detect major outside-in stratum corneum defects, Lucifer yellow (Lucifer Yellow CH dipotassium salt; L0144, Sigma-Aldrich) can be topically applied (1 mM in H_2O; 5 μL drop) onto the stratum corneum of the HEEs during the culture prior to harvesting. In case of a defect, the fluorescent dye will penetrate the stratum corneum and epidermis after 2 h of incubation (Fig. 2). Lucifer yellow can be visualized at 488 nm wavelength in formalin-fixed paraffin embedded HEEs.

To visualize the inside-out barrier and the position of tight junctions, EZ-Link Sulfo-NHS-LC-LC-Biotin (21335, Thermo Scientific) can be applied on the filter-side of the HEEs. Therefore the inserts are flipped upside-down and the tracer is applied on the bottom of the filter for 60 min. The penetration of this dye is restricted by tight junctions in the stratum granulosum. EZ-Link Sulfo-NHS-LC-LC-Biotin can be visualized with streptavidin Alexa Fluor (e.g., Alexa Fluor 594) in formalin-fixed paraffin embedded HEEs (Fig. 3).

Transepidermal Water Loss

The inside-out barrier function can be monitored by transepidermal water loss (TEWL) measurement. TEWL is a widely accepted parameter of skin barrier function in vivo, and patients suffering

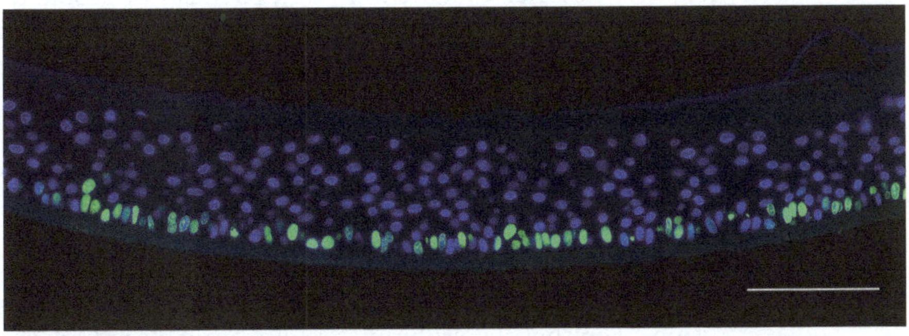

Fig. 1 Basal layer keratinocyte proliferation in HEE visualized by EdU incorporation. EdU is being incorporated into the proliferating keratinocytes of the stratum basale for 24 h (from day 7 to 8) and visualized with Alexa Fluor 488 (green nuclei). All nuclei are stained with DAPI (Blue nuclei). Scale bar = 100 μm

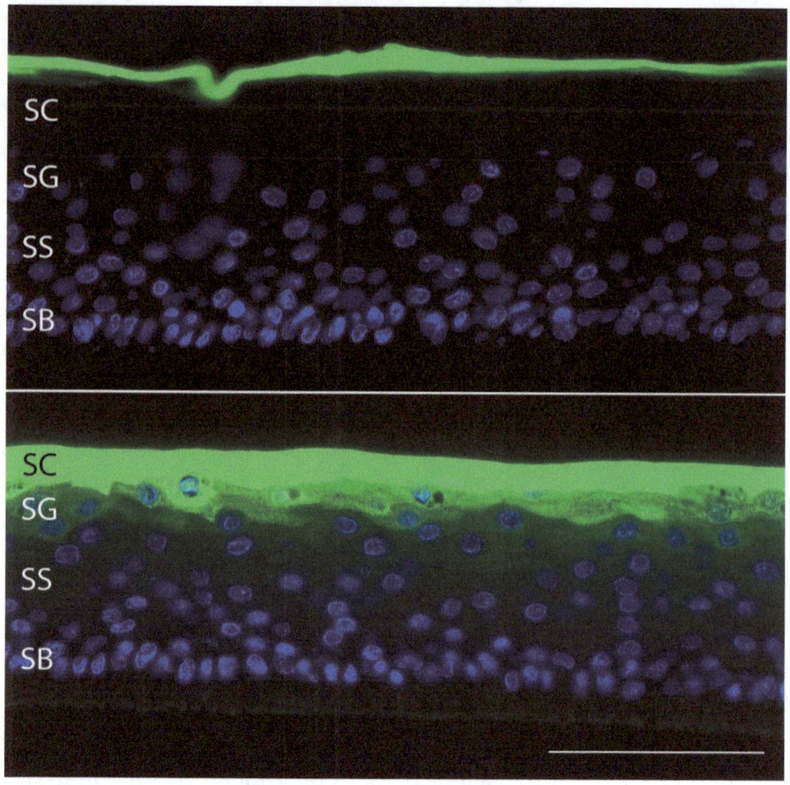

Fig. 2 Outside-in Lucifer yellow dye penetration. Topically applied Lucifer yellow (green fluorescent dye) remains on the uppermost stratum corneum layer of a fully differentiated HEE with a proper epidermal barrier (upper photograph). Major disruption (e.g. by chemical application) or absence of a stratum corneum (lower photograph) leads to extended dye permeation into the stratum corneum and into the viable layers of the epidermis. For quantitative analysis of skin barrier function, diffusion studies are considered the optimal experimental assay. *SC* stratum corneum, *SG* stratum granulosum, *SS* stratum spinosum, *SB* stratum basale. Nuclei are stained with DAPI (blue color). Scale bar = 100 μm

from skin barrier dysfunction (e.g., psoriasis or atopic dermatitis) have higher TEWL than healthy skin. For organotypic skin models, TEWL can be measured (e.g., using the AquaFlux Model AF200 closed chamber with in vitro adapter) after a sufficient acclimatization period outside the incubator to prevent biased measurements due to fluid evaporation. Note that in vitro TEWL values are, in general, higher than that for in vivo skin.

Transepithelial Electrical Resistance

Similar to in vitro studies on internal barrier organs (e.g., lung or gut), transepithelial electrical resistance (TEER) can be measured using an epithelial voltohmmeter (also available in high throughput formats) to follow the HEE formation in real time, and longitudinally study barrier function and repair. Important to note is that

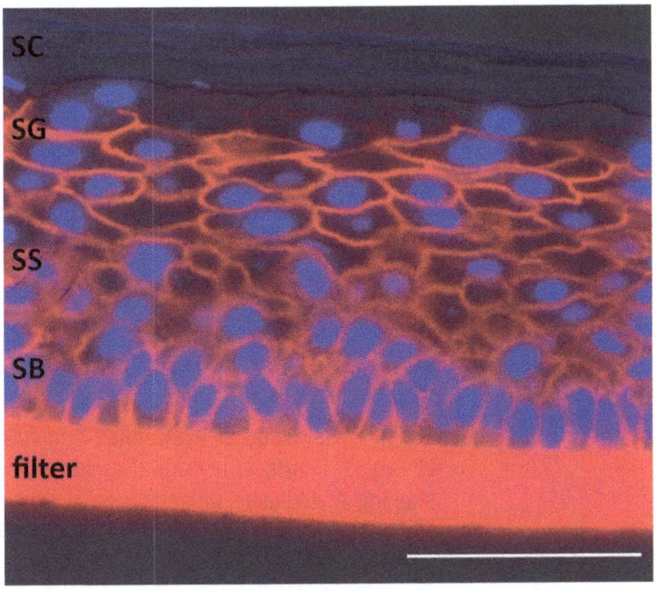

Fig. 3 Inside-out EZ-Link Sulfo-NHS-LC-LC-Biotin penetration. EZ-Link Sulfo-NHS-LC-LC-Biotin is applied at the basal layer side of the HEE and intercellularly through the epidermis (red color). Diffusion towards the stratum corneum is withdrawn by the tight junctions of the stratum granulosum. Photograph shows overlay picture of brightfield and fluorescence image. SC stratum corneum, SG stratum granulosum, SS stratum spinosum, SB stratum basale. Nuclei are stained with DAPI (blue color). Scale bar = 100 μm

besides tight junction formation, the thickness of the stratum corneum also contributes to rising TEER values, meaning that during HEE culture the TEER values will continue to rise with increasing stratum corneum formation.

Tissue Morphology, Gene and Protein Expression

The HEE biopsy can be harvested for multiple purposes: for formalin fixation and paraffin embedding, take a 3 mm punch biopsy and divide the rest of the HEE for, for example, gene expression analysis and western blotting. Of course, this can be adjusted depending on the required analyses.

Sections of formalin-fixed, paraffin-embedded punch biopsies can be stained with hematoxylin eosin for morphology assessment (Fig. 4) and by indirect immunoperoxidase technique with avidin–biotin complex enhancement (Vectastain Laboratories) or immunofluorescence for protein expression.

For gene expression analysis, the HEE can be directly placed in a designated RNA lysis buffer (including filter) and stored at −80 °C until further processing. Prior to RNA isolation (using a standard total RNA extraction kit according to manufactures instruction), take the plastic filter out of the lysis buffer and mechanically dissociate the HEE (using a micro pestle or needle and syringe) to maximize RNA yield.

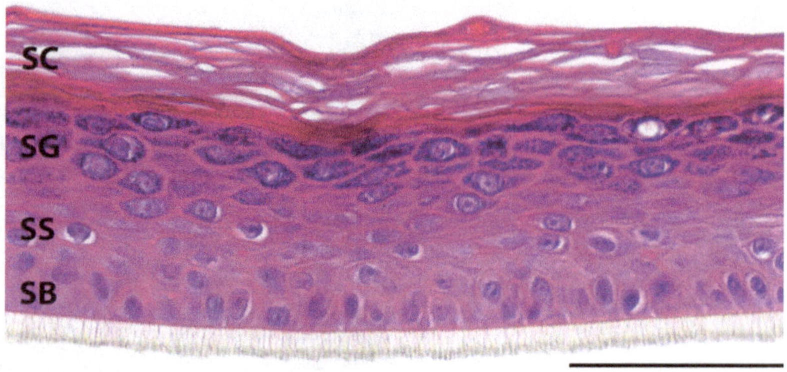

Fig. 4 Morphology of a stratified and fully differentiated HEE after 8 days of air–liquid interface culture. Hematoxylin (nuclei stained in blue) and eosin staining (cytoplasm stained in pink) reveals all layers of the epidermis, from the proliferating basal cells attached to the porous filter membrane up to the cornified cells of the stratum corneum. *SC* stratum corneum, *SG* stratum granulosum, *SS* stratum spinosum, *SB* stratum basale. Scale bar = 100 μm

For western blotting, the HEE can be directly placed in a designated cell lysis buffer (including filter). The use of protease and/or phosphatase inhibitors is recommended to prevent protein breakdown or dephosphorylation. The HEE will readily detach from the plastic filter, and the filter can be removed for further processing of the tissue lysate. Sonication may improve detection of cross-linked stratum corneum proteins.

After harvesting the HEEs, the conditioned culture medium may be stored at −20°C for protein analysis by ELISA, Luminex, or equivalent assays and/or directly plated on blood agar plates to check for possible infections.

4 Notes

1. Alternatively to gamma irradiation 3T3 cells can also be treated with 10 μg/mL Mitomycin C for 2 h upon confluency to cause growth-arrest and then be seeded as feeder cells.
2. Feeder cells can also be stored at −20 °C for later use. Not the living cells but the extracellular matrix proteins and factors secreted into the medium are essential for keratinocyte growth. Separately freeze the conditioned medium and the feeder-coated culture flasks and store until needed. Before use, thaw the flask and wash very stringently with PBS to eliminate all feeder cells attached to the flask surface. Add thawed conditioned medium and proceed with seeding the isolated keratinocytes (**step 5**).

3. For isolation of keratinocytes from specific patients or volunteers, take a 4 mm punch biopsy from the upper buttock and directly proceed with **step 4** of the protocol. Given the low cell yield, prepare feeder cells in a 6-well plate for seeding of isolated keratinocytes. Seed at a minimum of 50,000 cells/cm^2. Proceed as described and harvest cells for a second passage when (almost) confluent. Reseed keratinocytes onto new feeder cells at a density of 5000 cells/cm^2, for example, about two million cells per well are passaged to three T125 flasks. Harvest this P1 cell population for storage in liquid nitrogen as described.
4. Estimated, from 40 cm^2 skin tissue about 100 biopsies of 6 mm diameter can be gained when punches are collected tightly next to another.
5. Do not scrape too vigorously to prevent fibroblast from the dermal part of the biopsy to contaminate the keratinocyte suspension and cultures.
6. Critically evaluate the quality of the keratinocytes and only count those ones that are small, evenly rounded and not stained by trypan blue. Larger cells with atypical morphology (feeder remnants or differentiated keratinocytes) should be neglected for cell count. Per 6 mm punch biopsy on average 0.5–1 million keratinocytes can be isolated.
7. Phenol-red in the medium (included in DMEM) helps to monitor the pH of the medium. Lower pH (orange-yellow color) indicates that medium needs to be refreshed.
8. In case keratinocytes start to differentiate (reduced nucleus to cytoplasm ratio, more elongated cell shape and multilayering of keratinocytes) cells should be harvested even in case of preterm confluence rates. Keratinocyte differentiation must be prevented to retain proliferative capacity of the cells for later use, for example, in 3D epidermal equivalent cultures.
9. Patiently wait until cells have detached, do not vigorously spray off the cells as this may cause sheering and reduced viability.
10. Keep collagen cold to prevent coagulation. Prevent bubbles when pipetting to ensure evenly coating.
11. Final concentration of one million cells per mL is required, therefore, resuspend in a low volume.
12. For prolonged period before seeding, keep cell suspension on ice.
13. Make sure not to damage the filter with your pipette tip when removing surplus collagen solution and PBS from the bottom and the border of the insert.

14. The last drop may stick to the pipette tip. Release it by touching the liquid in the insert. Prevent the formation of air bubbles.
15. Prevent the formation of air bubbles inside the well and below the insert. Lifting the carrier plate should help removing the air bubbles. The incubator should be leveled and the inserts should be carefully placed in the 24 wells plate to ensure homogeneous distribution of cells across the insert area.
16. The medium level should exactly reach the bottom of the insert to ensure nutrition of the keratinocytes. Check daily and adjust if needed. Remove any air bubbles if present. The cultures should not be left unrefreshed during the weekend. Adhere to the protocol and refresh medium every other day to prevent low nutrient levels and pH alterations (indicated by orange color of the medium).
17. After a few biopsies, the punch biopsy can become blunt and should be replaced. Alternatively, a scalpel can be used for cutting out the filter membrane. However, this is not recommended as the HEE can more easily be damaged and come loose from the filter.
18. The following materials are suggested for 12-well inserts:
    - 12-Well cell culture plate (any supplier).
    - 12-Well ThinCert plate, deep wells (665110, Greiner Bio-One).

    The following changes in the protocol are required:
    - Step 1: Pipet 300 μL of collagen solution into the insert.
    - Step 9: Seed 300,000 cells (in 300 μL).
    - Steps 6 and 10: Pipet in total 1.2 mL medium inside the 12-well (400 μL + 800 μL) during the submerged phase.
    - Step 12: Use the ThinCert deep well plate for the air–liquid interface. Pipet 3.8 mL of 3D differentiation medium inside the well.
    - Step 14: Punch out the middle of the filter/HEE using a 10 mm biopsy punch.
19. The following changes to the protocol are required for N/TERT cells:
    - Steps 3, 5, 6, and 10: Keratinocyte Serum Free Medium (KSFM; 17005-034, Thermo Scientific) is used for seeding of the cells and during the first 48 h of the submerged phase, instead of CnT Prime.
    - Step 13: N/TERT cells generally need a few additional days at the air–liquid interface to develop into a fully stratified HEE. Harvesting is recommended from day 10 onward.

5 Troubleshooting

T1: Excessive scraping can cause release of dermal fibroblasts from the punch biopsies. Due to their high growth rate these can overgrow the keratinocytes during culture and thereby cause contamination of the keratinocyte cell pool. If fibroblasts grow out during cell culture, spray the cell layer with trypsin–EDTA regularly during a few minutes to detach the fibroblasts. Regularly check under the microscope for detachment of fibroblast and when keratinocytes start rounding, directly stop the enzymatic reaction with FBS, wash the cell layer and refresh the keratinocyte growth medium. Place flasks back in the incubator for further keratinocyte proliferation.

T2: Yellow and cloudy medium can indicate microbial contamination. In case of contamination: prevent spreading of contamination to neighboring inserts by carefully sealing the infected inserts with Parafilm and bring the other inserts over to a new plate with fresh medium. Contaminated medium can be plated on bacteria growth agar plates to check for type of infection. Figure 5 depicts the tissue morphology of a contaminated culture.

T3: During air–liquid interface culture and the process of cornification, the HEE surface should become dry. However, during the first few days of air–liquid interface culture, some moisture may remain at the rim of the insert, which should be carefully removed with a small volume pipette tip. If HEEs remain moist after day 4 or 5 of the air–liquid interface cultures, this may indicate improper epidermal development and lack of cornification.

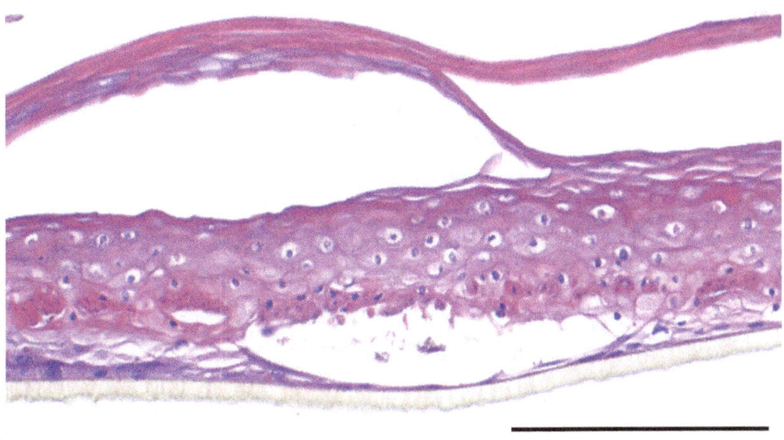

Fig. 5 Detrimental morphologic alterations by bacterial contamination. Hematoxylin and eosin staining shows drastic epidermal defects like structural damage and apoptotic cells indicated by bright pink cytoplasm and small pyknotic nuclei. Scale bar = 100 μm

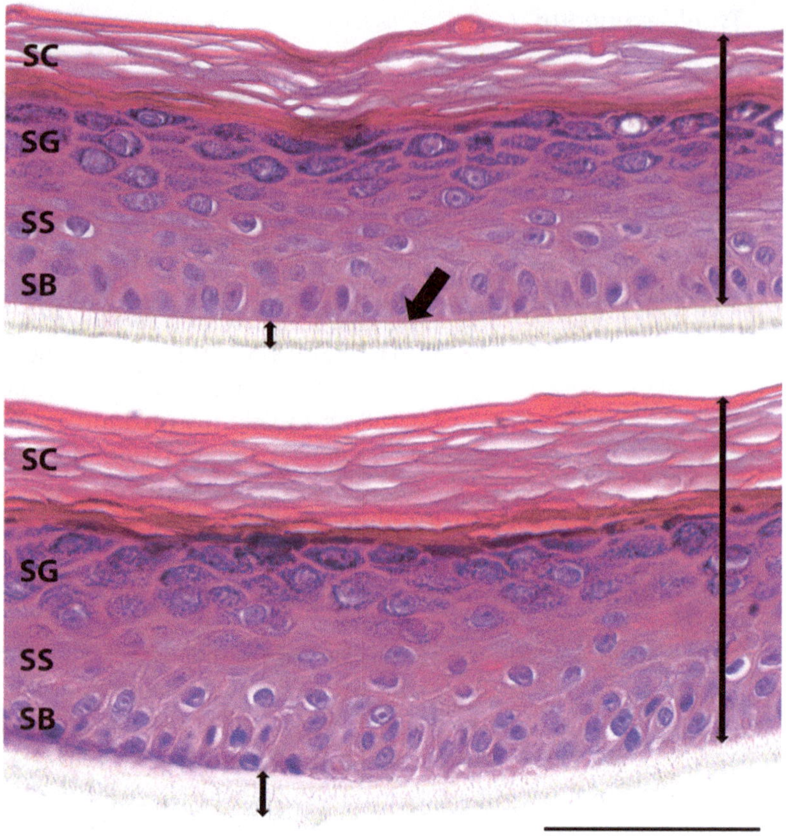

Fig. 6 Orientation of HEE sectioning influences morphological assessment. Both photographs are taken from the same HEE section. Top: true transverse section indicated by the straight vertical line pattern and even thickness of the porous filter membrane (small double headed arrow), as well as the straight line separating the basal layer and the filter (one-headed arrow). Bottom: oblique section indicated by the increased filter thickness (small double headed arrow) and disordered structure of the filter and basal layer. This imperfect orientation results in visually elongated cells in the basal layer and an overall thicker epidermis (indicated by the large double headed arrow). Scale bar = 100 μm

T4: Due to fixation and tissue processing, dehydration of the tissue occurs leading to curled or folded edges of the HEE. Do not try to even out but embed the HEE as it is. Ensure a practical positioning of the HEE for transverse sectioning. Oblique sectioning may influence the tissue morphology and interpretation of results. Figure 6 shows the difference between correct (top) and improper (bottom) orientation of the HEE. The structure and thickness of the filter is a good measure for this quality assessment.

References

1. Rheinwald JG, Green H (1975) Serial cultivation of strains of human epidermal keratinocytes: the formation of keratinizing colonies from single cells. Cell 6(3):331–343
2. Regnier MP, Prunieras M, Woodley D (1981) Growth and differentiation of adult human epidermal cells on dermal substrates. Front Matrix Biol 9:4–35
3. Riebeling C, Luch A, Tralau T (2018) Skin toxicology and 3Rs-Current challenges for public health protection. Exp Dermatol 27 (5):526–536. https://doi.org/10.1111/exd.13536
4. Niehues H, Bouwstra JA, El Ghalbzouri A, Brandner JM, Zeeuwen P, van den Bogaard EH (2018) 3D skin models for 3R research: the potential of 3D reconstructed skin models to study skin barrier function. Exp Dermatol 27(5):501–511. https://doi.org/10.1111/exd.13531
5. Tjabringa G, Bergers M, van Rens D, de Boer R, Lamme E, Schalkwijk J (2008) Development and validation of human psoriatic skin equivalents. Am J Pathol 173(3):815–823. https://doi.org/10.2353/ajpath.2008.080173
6. Dickson MA, Hahn WC, Ino Y, Ronfard V, Wu JY, Weinberg RA, Louis DN, Li FP, Rheinwald JG (2000) Human keratinocytes that express hTERT and also bypass a p16(INK4a)-enforced mechanism that limits life span become immortal yet retain normal growth and differentiation characteristics. Mol Cell Biol 20(4):1436–1447. https://doi.org/10.1128/mcb.20.4.1436-1447.2000
7. Smits JPH, Niehues H, Rikken G, van Vlijmen-Willems I, van de Zande G, Zeeuwen P, Schalkwijk J, van den Bogaard EH (2017) Immortalized N/TERT keratinocytes as an alternative cell source in 3D human epidermal models. Sci Rep 7(1):11838. https://doi.org/10.1038/s41598-017-12041-y
8. van Drongelen V, Danso MO, Mulder A, Mieremet A, van Smeden J, Bouwstra JA, El Ghalbzouri A (2014) Barrier properties of an N/TERT-based human skin equivalent. Tissue Eng Part A 20(21–22):3041–3049. https://doi.org/10.1089/ten.TEA.2014.0011
9. Maas-Szabowski N, Starker A, Fusenig NE (2003) Epidermal tissue regeneration and stromal interaction in HaCaT cells is initiated by TGF-alpha. J Cell Sci 116(Pt 14):2937–2948. https://doi.org/10.1242/jcs.00474
10. Nygaard UH, Niehues H, Rikken G, Rodijk-Olthuis D, Schalkwijk J, van den Bogaard EH (2015) Antibiotics in cell culture: friend or foe? Suppression of keratinocyte growth and differentiation in monolayer cultures and 3D skin models. Exp Dermatol 24(12):964–965. https://doi.org/10.1111/exd.12834
11. Buck SB, Bradford J, Gee KR, Agnew BJ, Clarke ST, Salic A (2008) Detection of S-phase cell cycle progression using 5-ethynyl-2′-deoxyuridine incorporation with click chemistry, an alternative to using 5-bromo-2-′-deoxyuridine antibodies. Biotechniques 44 (7):927–929. https://doi.org/10.2144/000112812

Chapter 6

Purification of Extracellular Microvesicles Secreted by Dermal Fibroblasts

Isabelle Petit, Ayelet Levy, and Daniel Aberdam

Abstract

Extracellular vesicles (EVs) secreted by all cells are key players in information transfer within a tissue or organism. With their highly cell-specific protein and RNA content, EVs can propagate cellular signals and modulate distant cells' behavior. Dermal fibroblasts are supportive cells for all skin cells and the roles of their EVs start to come to light only recently. In this chapter, we describe a protocol to isolate small EVs from primary human fibroblast culture using classical differential centrifugation methodology.

Key words Extracellular vesicles, Microvesicles, Exosomes, Ultracentrifugation, Dermal fibroblasts

1 Introduction

Skin homeostasis and repair rely on intricate communication between the different cell types residing in the cutaneous tissue: fibroblasts in the dermis, keratinocytes in the epidermis, melanocytes, specialized cells forming the appendages as well as blood vessels and immune cells. Extracellular vesicles (EVs) are secreted bilayered vesicles of various sizes that can be taken up by or activate proximal neighbor cells within the tissue or distant cells through the circulation. EV composition and cargo is highly cell specific and include proteins, RNAs, and miRNAs [1, 2]. While secretion of bilayered vesicles in the extracellular environment has been reported in various normal and pathological systems, few studies have questioned the composition and role of EVs secreted by skin cells. A pioneering article by Raposo and co-workers has documented the role of keratinocyte EVs in skin pigmentation revealing specific UV-induced miRNA incorporation into EVs [3]. Recently, EVs secreted by dermal fibroblasts (DF) have been reported to modulate keratinocyte migration [4] and we have shown that EVs from activated DF increase hair growth through dermal papilla activation [5]. Other studies have focused on the regenerative

Natalia V. Botchkareva and Gillian E. Westgate (eds.), *Molecular Dermatology: Methods and Protocols*,
Methods in Molecular Biology, vol. 2154, https://doi.org/10.1007/978-1-0716-0648-3_6,

© Springer Science+Business Media, LLC, part of Springer Nature 2020

potential of EVs, mostly isolated from various types of cultured mesenchymal stem cells (MSC) cells, on skin rejuvenation and wound healing [6]. Cells shown to have therapeutic potential such as MSC are secreting EVs that exert similar effects suggesting that EVs could replace cells in therapies [7].

In order to investigate the functions of EVs in skin homeostasis, EVs must be isolated from cultured cells. Primary fibroblasts can easily be isolated and amplified in vitro from human dermal biopsies using well-established protocols [8]. In this chapter, we describe differential centrifugation, a common EV isolation technique when dealing with large volume of cell medium to process. It consists of successive centrifugations to eliminate cell debris and large EVs and ultracentrifugation (UC) to pellet the remaining particles with size ranging from 30 to 150 nm (termed small EVs) comprising microvesicles shed by budding of the membrane (ectosomes) or released by fusion of endosomal multivesicular bodies (named exosomes). However, it is important to be aware that UC not only leads to the isolation of nanoparticles that contains a mixture of small EVs of various sizes, contents, and origins (small ectosomes and exosomes indistinctly) but also protein aggregates, soluble proteins, and damaged EVs that can account for bioactivity in functional assays. In an effort to further purify EVs UC can be followed by additional separation techniques such as size-exclusion chromatography, density gradient (both techniques allow for elimination of protein contaminants), or immunoisolation (the latter method, however, uses antibodies against common EV markers such as CD81 and CD63, and therefore isolates only a specific EV population) [9]. In a seminal paper, the *International Society of Extracellular Vesicles* (ISEV) proposes extensive guidelines and invaluable recommendations about EV isolation, quantification, and characterization techniques and functional analyses as well as provides minimal requirements for EV-related publications [10]. In this chapter, we describe only basic DF-EV isolation by UC and refer the readers to additional techniques when higher EV purity is required and for EV quantification and characterization [10–12]. The protocol described here has been adapted from Théry et al. [13] and also includes washing and sterilization procedures if using reusable UC tubes.

Of note, dermal fibroblasts are relatively big cells and consequently a large cell amplification step is required in order to obtain enough EVs. This characteristic implies that one must be aware that incubator space, medium quantities, time, and workload are all scaled up when compared to other cell types that are smaller and/or producing more EVs.

2 Materials

2.1 Cell Culture

1. Dermal fibroblast medium: DMEM high glucose (4.5 g/L) supplemented with 10% fetal bovine serum (FBS) and 2 mM glutamine.
2. EV-depleted dermal fibroblast medium (*see* Subheading 3.1).
3. Dermal fibroblasts isolated from primary human adult skin biopsy [8]. Use passage less than 10.
4. 150 mm plastic culture dishes.

2.2 sEV Purification

1. Dulbecco's phosphate buffered Saline (DPBS, without calcium and magnesium) (*see* **Note 1**).
2. 50 ml conical tubes.
3. 25 ml pipettes.
4. 65 ml reusable thick-wall polycarbonate tubes (Beckman, cat. 355622) with aluminum lids and adapters if using 45Ti rotor (*see* **Note 2**). Alternatively, 30 polypropylene tubes can be used if using swing 32Ti.
5. Tube racks.
6. Siliconized or low-binding Eppendorf tubes (1 and 0.65 ml).

2.3 Tube Washes and Sterilization

1. Detergent (Nalgene L900).
2. Hydrogen peroxide (H_2O_2).
3. Ethanol.
4. 2 L beakers and 0.5 L beakers.
5. Distilled water (6 L) and sterile dH_2O (6 L).
6. Sealable bags such as autoclave bags.

2.4 Equipment

1. Refrigerated tabletop centrifuge with fixed angled rotor for 50 ml tubes and reaching 10,000 × *g* (Eppendorf 5804R/5810R with rotor F-34-6-38 or equivalent) (*see* **Note 3**).
2. Ultracentrifuge Beckman Coulter Optima LE-80K or equivalent.
3. Fixed rotor 45Ti or swing 32Ti (Beckman Coulter) or equivalent (*see* **Note 4**).
4. Weight scale.

3 Methods

The time line of the procedure we used is presented in Fig. 1a. Details such as medium volumes and numbers of UC runs are indicated in Fig. 1b. All steps should be performed in sterile

A

Day	Actions
-7 *(Wed)*	Seeding 0.5-1x10^{6} cells/150mm x8-13 plates
	100K runs for serum depletion
0 *(Wed)*	Seeding 0.5-1x10^{6} cells/150cm^{2} x40-80 plates
5 *(Mon)*	Medium change to 10% EV-depleted
7 *(Wed)*	EV-containing medium collection 0.3K, 2K, 10K, 1st 100K
8 *(Thur)*	100K
9 *(Fri)*	2nd 100K (wash) Aliquots

B

		Serum EV depletion			EV isolation		
Plates	Vol/plate	Total Vol 10% depl. required	20% depl. prepared	Number of depletion runs (ON)	0,3K, 2K, 10K each:	Number of runs 100K-1st	Number of runs 100K-2nd (wash)
50	20 ml	1000 ml	600 ml	2 (45Ti) or 4 (32Ti)	20x 50 ml-tubes	3 (45Ti) or 6 (32Ti)	1

Fig. 1 EV isolation from dermal fibroblasts. (**a**) Time outline of the experimental procedure describing days and actions performed. (**b**) Details of the experiments including volume of medium and number of UC runs needed for the purification of EVs from 50 × 150 mm culture plates of DF cells

conditions in a laminar flow hood in order to obtain sterile EVs. All centrifugations are performed at 4 °C in precooled centrifuges.

3.1 Depletion of Serum EVs

1. Fill UC tubes with DMEM with 20% serum (without glutamine) up to 65 ml. Weigh tubes and equalize weight using DMEM 20% serum (so that there is less than 0.03 g difference).
2. Ultracentrifuge 18 h at 100,000 × *g* (100K) (*see* **Note 5**).
3. Remove with caution the lids from all the tubes and collect first 25–30 ml while tubes are still in the rack and transfer supernatants to sterile bottle. Hold with one hand the first tubes tilting on the side where the pellet is supposed to be (pellet facing down). Slowly collect the rest of the supernatant with a

25 ml pipette placed on the opposite side of the pellet location leaving 3–5 ml. Repeat the procedure for all tubes. Tubes and lids are washed immediately (*see* Subheading 3.4).

4. EV-depleted 20% serum is kept at 4 °C and the procedure is repeated until the needed volume is reached.
5. EV-depleted 10% medium is prepared by diluting the 20% medium with DMEM. Glutamine (2 mM final) is added and the medium is filtered through 0.22 mm membrane. EV-depleted medium can be kept at 4 °C for up to 3 weeks.

3.2 Cell Amplification

1. Grow DF cells in high glucose DMEM supplemented with 10% FBS and glutamine in 150 mm dishes by seeding 0.5–1.10^{6} cells every 7 days. We recommend a minimum of 40 dishes per EV purification experiment (range 40–80 plates).
2. Medium is changed to 10% EV-depleted 48 h before the day of collection (*see* timeline Fig. 1a); aspirate medium and add 7 ml PBS, swirl the dish.
3. Aspirate PBS and add 20 ml 10% EV-depleted medium.
4. Return to incubator for 48 h (*see* **Note 6**).

3.3 EV Purification

1. Collect DF conditioned medium in 50 ml tubes and discard the cells except three dishes replenished with new DF medium for cell count later on.
2. Centrifuge 300 × *g* for 10 min at 4 °C.
3. Collect carefully supernatant by holding tubes straight up leaving 1 ml of medium above pellet and transfer to new 50 ml tubes.
4. Centrifuge 2000 × *g* for 20 min at 4 °C.
5. Collect carefully supernatant by holding tubes straight up leaving 1 ml of medium and transfer to new 50 ml tubes. Mark tubes at approximate location where the pellet will be.
6. Centrifuge 10,000 × *g* (10K) for 30 min at 4 °C in fixed-angle rotor table centrifuge (*see* **Note 3**).
7. Remove with caution the lids from all the tubes and collect first 25–30 ml while tubes are still in the rack and transfer supernatants to sterile bottle. Hold with one hand the first tubes tilting on the side where the pellet is supposed to be (pellet facing down). Slowly collect the rest of the supernatant with a 25 ml pipette placed on the opposite side of the pellet location. Collect carefully 10K supernatant while the tube tilting on pellet side and placing the pipette opposite side of the pellet leaving 1–2 ml. Transfer to UC tubes. Close with appropriate lids, weight tubes and equilibrate (less than 0.03 g differences) (*see* **Note 7**).

8. Place in cooled 45Ti rotor and note the position of the external side where the pellet will be located. Store the rest of the collected 10K supernatants at 4 °C in UC tubes or 50 ml tubes.
9. Ultracentrifuge 90 min at 100,000 × *g* at 4 °C.
10. Remove cells from the three dishes kept aside using trypsin solution. Wash with DF medium and count independently. Note the average number of cells per plate and the total number of producing cells for the record.
11. Remove with caution the lids and collect first 25–30 ml while tubes are still in the rack, transfer supernatants to waste bottle. Hold with one hand the tube tilting on the opposite side where the pellet is supposed to be (pellet facing up). Slowly collect the rest of the supernatant with a 25 ml pipette placed on the opposite side of the pellet location. Discard the supernatant. Still holding the tube, then use a 1 ml micropipette to remove the last drop of liquid. Put back the tube in the rack. Repeat the procedure for the other tubes.
12. Add 3 ml of cold PBS in every tube. Take one tube, resuspend thoroughly with a 1 ml micropipette by gently flushing on the area where the pellet is located (*see* **Note 8**). With fixed-angle 45Ti rotor the pellet is located where the side and the rounded bottom meet. DF EV pellet can hardly be seen and appears translucent. Pool together all washes to one of the tubes.
13. Repeat this step with another 3 ml of cold PBS. Transfer to the collecting tube. If reusable, place the empty tubes in a water-containing beaker until the washing procedure or directly in the detergent-containing beaker. Place the lids and adapters in distilled water-containing beaker (*see* Subheading 3.4).
14. Keep the collecting tube containing EVs at 4 °C and repeat ultracentrifugation/EV pellet collecting rounds with the rest of the 10K supernatant.
15. The collecting tube(s) is/are filled with cold PBS, equilibrated and centrifuged 90 min at 100,000 × *g* at 4 °C.
16. Remove the lid(s) carefully. Collect 25–30 ml while the tube (s) is/are still in the rack, transfer supernatant to waste bottle. Hold with one hand the tube tilting on the opposite side where the pellet is supposed to be (pellet facing up). The EV pellet can usually be seen at this step and appear translucent. Collect carefully the remaining last drop with a 1 ml pipette.
17. Resuspend the EV pellet carefully with 200–400 μl ice-cold PBS by extensively flushing the area where the pellet is located (*see* **Note 8**). Transfer to a 1 ml low binding tube. Repeat with an additional volume of 200–400 μl PBS. Note the final volume for record.

A

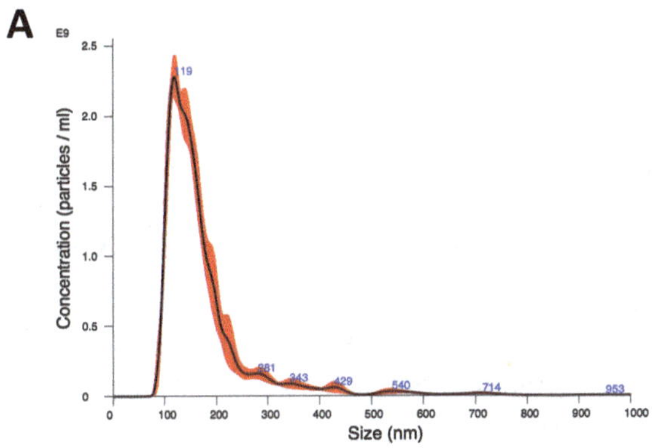

Median size (nm)	156.6±15.2
Median mode (nm)	104.9±13.4
Total EV number (for 50 plates)	2.75±0.7 x10^{11}
EV number/cell	1115±616

B

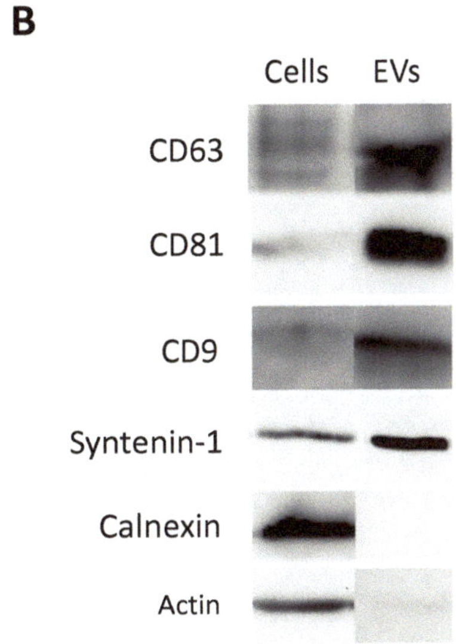

Fig. 2 DF-EV quantification and characterization. (**a**) Quantification of DF-EVs by NTA analysis. Particle size was determined by automatic tracking of particle diffusion due to Brownian motion using NanoSight LM14 instrument (Malvern). NTA graph was obtained by NTA3.1 software averaging data from five videos of 60 s for each of two different dilutions of DF-EVs. The table below presents the average median size and mode as well as total number of particles and number of particles per cell from 12 independent purification experiments. (**b**) Characterization of DF-EVs by SDS-PAGE western blot analysis. DF-EVs (typically 10–20 μl or 1-- 5×10^9 particles) and DF cell lysate (20 μg) were incubated with SDS-PAGE sample buffer containing reducing agent or not (antibodies for tetraspanin proteins CD63, CD9, and CD81 often require nonreducing conditions), boiled for 5 min at 95 °C and run on 10% SDS-PAGE gel. Antibodies used are: CD63 (BD Biosciences, #556019), CD81 (Santa Cruz Biotechnology, #166029), CD9 (Millipore, #CBL-162), syntenin-1 (Abcam, #133267), calnexin (Santa Cruz Biotechnology, sc-11397), and actin (Santa Cruz Biotechnology, #1615). DF-EVs are enriched in classical EV markers CD63, CD81, CD9, and syntenin-1 when compared to cell lysate, while cytosolic proteins actin and calnexin are absent in EVs

18. Aliquot EV preparation in 0.65 ml low binding tubes with convenient ready-to-use 10–50 μl volumes to avoid freezing–thawing rounds and store at −80 °C.

DF-EVs can be quantified with several methods [14]. We used nanoparticles tracking analysis (NTA) based on light scattering that provides both particle concentration and size distribution (detailed protocol in [15]). Figure 2a shows typical NTA analysis of DF-EVs revealing that the majority of particles are about 100 nm of size. The average total number of DF-EVs we obtained from 50×150 mm plates was $2.75 \pm 0.7 \times 10^{11}$ particles. A molecular protein characterization by SDS-PAGE blot analysis (method described in [13]) revealed the presence of classical EV markers

CD63, CD81, CD9, and syntenin-1 in DF-EVs, while cytosolic calnexin and actin are absent (Fig. 2b).

3.4 Tube Cleaning and Sterilization

Reusable polycarbonate UC tubes can be washed thoroughly with mild detergent and sterilized with H_2O_2 (*see* **Note 9**). Washing steps are performed in nonsterile conditions while sterilization procedure is carried under a laminar flow cabinet (*see* **Note 10**).

1. Prepare four large beakers with 1.5 L distilled water. In the first beaker add 6 ml of detergent for EV depletion tubes or 4 ml of detergent for EV isolation tubes. Cover the beakers with aluminum foil with labels.
2. Immediately after use tubes are quickly rinsed with tap water over the sink and soaked in the detergent-containing beaker for 10 min (*see* **Note 11**). Tubes are manipulated with large plastic forceps and it is important that tubes are fully filled with liquid as they have a tendency to float.
3. Using a bottlebrush or a toothbrush rub gently the pellet location. Rinse well under running tap water. Repeat for all tubes.
4. Soak the tubes successively in the three beakers containing 1.5 L distilled water for at least 10 min each wash.
5. Tubes are left on absorbing paper until dry (at least overnight).
6. Sterilization procedure is only carried when tubes are fully dry. Under laminar flow cabinet prepare four large beakers with 1.5 L *sterile* distilled water. In the first beaker add H_2O_2 to a final concentration of 10%. Cover the beakers with aluminum foil with labels. H_2O_2 is light sensitive, cover entirely the H_2O_2-containing beaker with aluminum foil and perform the sterilization procedure without direct light from the cabinet (*see* **Note 12**).
7. Soak the tubes in the H_2O_2 beaker for 10 min then successively in the three beakers containing 1.5 L distilled water for at least 10 min each wash. Do not leave the tubes in the H_2O_2 solution more than 10 min.
8. After the last wash tubes are left on absorbing paper under the cabinet until completely dry (up to overnight).
9. Place the tubes in sealable bags and remove from the cabinet after folding the open extremity to avoid contact with air. Bags are sealed and stored.
10. UC lids and adapters are washed immediately after use under the laminar flow in a 0.5 L beaker containing distilled water for 15 min. Manipulate the lids and adapters with small sterile forceps.

11. Transfer the lids and adapters into 70% ethanol-containing 0.5 L beaker (prepare 70% ethanol with sterile water) for 15 min for sterilization.
12. Place the lids and adapters on absorbing paper under the cabinet until completely dry.
13. Store the lids and adapters in a sterile beaker well covered with aluminum fold.

4 Notes

1. It is critical to use PBS devoid of nanoparticle contamination. We use only unopened commercial 1× PBS.
2. We keep two distinct sets of UC tubes, one for serum depletion and the other one devoted to EV isolation.
3. Alternatively, 10,000 × *g* centrifugation can be performed in the ultracentrifuge.
4. Fixed rotor 45Ti is preferable over alternative swing rotors. It also contains larger volume.
5. After UC runs it is important to manipulate the tubes with extreme caution, to use appropriate tube racks that do not cause tube movement and to avoid unnecessary manipulation that could disturb the EV pellet.
6. Allow at least 24 h before conditioned medium collection. We routinely performed collection 48 h post medium change. It is critical that cells are in healthy condition with limited cell death.
7. EV isolation can be paused at this step and EV-containing medium stored at 4 °C until ultracentrifugation rounds are performed (usually completed in 2–3 days, *see* Fig. 1).
8. This step is critical for optimal recovery. Extensive flushing is required. Always use filter tips when using micropipette to avoid contamination.
9. Alternative protocols are available for UC tube sterilization depending on tube material. Refer to tube datasheet for appropriate procedures. For polycarbonate tubes we found that tube life is significantly extended when using H_2O_2 over autoclaving. However reusable UC tubes will eventually show signs of weakness such as cracks that will potentially lead to leakage or errors during the UC run. It is therefore important to thoroughly examine each tube for integrity before reusing and discard cracked tubes.
10. Washing beakers can be kept for 48 h and reused. Do not use the same washing beakers for serum depletion tubes (those contain a large serum EV pellet) and EV isolation tubes.

11. Do not soak in detergent more than 10 min. Alternatively, tubes can be stored in a water-containing beaker until one can proceed with the washing procedure.
12. Caution must be taken when manipulating the H_2O_2 solution. Sterilization beakers can be kept and reused for 48 h only. Depletion tubes and isolation tubes can be sterilized with the same solutions if starting first with the EV isolation tubes.

Acknowledgments

This work was supported by grants from Fondation pour la Recherche Médicale (FRM team 2014), Société Française de Dermatologie (SFD), and INSERM.

References

1. Tkach M, Théry C (2016) Communication by extracellular vesicles: where we are and where we need to go. Cell 164:1226–1232
2. Stahl PD, Raposo G (2019) Extracellular vesicles: exosomes and microvesicles, integrators of homeostasis. Physiology (Bethesda) 34:169–177
3. Cicero Lo A, Delevoye CED, Gilles-Marsens F et al (2015) Exosomes released by keratinocytes modulate melanocyte pigmentation. Nat Commun 6:7506
4. Terlecki-Zaniewicz L, Pils V, Bobbili MR et al (2019) Extracellular vesicles in human skin: cross-talk from senescent fibroblasts to keratinocytes by miRNAs. J Invest Dermatol 139:2425. https://doi.org/10.1016/j.jid.2019.05.015
5. le Riche A, Aberdam E, Marchand L et al (2019) Extracellular vesicles from activated dermal fibroblasts stimulate hair follicle growth through dermal papilla-secreted norrin. Stem Cells 27:247. https://doi.org/10.1002/stem.3043
6. Rani S, Ritter T (2015) The exosome – a naturally secreted nanoparticle and its application to wound healing. Adv Mater 28:5542–5552
7. Pelizzo G, Avanzini MA, Icaro Cornaglia A et al (2018) Extracellular vesicles derived from mesenchymal cells: perspective treatment for cutaneous wound healing in pediatrics. Regen Med 13:385–394
8. Rittié L, Fisher GJ (2005) Isolation and culture of skin fibroblasts. Methods Mol Med 117:83–98
9. Kowal J, Arras G, Colombo M et al (2016) Proteomic comparison defines novel markers to characterize heterogeneous populations of extracellular vesicle subtypes. Proc Natl Acad Sci U S A 113:E968–E977
10. Théry C, Witwer KW, Aikawa E et al (2018) Minimal information for studies of extracellular vesicles 2018 (MISEV2018): a position statement of the International Society for Extracellular Vesicles and update of the MISEV2014 guidelines. J Extracell Vesicles 7:1535750
11. Lane RE, Korbie D, Trau M, Hill MM (2017) Purification protocols for extracellular vesicles. Methods Mol Biol 1660:111–130
12. Hartjes TA, Mytnyk S, Jenster GW et al (2019) Extracellular vesicle quantification and characterization: common methods and emerging approaches. Bioengineering (Basel) 6:7
13. Théry C, Amigorena S, Raposo G, Clayton A (2006) Isolation and characterization of exosomes from cell culture supernatants and biological fluids. Curr Protoc Cell Biol. Chapter 3:Unit 3.22–3.22.29
14. Shao H, Im H, Castro CM et al (2018) New technologies for analysis of extracellular vesicles. Chem Rev 118:1917–1950
15. Gardiner C, Ferreira YJ, Dragovic RA et al (2013) Extracellular vesicle sizing and enumeration by nanoparticle tracking analysis. J Extracell Vesicles 2:19671

Chapter 7

A Method to Investigate the Epidermal Permeability Barrier In Vitro

Yoshikazu Uchida and Anna Celli

Abstract

The epidermal permeability barrier serves as a multifunctional partition to protect its host from the external environment. Most epidermal permeability barrier studies have been conducted using in vivo human and experimental animals, although some studies have used in vitro cultured cells. There currently is an increased demand for these cultured models, thus avoiding the use of laboratory animals. Here, we first summarize required features that need to be recaptured in cultured keratinocytes for an epidermal permeability barrier study and second, we describe a method for culturing these cells. We also introduce methods to analyze epidermal permeability barrier function using cultured keratinocytes.

Key words Cultured human keratinocytes, Epidermal permeability barrier, Lipid analysis, Reconstituted three-dimensional organotypic epidermal equivalents, Submerged cultured keratinocytes, TEER assay, Transepidermal water loss

1 Introduction

1.1 Overview of the Epidermal Permeability Barrier

The skin consists of multiple barriers: mechanical, thermal, antimicrobial, irradiation, antioxidant, and permeability [1, 2]. Each of the skin constituents—epidermis, dermis, sebaceous glands, sweat glands, hair and adipose tissue—contribute to making all these barriers competent to maintain proper biological functions. Among the skin constituents, the epidermis (thee outermost layer of skin) is the first line of defense, so it is largely responsible for skin barrier functions [1, 2]. The permeability barrier performs many tasks, that is, preventing penetration/invasion of small and large exogenous chemicals, particles, and microorganisms, protecting against excess water evaporation (transepidermal water loss, TEWL); and preventing loss of endogenous compounds, such as ions, nucleotides, proteins, lipids, and carbohydrates [1–3]. Epidermal permeability barrier deficiencies can lead to lethal consequences, indicating that this barrier functioning correctly is

Natalia V. Botchkareva and Gillian E. Westgate (eds.), *Molecular Dermatology: Methods and Protocols*,
Methods in Molecular Biology, vol. 2154, https://doi.org/10.1007/978-1-0716-0648-3_7,
© Springer Science+Business Media, LLC, part of Springer Nature 2020

mandatory for mammalian survival in a terrestrial environment. Lipid-dominant extracellular lamellar membranes and tight junctions play major roles in the permeability barrier. Extracellular lamellar membranes are a localized apical phase of stratified keratinocytes in the stratum corneum, while tight junctions are distributed in the lateral phase of keratinocytes in the nucleated layers of epidermis [1–3]. In this chapter, we focus on one method of investigating the epidermal permeability barrier in the stratum corneum.

1.2 Using Cultured Keratinocytes for Investigating the Epidermal Permeability Barrier

Permeability barrier function has long been investigated in in vivo skin (experimental animals and human), as well as in cultured keratinocytes. Additionally, recent advancements in technology, that is, the Raman spectrometer, confocal laser microscopy, optical ultrasound image, and quantitative RT-PCR using tape-stripped stratum corneum, have allowed investigators to conduct certain minimally invasive in vivo biological studies. However, one advantage for using in vitro cultured keratinocyte studies is that it is easier to modify gene expression using siRNA and CRISPR techniques than to explore in vivo models; and in vitro studies are still the essential experimental method used to characterize precise biological features of barrier formation.

1.3 Features Needed in Cultured Keratinocytes for Investigating Epidermal Permeability Barrier

Lipids are essential components of extracellular lamellar membranes, which are responsible for epidermal permeability barrier function in the stratum corneum [4]. In addition, the cornified membranes, consisting of cross linked proteins, replace the plasma membrane of keratinocytes at the transition from the granular layer to the stratum corneum and provide mechanical strength, becoming the foundation of a scaffold formation termed the *c*orneocyte lipid *e*nvelope (CLE) [1–3, 5]. Moreover, although neosynthesis of biological components does not occur in the stratum corneum, presynthesized cellular components are catabolized in the stratum corneum by β-glucocerebrosidase and sphingomyelinase, that convert glucosylceramide and sphingomyelin to ceramide, respectively [6]. Ceramides are further hydrolyzed to fatty acids and sphingoid bases by ceramidase [1–3], and cholesterol sulfate is hydrolyzed to cholesterol by cholesterol sulfatase. Acidic and neutral proteases contribute to desquamation [3, 7–10]. In addition, amino acids generated from filaggrin become major components of natural moisturizing factor (NMF) [11]. These enzyme-driven hydrolytic processes occur during differentiation, in particular its late stage, leading to formation of the epidermal permeability barrier and, therefore, are required features to be recaptured in cultured keratinocytes in order to properly study the permeability barrier.

1.4 Submerged Cultured Keratinocytes to Recapture Proliferating Keratinocytes and Early and Late Stages of Differentiated Keratinocytes

Reconstituted three-dimensional organotypic epidermal equivalents, which mimic epidermis, are commercially available. Because these keratinocytes are grown at the air-liquid interface, cells are carrier-supported to form a stratum corneum [12]. Transwells designed for lifting cultures are also commercially available. However, for most biochemical studies, it is not convenient or cost-effective, to obtain the large amounts of cells needed from such organotypic cultures. In addition, although transparent engineered epidermal membranes are available, layered cells still cannot be observed using light microscopy. Utilization of submerged cultured keratinocytes showing late stages of differentiation is an alternative method to use to investigate differentiated keratinocyte features. Since proliferative and different stages of cell differentiation can be obtained by this submerged cultured system (developed here in our laboratory), our system is convenient for comparing proliferating, early and late stages of differentiated keratinocytes in the same experiment [13].

1.5 Organotypic Epidermal Equivalents

Reconstituted three-dimensional organotypic epidermal equivalents have been widely used as valuable in vitro models to study epidermal barrier development and function. Such models need to recapitulate morphological, biochemical, and functional barrier properties of in vivo epidermis, including pH and other ionic gradients necessary for adequate lipid synthesis, secretion, and processing.

While many agents such as extracellular calcium, thyroid hormone, corticosteroids, estrogen, vitamins, essential fatty acids, have been shown to accelerate epidermal barrier formation, air exposure of confluent cultured human or murine keratinocytes is crucial for the development of a competent permeability barrier [14–16]. Moreover, similar to what is observed in infant development, exposure to a reduced-humidity environment has been linked to improved barrier function in in vitro organotypic epidermal equivalents [12].

The following summarizes the protocol developed in our laboratory [12] for the culturing of organotypic epidermal equivalents specifically suited for human epidermal permeability barrier function studies.

1.6 Evaluation of Epidermal Permeability Barrier Constituents

As above (Subheading 1.3), proteins, which form cornified envelopes and hydrolytic enzymes required for desquamation and processing of barrier lipids and for producing NMF in the stratum corneum, are important components of epidermal permeability. Generation of these proteins and their catabolites and enzymes are assessed by immunohistochemistry and western blot analysis. Unspecific binding often occurs in the stratum corneum between components and some antibodies (*see* **Note 1**).

1.7 Thin-Layer Chromatography

Lipids are largely responsible for forming the epidermal permeability barrier in the stratum corneum. Liquid chromatography–tandem mass spectrometry (LC-MS/MS) analysis, established well over 10 years ago, allows us to precisely characterize lipid species using its analytical capability to detect heterogeneous mixtures of ceramide and glucosylceramide. However, LC-MS/MS analysis requires expensive equipment, as well as skilled/trained personnel. Thin-layer chromatography (TLC) is an alternative method for assessing lipid profiles in keratinocytes. Although analysis of multiple lipid components such as ceramide, glucosylceramide, cholesterol, fatty acids, and triglyceride using LC-MS/MS is required for different elution solvent systems, only one or two TLC analyses are needed to provide certain levels of lipid profiles. Here we describe a method to conduct lipid analysis using TLC.

1.8 Studies of Physiological Epidermal Barrier Function

1.8.1 TEER Assay

Transepithelial electrical resistance (TEER) is a noninvasive measurement that can be used to assess epidermal permeability function in human epidermal equivalent models. Generally, two TEER measurement methods (resistance-based for model systems and impedance-based for whole tissue measurements) have been established.

Resistance-based TEER measurements are particularly indicated to monitor the development of in vitro model systems and their ability to form a competent permeability barrier. These measurements are generally performed using a commercially available instrument called epithelial voltohmmeter (EVOM). EVOMs use a pair of electrodes commonly known as "chopsticks" or STX2 electrodes to apply an alternating square wave voltage to the membrane, reading the resulting current, and calculating the electrical resistance from Ohm's first law [17, 18]. STX2 electrodes are composed of parallel semiflexible electrodes of different lengths held at a fixed distance from each other. When measuring TEER, the longer electrode is placed in the outer cell culture compartment while the shorter electrode is carefully placed in the inner compartment of the insert so as to not puncture or damage the membrane. Since electrical resistance readings are dependent on temperature, distance between the electrodes and electrode position respect to the membrane, care should be taken to ensure standard measurement practices across samples, and multiple measurements should be performed for each epidermal equivalent membrane. For these measurements, 2D (monolayers) or 3D (epidermal equivalents) cell layers are generally seeded on commercially available transwell inserts on a semipermeable cell culture grade membrane and grown until confluent.

1.8.2 Epidermal Permeability Barrier Recovery Assay by TEER Measurement

TEER measurements can be used to easily monitor epidermal barrier recovery after barrier disruption in three-dimensional in vitro epidermal equivalent models.

This type of assay, also referred to as treadmill test, is commonly used to assess barrier homeostatic ability which can be compromised even if basal barrier function is normal.

For example, acidic sphingomyelinase deficient transgenic mice (a model of Pick disease type A and B) show normal TEWL, while barrier recovery following acute barrier disruption by tape stripping or acetone treatment (removing upper layer of stratum corneum or extracting lipids in the stratum corneum, respectively) is delayed compared with normal control mice [19]. Barrier homeostatic function declines in some populations of people (psychological/ sociological stress, ageing, atopic dermatitis). Barrier recovery from acute barrier disturbance assesses barrier homeostatic function. Three-dimensional organotypic epidermal equivalents with a fully competent permeability barrier can be used to assess homeostatic barrier function in vitro using a TEER assay [12].

1.8.3 TEWL Measurement

Water passively evaporates through the skin due to the difference in water vapor pressure across the epidermal permeability barrier. The amount of water that is lost through the skin via passive evaporation is commonly referred to as transepidermal water loss (TEWL) and it is a good indicator of epidermal permeability barrier function [20].

Due to the passive evaporation through the skin of water from the body to ambient air, a water vapor gradient proportional to the epidermal water loss exists around the skin to about10 mm above the surface [21, 22]. TEWL measurements are widely used to assess the permeability barrier status of humans and rodents and can also be used to monitor the permeability barrier function of in vitro systems.

TEWL is indirectly measured by measuring the change in water vapor density from the proximity of the skin to a more distant position [23]. Three different types of instruments can be used to measure TEWL: open chamber sensors, closed chamber sensors, and condenser chamber sensors.

The open chamber sensor consists of a hollow cylinder, which is placed in close proximity with the skin or in vitro epidermal equivalent systems with two sensors for temperature and for humidity placed at different distances from the sensor's openings. The TEWL is inferred by the difference in water vapor density between the two sensors. This type of sensor has the advantage of not occluding the skin, thus reducing occlusion derived changes in barrier function. Moreover, since humidity does not accumulate in a closed chamber, these sensors allow for continuous TEWL measurements. The drawback of these devices is that the measurements are significantly influenced by ambient ventilation and humidity, which need to be controlled.

The closed chamber used in closed chambered sensors protects the measurements from ambient ventilation, but measurements with these systems require the sensor to be lifted from the skin periodically to avoid accumulation of water vapor.

The condenser chamber sensors utilize the same principle as the open chambered ones but use a cooled sensor cover to eliminate water vapor accumulation by freezing it to ice. This strategy allows the measurements to be protected from ambient air currents and disturbances [24, 25].

Recently, an ad hoc sensor array based on the open chamber design has been developed to allow the continuous monitoring of TEWL in in vitro organotypic epidermal equivalents and other systems directly in their 24-well cell culture plates (Tewitro TW24; Courage + Khazaka Electronic, Cologne, Germany; http://www.courage-khazaka.de/index.php/en/products/scientific/382-tewitro-e). However, TEWL readings of in vitro systems can also be performed with the other sensor types [12, 25, 26].

Due to the sensitivity to ambient condition and air fluxes of the TEWL sensors, care should be taken to ensure that measurements are conducted in areas protected from air fluxes and under standard temperature and humidity conditions. Moreover, the samples should be allowed to equilibrate at room temperature and humidity for about 30 min prior to measurements. Due to the variability in TEWL readings, multiple readings should be performed and averaged per sample. Also, organotypic epidermal equivalents show batch to batch variation, making comparisons across separate experiments challenging, so care should be taken to have an internal control for each batch.

2 Materials

2.1 Preparation of Primary Cultured Human Keratinocytes

1. Hank's Balanced Salt Solution (HBSS) with Ca^{2+} and Mg^{2+}.
2. 10% Hibiclens™ in sterile water or equivalent antiseptic solution, HBSS Ca^{2+} and Mg^{2+} free (HBSS-CMF), Penicillin/Streptomycin (10,000 I.U./ml).
3. 25 U/ml dispase in HBSS-CMF; 0.05 mg/ml gentamycin in HBSS-CMF.
4. 0.05% trypsin–EDTA made up using HBSS-CMF (final 0.53 mM).
5. Trypsin Neutralizing Solution (TNS): HBSS–Chelexed FCS, 20:1 (v/v) stirred for 60 min at room temperature followed by sterilizing using filter (0.45 μm).
6. Serum-free keratinocyte growth medium (KGM) supplemented with 0.07 mM $CaCl_2$ and bovine pituitary extracts (BPE).
7. KGM supplemented with 1.2 mM CaCl2 and BPE.
8. Dulbecco's Modified Eagle's Medium (DMEM) and Ham F-12 (2:1, vol/vol), containing 1.2 mM $CaCl_2$ supplemented

with 10% fetal bovine serum, insulin (10 μg/ml), hydrocortisone (0.4 μg/ml), and ascorbic acid (50 μg/ml) (*see* **Note 2**).

2.2 Organotypic Epidermal Equivalents

1. Hanging polyethylene terephthalate inserts with 0.4 μm pore size (Millipore Sigma).
2. Membrane coating solution: CELLstart (Thermofisher) diluted 1:50 in Dulbecco's phosphate-buffered saline (DPBS) with calcium (0.9 mM) and magnesium (0.49 mM).
3. CnT Prime 07 culture medium (CELLnTEC, CELLnTEC Advanced Cell Systems AG).
4. CnT Prime 3D Barrier (CELLnTEC).

2.3 Lipid Extraction from Cultured Cells and Saponification

1. Glass tube (16 mm × 150 mm) with Teflon seal inner.
2. Chloroform ($CHCl_3$).
3. Methanol (MeOH).
4. Second Washing Solution: $CHCl_3$–MeOH saturated water is prepared by mixing $CHCl_3$–MeOH–water (1:1:0.9) and standing overnight. Upper phase is $CHCl_3$–MeOH saturated water used as Second Washing Solution. Lower phase can be used for rinsing tubes.
5. Saponification solution: Dissolved 1 g NaOH 2.5 ml water and then add 22.5 ml MeOH (1 M NaOH solution).
6. Universal pH test paper.

2.4 Thin-Layer Chromatography

1. HPTLC plate 10 × 20 cm (Merck) (*see* **Note 3**).
2. Acetic acid.
3. Acetone.
4. Chloroform.
5. Diethyl ether.
6. Ethyl acetate.
7. n-Hexane.
8. Hexyl acetate.
9. Methanol.
10. Lipid standards: cholesterol (0.1–0.5 μg); cholesterol palmitate (palmitate or other fatty acyl) and triglyceride (1–2 μg); ceramide, glucosylceramide, and phospholipids (0.5–1 μg).
11. Charring solution: 1.5% cupric sulfate in acetic acid–sulfuric acid–orthophosphoric acid–water (50:10:10:30, v/v).

2.5 Equipment

1. Lamina flow food.
2. Scalpel.
3. Forceps.

4. Benchtop centrifuge.
5. Refrigerator.
6. CO_2 incubator.
7. Inverted microscope.
8. TLC development chamber.
9. Hot plate.
10. Probe type sonicator (*see* **Note 4**).
11. Nitrogen gas evaporator (*see* **Note 5**).
12. Dry oven (*see* **Note 6**).
13. Epithelial voltohmmeter (EVOM).
14. Tewameter (Courage + Khazaka, TM300).

3 Method

3.1 Primary Culture of Human Keratinocytes

Our lab (Department of Dermatology, UC San Francisco and VA Health Care System) has followed the Pittelkow and Scott [27] primary keratinocyte culture method with some modifications.

3.1.1 Skin

Human keratinocytes are isolated from newborn human foreskins. Skin samples are kept in HBSS at 4 °C (*see* **Note 7**).

1. Wash the skin in 10% Hibiclens™ in sterile water for 10 s.
2. Rinse the skin two times with 25 ml of Hibiclens solution for 10 s.
3. Rinse the skin with 25 ml of HBSS-CMF containing Penicillin/Streptomycin (10,000 I.U./ml).
4. Section the skin into smaller (approximately 0.25 mm^2) pieces to facilitate exposure to the dissociating agent.
5. Incubate the skin with 4 ml of HBSS-CMF containing dispase (final 25 U/ml) and gentamycin (final 0.05 mg/ml) in 60 mm petri dish overnight in 4 °C.
6. Using a scalpel, hold one side of the tissue and with forceps, hold the epidermis sheet and pull it off slowly and smoothly.
7. Transfer the epidermal sheet into 15 ml of sterile centrifuge tube with 4 ml 0.05% trypsin–EDTA in HBSS-CMF and incubate for 10–15 min at 37 °C.
8. Add 6 ml of TNS and gently mix by pipette up and down for five times. Continue on to the next piece of tissue.
9. Centrifuge the tubes at 1000 rpm for 5 min.
10. Aspirate supernatant and re-suspend the cell pellet in 5 ml of serum free KGM supplemented with 0.07 mM $CaCl_2$ and BPE.

11. Plate the cells in the T-175 flasks containing 20 ml of KGM supplemented with 0.07 mM $CaCl_2$ and BPE.
12. Incubate cells in 5% CO_2 at 37 °C.
13. Change the media every other day.
14. Cells are passaged when 70–80% confluency is reached (*see* **Note 8**).

3.2 Submerged Cultured Keratinocytes to Recapture Proliferating Keratinocytes and Early and Late Stages of Differentiated Keratinocytes

1. Culture keratinocytes (2nd–4th passage) in serum free KGM supplemented with 0.07 mM $CaCl_2$ and BPE until 70–80% confluence. Changed the medium every other day (*see* **Note 9**).
2. After keratinocytes reach 70–80% confluence, culture them with serum-free KGM containing 1.2 mM $CaCl_2$ for 2 days (*see* **Note 10**).
3. Replace the medium with DMEM and Hams F-12 (2:1, vol/-vol), containing 1.2 mM Ca^{2+}, supplemented with 10% fetal bovine serum, insulin (10 μg/ml), hydrocortisone (0.4 μg/ml), and ascorbic acid (50 μg/ml).
4. Culture the cells for 9–12 days with medium changes every other day (*see* **Note 11**).

3.3 Organotypic Epidermal Equivalents

1. Coat the hanging polyethylene terephthalate inserts (0.4 μm pore size (Millipore Sigma)) with CELLstart (Thermofisher) diluted 1:50 in Dulbecco's PBS with calcium and magnesium for 2 h at 37 °C.
2. Aspirate the excess of coating solution and wash the inserts in cell culture medium once.
3. Add 1.5 ml of growth medium to each well of a 12-well plate to the outside compartment.
4. Seed first passage of the keratinocytes at a density of 2.2×10^5 cells/cm^2 (5×10^5 cells in 0.5 ml medium for a 12-well plate insert) in supplemented CnT Prime 07 culture medium (CELLnTEC) and incubate in a humidified atmosphere of 5% CO_2 95% air at 37 °C for 72 h.
5. Aspirate the growth medium from the outer and inner compartments, and replace with CELLnTEC at 37 °C for 16–20 h.
6. Aspirate the medium from both compartments and airlift the cells, and add CnT Prime 3D Barrier, only from the outer compartment.
7. Change medium of outer compartment daily with CnT Prime 3D Barrier and culture in a humidified atmosphere of 5% CO_2 95% air at 37 °C incubator for 12 days.
8. Move cultures to a dry incubator (below 50% RH) for 48 h to complete the formation of a competent epidermal barrier as assessed by morphological and functional measurements [12].

3.4 Evaluation of Epidermal Permeability Barrier Constituents

3.4.1 Unbound Lipid Extraction from Cultured Cells

1. Rinsing the extraction tube: Put 2 ml of chloroform ($CHCl_3$)–methanol (MeOH) (1:1, v/v) in tube and shake by hand. Repeat three times (Check solvent has not leaked).
2. Suspend the whole harvested cells into 1.6 ml of PBS and homogenize by probe type sonicator (on ice). Alternatively, glass homogenizer could be used.
3. Add 2 ml of $CHCl_3$ and 4 ml MeOH to 1.4 ml of cell homogenates and shake vigorously for 5 min by hand and leave overnight at room temperature.
4. Shake tubes by hand for 5 min (vigorously).
5. Centrifuge tubes at 1500–2500 × *g* for 5 min (cells/debris is precipitated).
6. Transfer the supernatant extracts to a new cleaned glass (rinsed tube, as above 1).
7. Add 2 ml of $CHCl_3$ and 2 ml of water to the extracts.
8. Shake tubes for 5 min by hand.
9. Centrifuge tubes at 1500–2500 × *g* for 5 min.
10. Discard the upper phase very carefully (*see* **Note 12**).
11. Add 6 ml of Second Washing Solution to the lower phase.
12. Shake tubes by hand for 5 min.
13. Centrifuge tubes at 1500–2500 × *g* for 5 min.
14. Discard the upper phase from the lower fraction phase.
15. Dry up lipid extract using N_2 in a fume hood.
16. Dissolve lipids in $CHCl_3$–MeOH (1:1, v/v) and store at −20 °C.

3.4.2 Thin-Layer Chromatography

Separation of Triglyceride, Free Fatty Acid, Cholesterol, Ceramide, and Glucosylceramide

1. Develop (*see* **Note 13**) HPTLC plate with $CHCl_3$–MeOH–H_2O (60:30:4, v/v) to top of plate prior to use in order to clean the HPTLC plate to minimize background color following charring.
2. Dry HPTLC plate on a hot plate at 40–60 °C.
3. Store HPTLC plate in a clean place.
4. Heat the HPTLC plate in the oven or on the hot plate at 100 °C for 10–15 min to dry and cool down to room temperature immediately before use.
5. Mark out application line (1 cm from bottom) by pencil (2, 5, and 8.5 cm lines from applied position).
6. Mark out sample application position (1.5 cm from left).
7. Apply sample to a 4 mm in length. Leave 6–8 mm distance between the sample applied areas.
8. Apply lipid standards on both sides of the TLC plates.

9. Insert filter paper into developing chamber to cover glass. Pour solvent 10–15 min before putting TLC plate to saturate chamber with solvent.
10. Develop with $CHCl_3$–MeOH–H_2O (40:10:1, v/v) to 2 cm line from applied position.
11. Dry on hot plate at 60 °C.
12. Develop with $CHCl_3$–MeOH–H_2O (40:10:1, v/v) to 5 cm line from applied position.
13. Dry on hot plate 60 °C.
14. Develop with $CHCl_3$–MeOH–acetic acid (47:2:0.5, v/v) to 8.5 cm line from applied position.
15. Dry on hot plate 60 °C.
16. Develop with n-hexane–diethyl ether–acetic acid (30:15:0.5, v/v) develop 8.5 cm line from applied position or top of HPTLC plate.
17. Dry on hot plate 60 °C.

Separation of Glucosylceramide

1. Develop with $CHCl_3$–MeOH–H_2O (40:10:1, v/v) to 1.0 cm.
2. Dry on hot plate (60 °C).
3. Develop with $CHCl_3$–MeOH–H_2O (40:10:1, v/v) to 7.5 cm.
4. Dry on hot plate (60 °C).
5. Develop by n-Hexane–diethyl ether–acetic acid (65:35:1, v/v) to the top of the plate.
6. Dry on hot plate (60 °C).

Development for Separation of Ceramide

If glucosylceramide and ceramide are the focus of study, the following solvent system (modification of Ponec et al. method [28]) is appropriate.

1. Develop the plate with $CHCl_3$ to 1.5 cm.
2. Dry on hot plate (60 °C).
3. Develop with $CHCl_3$–MeOH–acetone (76:16:8, v/v) to 1.0 cm.
4. Dry on hot plate (60 °C).
5. Develop with $CHCl_3$–MeOH–hexyl acetate–acetone (86:4:1:10, v/v) to 7.0 cm.
6. Dry on hot plate (60 °C).
7. Develop with $CHCl_3$–MeOH–acetone (76:20:4, v/v) to 2.0 cm.
8. Dry on hot plate (60 °C).
9. Develop with $CHCl_3$–MeOH–diethyl ether–hexyl acetate–ethyl acetate–acetone (72:4:4:1:4:16, v/v) to 7.5 cm.

10. Dry on hot plate (60 °C).
11. Develop with n-hexane–diethyl ether–ethyl acetate (80:16:4, v/v) to the top of the plate.
12. Dry on hot plate (60 °C).
13. Develop with n-hexane–diethyl ether–acetic acid (65:35:1, v/v) to the top of the plate.

Separation for Phospholipids

Develop with $CHCl_3$–MeOH–28% NH_4OH (32:12:2.5, v/v) and develop to 8.5 cm.

Visualizing Lipids on TLC Plate

1. Dip fully developed and dry HPTLC plate into charring solution and then dry on hot plate at 40–60 °C.
2. Gradually increase hot plate temperature to 90 °C to fully dry charring solution. Avoid placing HPTLC plate directly next to 90 °C hot plate, as this will bend the glass slightly which may influence scanning.
3. Put into 160 °C oven for 20 min.
4. If you need to keep the HPTLC plates, cover with clean glass, wrap with aluminum foil and store at −20 °C.

3.4.3 Corneocyte Lipid Envelope (CLE) Analysis

After unbound lipid extraction from cells as above, some unbound lipids are still retained in cells (delipidated cells). To assure that no unbound lipids are present in cells, cells are further treated with $CHCl_3$–MeOH.

1. Dry delipidated cells (dl-cells) following unbound lipid extraction using nitrogen gas evaporator (*see* **Note 5**).
2. Add $CHCl_3$ 4 ml, MeOH 2 ml to dl-cells.
3. Shake tubes by hand for 5 min (vigorously) and leave for 2 h.
4. Centrifuge tubes at 1500–2500 × *g* for 5 min.
5. Add 3 ml $CHCl_3$ and 3 ml MeOH to dl-cells.
6. Shake for 5 min and leave for 2 h.
7. Centrifuge tubes at 1500–2500 × *g* for 5 min.
8. Transfer the extracts to a new cleaned glass tube.
9. Add 4 ml $CHCl_3$ and 2 ml MeOH to dl-cells in the glass tube.
10. Shake tubes by hand for 5 min (vigorously) and leave for 2 h.
11. Centrifuge tubes at 1500–2500 × *g* for 5 min.
12. Transfer the extracts to a new cleaned glass tube.
13. Add 4 ml $CHCl_3$ and 2 ml MeOH to dl-cells in the glass tube.
14. Shake tube by hand for 5 min (vigorously) and leave for 2 h.
15. Centrifuge tubes at 1500–2500 × *g* for 5 min.
16. Transfer the extracts to a new cleaned glass tube (EXT A).

17. Add 3 ml $CHCl_3$ and 3 ml MeOH to remaining dl-cells pellet.
18. Shake tubes by hand for 5 min (vigorously) and leave for 2 h.
19. Centrifuge tubes at 1500–2500 × *g* for 5 min.
20. Transfer the extracts to a new cleaned glass tube (EXT B).
21. Add 4 ml $CHCl_3$ and 2 ml MeOH to remaining dl-cells pellet.
22. Shake tubes by hand for 5 min (vigorously) and leave for 2 h.
23. Centrifuge tubes at 1500–2500 × *g* for 5 min.
24. Transfer the extracts to a new cleaned glass tube (EXT C).
25. Combine Ext A, B, and C, and dry down under a stream of N2, followed by dissolving the extract in 0.1 ml of $CHCl_3$–MeOH, 2:1 (v/v). NB. dl cell extracts should be applied to TLC plate and developed as above, to confirm absence of lipids before saponification of bound lipids is performed as described in **steps 26–43** below).
26. Turn on water bath (set 60 °C).
27. Dry delipidated cells (dl-cells) using nitrogen gas evaporator (*see* **Note 5**).
28. Add saponification solution (1 M NaOH in MeOH–H_2O) (9:1, v/v) into dried delipidated cells and mix using vortex mixer.
29. Incubate at 60 °C for 1 h.
30. Mix the sample every 10 min during incubation using vortex mixer.
31. Cool down tube to room temperature and add 0.5 ml of 2 N HCl to check if solution is at pH 4 using universal pH test paper.
32. If pH is still not about 4, add a small of volume (<50 μl) of 2 N HCl until pH reaches about 4.
33. Add $CHCl_3$, MeOH and water (final 2–2-2 ratio, v/v). Record total volume.
34. Shake tubes by hand for 5 min (vigorously).
35. Centrifuge tubes at 1500–2500 × *g* for 5 min.
36. Discard upper phase and add second wash 1/2 vol of initial $CHCl_3$, MeOH, and water (final 2:2:2 ratio, v/v) solution (above).
37. Shake tubes by hand for 5 min (vigorously).
38. Centrifuge tubes at 1500–2500 × *g* for 5 min.
39. Shake tubes by hand for 5 min (vigorously).
40. Centrifuge tubes at 1500–2500 × *g* for 5 min.
41. Dry up lipid extract using N_2 in the fume hood.

42. Dissolve lipids in $CHCl_3$–MeOH (1:1, v/v) and store at −20 °C.
43. Analyze bound by TLC (as above using solvent system described in Subheading "Separation of Triglyceride, Free Fatty Acid, Cholesterol, Ceramide, and Glucosylceramide".

3.5 Studies of Physiological Epidermal Barrier Function

3.5.1 TEER Assay

1. Immerse STX2 electrodes in 70% ethanol for 15 min to sterilize and wash with PBS or culture medium.
2. Remove the growth medium from both the inside and the outside of the insert and replace with fresh media.
3. Place STX2 electrodes to both the inside and outside compartment.
4. Measure resistance three or more times per transwell from different positions and obtain a spatially averaged epithelial resistance (R_{av}).
5. Measure resistance in an empty insert immersed in growth media to obtain the semipermeable membrane resistance (R_m).
6. The epithelium resistance (R_e) is calculated by subtracting R_m from the measured averaged epithelial resistance (R_{av}).
7. Since the resistance is inversely proportional to the membrane surface area, it is customary to multiply R_e by the membrane area to standardize measurements across different insert sizes. TEER measurements are usually expressed in units of Ω cm^2.

3.5.2 Epidermal Permeability Barrier Recovery Assay by TEER Measurement

1. Sterilize D-squames by 15 min UV exposure.
2. Cut the D-squames in strips ~3–5 mm wide.
3. You will already have prepared human keratinocyte organotypic epidermal equivalents, cultured as described above (Subheading 3.3) in 12-well or larger transwell inserts for 14 days, ready to be used for the assay.
4. A first TEER reading to assess basal TEER levels is performed as described above (Subheading 3.5.1).
5. After the basal TEER measurement, aspirate all media from the inner insert compartment, and return to the dry incubator (5% CO_2 at 37 °C) for 30 min to evaporate any leftover medium.
6. Fold the D-squame strips into an "L" shape with a longer and shorter arm.
7. Retrieve organotypic epidermal equivalents from the incubator and place in the culture hood.
8. Peel the backing from the D-squame strips using sterile forceps and apply the adhesive short 'L' side to the apical side of the organotypic epidermal equivalents. A pipette tip can be used to apply gentle pressure to the top of the D-squame to promote adhesion.

9. Organotypic epidermal equivalents are then tape stripped by peeling off the D-squame strips by pulling up the long "L" side with forceps.
10. Measure TEER as described above (*see* **Note 15**).
11. Inserts are returned to the dry incubator (5% CO_2 at 37 °C).
12. Barrier recovery is monitored at 3, 6, and 24 h after tape stripping by subsequent TEER measurements. Since basal TEER values usually keep increasing daily due to the absence of stratum corneum desquamation, non-tape stripped inserts rather than pre-tape strip basal TEER values are to be used as control.

3.5.3 TEWL Measurement

The following is an example of TEWL measurement of organotypic epidermal equivalents using TEWL (open chamber sensor, e.g., Tewitro TW24) (as described in [12]) (*see* **Note 16**).

1. The organotypic epidermal equivalents are generated as described above (Subheading 3.3).
2. When ready to perform TEWL, remove the plate from the incubator and keep at room temperature for 30 min for equilibration.
3. Place a 50 μl drop of culture media at room temperature on a Parafilm sheet.
4. Cut the organotypic culture out of its insert and place stratum corneum up on top of the media drop (*see* **Note 17**).
5. Hold the probe at approximately 1 mm above the stratum corneum surface, and take a reading. At least three separate measurements per samples must be done.

4 Notes

1. We recommend that western blot analysis is performed to assure that stratum lysate shows a band or bands with predicted molecular size of the target protein.
2. Ascorbic acid solution (10 mg/ml) is directly added into cultures and immediately agitated (since ascorbic acid is acidic, medium color at site of addition becomes yellow. Yellow color becomes red following immediate agitation).
3. Two types of TLC plates, thin-layer chromatography and high-performance thin-layer chromatography (HPTLC), are commercially available. The HPTLC plate is made of a smaller particle size silica than the TLC plate, and HPTLC shows better separation compared with TLC. Since epidermal glucosylceramide and ceramide consist of heterogeneous species,

HPTLC is better to use for these lipid analyses. Using our method, TLC is charred to visualize lipids, and HPTLC plate that does not contain a fluorescence agent is used.

4. Alternatively, glass homogenizer can be used.
5. Alternatively, nitrogen gas is directly blown into each tube, keeping the gas above the liquid layer and doing this in a fume hood.
6. Alternatively, a hot plate can be used.
7. The skin can be stored for about 24 h prior to isolating keratinocytes.
8. Cells generally reach 70–80% confluency 3–4 days after plating.
9. Keratins 5 and 14 are markers of proliferating basal keratinocytes [29]. The expression of these proteins is confirmed by RT-PCR, western blot analysis or immunohistochemistry.
10. Cell proliferation is increased for a while (approximately 12 h) after switching to high Ca^{2+} conditions and then is followed by decreasing DNA synthesis. These cells are defined as early stage differentiated keratinocytes and will express involucrin and low levels of loricrin [29].
11. After 7 days of culture, multilayer keratinocytes are observed under the microscope. The surface of differentiated cells displays rectangular shapes. Late stage differentiated keratinocytes synthesize loricrin [29] and heterogeneous ceramides, which form lamellar structures in the stratum corneum and also the immediate precursors of ceramide, heterogeneous glucosylceramide species [13] will be detected. In addition, cornified envelope bound ceramides are formed. Formation of lamellar membrane structures can be observed using electron microscopy [13].
12. There will be two distinct phases in the tube after centrifugation. *The lower phase* is the *lipid fraction*. The upper phase contains hydrophilic cell constituent (carbohydrates, proteins, salts, and polar lipids, including glycolipids (polyglycosylated lipids, sphingoid base, and certain lysophospholipids).
13. Development here means when the development solvent, added to the bottom of TLC chamber, travels up the plate by capillary action; the lipid species are separated on the plate, according to the solvent used.
14. If lipid(s) are present, repeat lipid extraction.
15. A resistance smaller than 300 Ω is not used for further measurements. After 7 days cultures, a resistance is approximately 750–1000 Ω. A single D-squame tape strip is usually sufficient to obtain a typical resistance reduction of 50–70%.

16. The organotypic culture surface area is required to be larger than the TEWL probe for these measurements.
17. This step is necessary to allow the TEWL probe to be in close proximity to the sample.

Acknowledgments

We thank Ms. Joan Wakefield for superb editorial assistance. This study was supported by UCSF grant #129594A, Leo Pharma sponsor "Full thickness skin models from human pluripotent stem cells for identification and test effectiveness of personalized therapies in atopic dermatitis" UCSF HDFCCC Laboratory for cell analysis (NIH P30CA082103 and S10OD021818-01) to A.C.

References

1. Uchida Y (2017) Skin lipids. In: Sakamoto K, Lochhead RY, Maibach H, Yamashita Y (eds) Cosmetic science and technology: theoretical principles and applications. Elsevier, UK, pp 685–698
2. Uchida Y, Park K (2016) Stratum corneum. Immunology of the skin. Springer, Tokyo, pp 15–30
3. Elias PM (2012) Structure and function of the stratum corneum extracellular matrix. J Invest Dermatol 132:2131–2133
4. Elias PM, Menon GK (1991) Structural and lipid biochemical correlates of the epidermal permeability barrier. Adv Lipid Res 24:1–26
5. Elias PM, Gruber R, Crumrine D, Menon G, Williams ML, Wakefield JS, Holleran WM, Uchida Y (2014) Formation and functions of the corneocyte lipid envelope (CLE). Biochim Biophys Acta 1841:314–318
6. Holleran WM, Takagim Y, Uchida Y (2006) Epidermal sphingolipids: metabolism, function, and role(s) in skin disorders. FEBS Lett 23:5456–5466
7. Horikoshi T, Igarashi S, Uchiwa H, Brysk H, Brysk MM (1999) Role of endogenous cathepsin D-like and chymotrypsin-like proteolysis in human epidermal desquamation. Br J Dermatol 141:453–459
8. Igarashi S, Takizawa T, Yasuda Y, Uchiwa H, Hayashi S, Brysk H, Robinson JM, Yamamoto K, Brysk MM, Horikoshi T (2004) Cathepsin D, but not cathepsin E, degrades desmosomes during epidermal desquamation. Br J Dermatol 151:355–361
9. Kishibe M (2019) Physiological and pathological roles of kallikrein-related peptidases in the epidermis. J Dermatol Sci. in press 95:50
10. Miyai M, Matsumoto Y, Yamanishi H, Yamamoto-Tanaka M, Tsuboi R, Hibino T (2014) Keratinocyte-specific mesotrypsin contributes to the desquamation process via kallikrein activation and LEKTI degradation. J Invest Dermatol 134:1665–1674
11. O'Regan GM, Kemperman PM, Sandilands A, Chen H, Campbell LE, Kroboth K, Watson R, Rowland M, Puppels GJ, McLean WH, Caspers PJ, Irvine AD (2010) Raman profiles of the stratum corneum define 3 filaggrin genotype-determined atopic dermatitis endophenotypes. J Allergy Clin Immunol 126:574–580.e571
12. Sun R, Celli A, Crumrine D, Hupe M, Adame LC, Pennypacker SD, Park K, Uchida Y, Feingold KR, Elias PM, Ilic D, Mauro TM (2014) Lowered humidity produces human epidermal equivalents with enhanced barrier properties. Tissue Eng Part C Methods 21:15–22
13. Uchida Y, Behne M, Quiec D, Elias PM, Holleran WM (2001) Vitamin C stimulates sphingolipid production and markers of barrier formation in submerged human keratinocyte cultures. J Invest Dermatol 117:1307–1313
14. Hanley K, Jiang Y, Elias PM, Feingold KR, Williams ML (1997) Acceleration of barrier ontogenesis in vitro through air exposure. Pediatr Res 41:293–299
15. Prunieras M, Regnier M, Woodley D (1983) Methods for cultivation of keratinocytes with an air-liquid interface. J Invest Dermatol 81:28s–33s

16. Williams ML, Hanley K, Elias PM, Feingold KR (1998) Ontogeny of the epidermal permeability barrier. J Invest Dermatol Symp Proc 3:75–79
17. Srinivasan B, Kolli AR, Esch MB, Abaci HE, Shuler ML, Hickman JJ (2015) TEER measurement techniques for in vitro barrier model systems. J Lab Autom 20:107–126
18. Chen S, Einspanier R, Schoen J (2015) Transepithelial electrical resistance (TEER): a functional parameter to monitor the quality of oviduct epithelial cells cultured on filter supports. Histochem Cell Biol 144:509–515
19. Schmuth M, Man MQ, Weber F, Gao W, Feingold KR, Fritsch P, Elias PM, Holleran WM (2000) Permeability barrier disorder in Niemann-Pick disease: sphingomyelin-ceramide processing required for normal barrier homeostasis. J Invest Dermatol 115:459–466
20. Honari G, Maibach H (2014) Skin structure and function. In: Dermatotoxicology, 8th edn. Elsevier, UK, pp 1–10
21. Pinnagoda J, Tupker RA, Agner T, Serup J (1990) Guidelines for transepidermal water loss (TEWL) measurement. A report from the Standardization Group of the European Society of Contact Dermatitis. Contact Dermatitis 22:164–178
22. Imhof RE, De Jesus ME, Xiao P, Ciortea LI, Berg EP (2009) Closed-chamber transepidermal water loss measurement: microclimate, calibration and performance. Int J Cosmet Sci 31:97–118
23. Crotty Alexander LE, Drummond CA, Hepokoski M, Mathew D, Moshensky A, Willeford A, Das S, Singh P, Yong Z, Lee JH, Vega K, Du A, Shin J, Javier C, Tian J, Brown JH, Breen EC (2018) Chronic inhalation of e-cigarette vapor containing nicotine disrupts airway barrier function and induces systemic inflammation and multiorgan fibrosis in mice. Am J Physiol Regul Integr Comp Physiol 314: R834–R847
24. Panisset F, Treffel P, Faivre B, Lecomte PB, Agache P (1992) Transepidermal water loss related to volar forearm sites in humans. Acta Derm Venereol 72:4–5
25. Hatano Y, Terashi H, Arakawa S, Katagiri K (2005) Interleukin-4 suppresses the enhancement of ceramide synthesis and cutaneous permeability barrier functions induced by tumor necrosis factor-alpha and interferon-gamma in human epidermis. J Invest Dermatol 124:786–792
26. Kuntsche J, Bunjes H, Fahr A, Pappinen S, Ronkko S, Suhonen M, Urtti A (2008) Interaction of lipid nanoparticles with human epidermis and an organotypic cell culture model. Int J Pharm 354:180–195
27. Pittelkow MR, Scott RE (1986) New techniques for the in vitro culture of human skin keratinocytes and perspectives on their use for grafting of patients with extensive burns. Mayo Clin Proc 61:771–777
28. Ponec M, Weerheim A, Kempenaar J, Mommaas AM, Nugteren DH (1988) Lipid composition of cultured human keratinocytes in relation to their differentiation. J Lipid Res 29:949–961
29. Shin KO, Choe SJ, Uchida Y, Kim I, Jeong Y, Park K (2018) Ginsenoside Rb1 enhances keratinocyte migration by a sphingosine-1-phosphate-dependent mechanism. J Med Food 21:1129–1136

Chapter 8

Isolating Dermal Papilla Cells from Human Hair Follicles Using Microdissection and Enzyme Digestion

Summik Limbu and Claire A. Higgins

Abstract

The dermal papilla (DP) is a cluster of mesenchymal cells located at the bottom of the hair follicle. Cells within the DP interact with numerous other cell types within the follicle, including epithelial stem cells, matrix cells, and melanocytes, regulating their function. The diameter of the DP is directly proportional to the width of the hair shaft, and a decrease in both cell number and DP size is observed in hair loss conditions such as androgenetic alopecia. Conversely, microdissected ex vivo DP can instruct growth of de novo hair follicles. The study of DP cells and their role in human hair growth is often hampered by the technical challenge of DP isolation and culture. Here we describe a method used within our research group for isolating DP from human hair follicles.

Key words Dermal papilla, Microdissection, Enzyme digestion, In vitro culture, Hair follicles, Human tissues

1 Introduction

The hair follicle is a dynamic mini-organ that undergoes continuous cycling and regeneration throughout its lifetime. A key component of the hair follicle that regulates this cycle is the dermal papilla (DP), a flame-shaped structure found at the base of the hair follicle. In human scalp follicles, there are on average 1000 cells per papilla [1], embedded in an extracellular matrix that is unique from that of the interfollicular dermis [2, 3] (Fig. 1a). These cells demonstrate heterogeneity from the apical tip to the base of the DP, and differentially signal to adjacent epithelial matrix cells controlling their proliferation, migration and differentiation into the epithelial layers of the hair follicle [4] (Fig. 1b, c). Both intact and cultured rodent vibrissae DP cells have been shown to induce formation of

Electronic supplementary material: The online version of this chapter (https://doi.org/10.1007/978-1-0716-0648-3_8) contains supplementary material, which is available to authorized users.

Natalia V. Botchkareva and Gillian E. Westgate (eds.), *Molecular Dermatology: Methods and Protocols*,
Methods in Molecular Biology, vol. 2154, https://doi.org/10.1007/978-1-0716-0648-3_8,
© Springer Science+Business Media, LLC, part of Springer Nature 2020

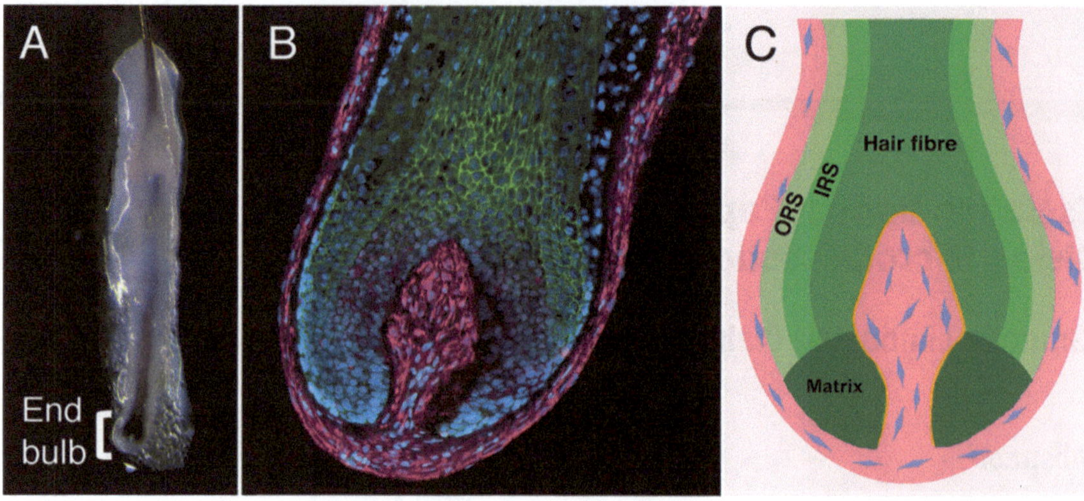

Fig. 1 Anatomy of a hair follicle. (**a**) The DP is located within the end bulb, at the bottom of the hair shaft. (**b**) The flame-shaped DP, immunostained with vimentin (pink), is surrounded by epithelial matrix cells (β-catenin, green). A dermal sheath also surrounds the whole hair follicle and is connected to the DP at its base, through the DP stalk (Blue = DAPI). (**c**) Illustration to show the structure of the end bulb and different layers of the hair follicle (*ORS* outer root sheath, *IRS* inner root sheath)

new hair follicles, making these cells an attractive therapy for regenerative medicine activities [5–8]. Additionally, cultured DP cells can differentiate into other cell types such as adipocytes, osteoblasts and chondrocytes [9]. This plasticity makes DP cells a useful tool for studying cell differentiation, and modeling both normal and disease processes in vitro [10].

The DP is separated from the epithelial compartment of the hair follicle by a thick basement membrane and therefore microdissection techniques can be to cleanly isolate. Culturing of rodent whisker DP cells after microsurgical isolation was first described in 1981 [11], while human scalp DP were successfully microdissected and cultured just a few years later, in 1984 [12]. In this book chapter, we present the method of microdissection of human hair follicles that is used within our laboratory, in a step by step approach.

In recent years, the use of collagenase type 4 has been described as a way to isolate DP [13, 14]; however, it is not as straightforward as a simple enzymatic digestion. Because of the unique extracellular matrix surrounding and encapsulating the human DP [2], the DP itself is not digestible with collagenase type 4. It is actually the tissue surrounding the DP that is digested, leaving behind an undigested DP which adheres to the culture dish as a sticky cluster of cells. This is not the case for murine DP which can be easily digested into a single cell suspension using collagenase and trypsin [15]. We assume that this is due to differences in the extracellular matrix makeup between human and mouse DP.

One of the difficulties in obtaining a good culture of human DP cells is in the first step after microdissection—attachment of the DP to the culture plate. If the DP does not attach to the culture plate, and remains suspended in the culture media, cells will not migrate out of the DP and no culture can be established. We have previously used a needle to attach DP to the culture dish, using it to pin the DP down so it remains adhered to the plastic enabling outgrowth of cells [16]. However, more recently we have adapted the collagenase digestion approach for isolating DP, which is described above, in order to improve anchoring of the DP to the culture dish. In this chapter, we describe both this "needle pinning" approach, and the enzymatic digestion approach for adhering the DP to the culture dish. Both techniques can be used to isolate human DP cells successfully and once established, there appears to be little difference in cell morphology.

2 Materials

Prepare growth, wash, and dissection media on the day of DP isolation. Prepare media in a cell culture hood at room temperature using aseptic technique. Store any remaining growth media in the fridge. Prepare wash and dissection media in 50 mL falcon tubes for convenience. These should be stored on ice next to the microscope while the microdissection is being performed.

2.1 Microdissection

1. Sterile 27 G × 3/4″ needles.
2. Sterile 1 mL syringes, without needles.
3. 100 mm petri dishes.
4. Binocular stereomicroscope.
5. Watchmaker forceps.
6. Noyes Spring Scissors.
7. 50 mL falcon tubes.
8. Wash media: 2× antibiotic-antimycotic (ABAM) made up from 100× stock in Dulbecco's Minimal Essential Media (DMEM) containing high glucose, GlutaMAX™, and pyruvate.
9. Dissection media: 1× ABAM in DMEM.
10. Laminar flow hood (referred to as a cell culture hood).
11. Sterile 200 μL pipette tips and pipette.
12. Sterile Pasteur pipettes.
13. Ice/Ice bucket.
14. Freshly isolated human Hair follicles.

2.2 Papilla Adherence and Growth

1. 27 G × 3/4″ needles.
2. 1 mL syringes, without needles.
3. 100 mm petri dishes.
4. 35 mm cell culture dishes.
5. Mixture of collagenase type 1 (1 mg/mL) and DNase 1 (12.8 U/mL) in DMEM with no serum. This is the working concentration. To make up the collagenase stock concentration of 10 mg/mL, weigh out 100 mg of collagenase type 1 and add 10 mL of sterile Hank's Balanced Salt Solution Modified with 10 mM HEPES in a cell culture hood. Filter the stock solution using a sterile 0.22 μm syringe filter and a 10 mL sterile syringe. Store the stock solution as 1 mL aliquots in sterile Eppendorf tubes at −20 °C until required. Make up collagenase solution by adding 1 ml stock with 9 ml DMEM. Add DNase 1 (at a final concentration 12.8 U/mL) to diluted collagenase solution on the day of isolation.
6. Growth media: 20% (v/v) fetal bovine serum (FBS) and 1× ABAM in DMEM.
7. Cell culture incubator—95% H_2O, 37 °C, 5% CO_2.

3 Methods

Ensure appropriate ethical approvals are in place before starting work since human tissue is to be used. Health and Safety forms relevant for your institution, for the use and disposal of human tissue and sharps should have also have been reviewed and approved before starting microdissection.

An ideal setup would be to have a stereomicroscope inside a cell culture hood; however, if this is not available it is possible to perform the microdissection on a laboratory workbench. Select a location in the corner of a room, where traffic and air movement is minimal. Spray the work surface with 70% ethanol before starting, and use sterile pipettes, petri dishes, tools, and needles throughout (Fig. 2a–d). Change needles regularly during the dissection procedure. We use petri dishes to place needles and tools into while they are in use, so that they do not touch the work surface.

All procedures are performed at room temperature, unless otherwise specified. Procedures described are for a right-handed person. Left-handed individuals may find it easier to reverse the needle handling instructions. All microdissection procedures are performed using the stereomicroscope. A Supplementary Video 1 has been produced which demonstrates all of the following methods.

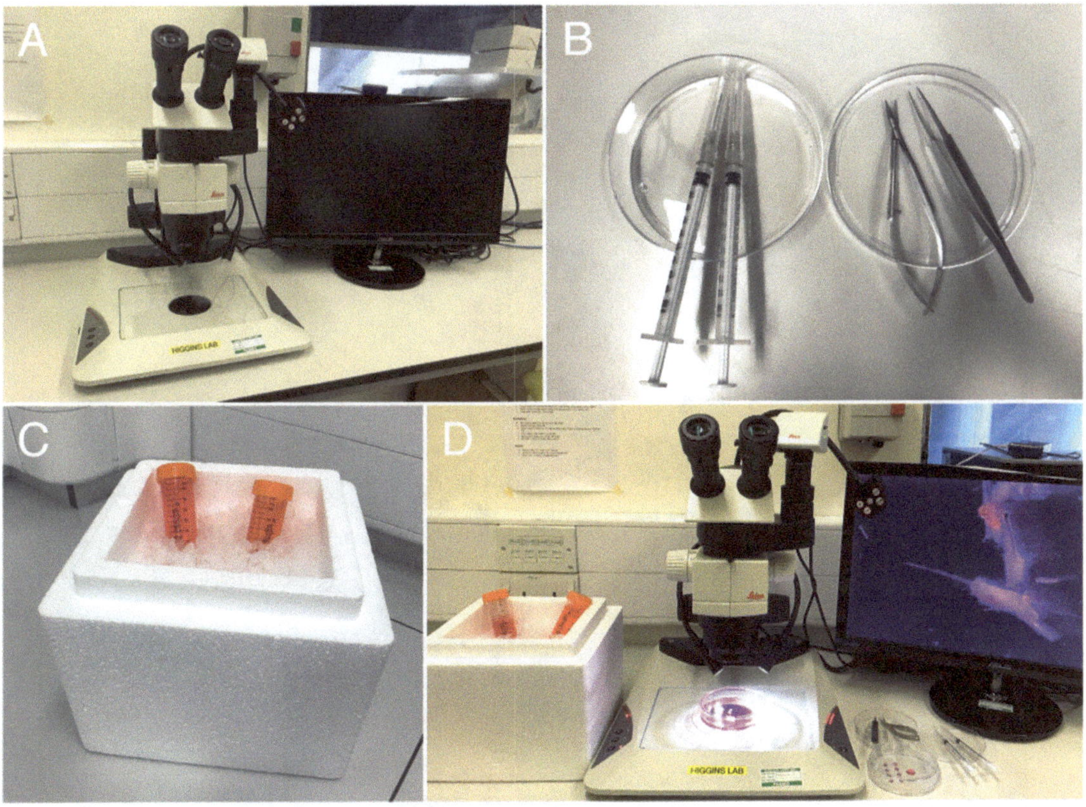

Fig. 2 Setting up the workstation for microdissection. (**a**) Place the binocular stereo microscope in the corner of the room with minimum foot traffic. (**b**) Place forceps, scissors and needles inside petri dishes to keep them clean during microdissection. (**c**) Place all media and samples on ice throughout the process. (**d**) Example of a workstation ready for microdissection

3.1 Microdissection of Hair Follicles

3.1.1 End Bulb Transection

1. Before isolation, prepare wash and dissection media in 50 mL Falcon tubes. Store unused media in an ice bucket.
2. Once hair follicles are obtained, use straightaway. If media is not prepared, keep the follicles on ice until all the reagents are ready.
3. If the follicles are transported from the clinic to the lab in saline or PBS, pour all of the hair follicles (including transport media) into an empty 100 mm dish. If follicles are wrapped in a damp gauze, skip straight to **step 4**.
4. Add 10 mL of wash media into a clean 100 mm petri dish and transfer the hair follicles into the wash media using forceps. Leave follicles in the wash media for 30 min on ice.
5. After 30 min, use forceps to transfer the hair follicles from the wash media into a new 100 mm petri dish containing dissection media. Ensure that all of the hair follicles have been moved into the dissection media.

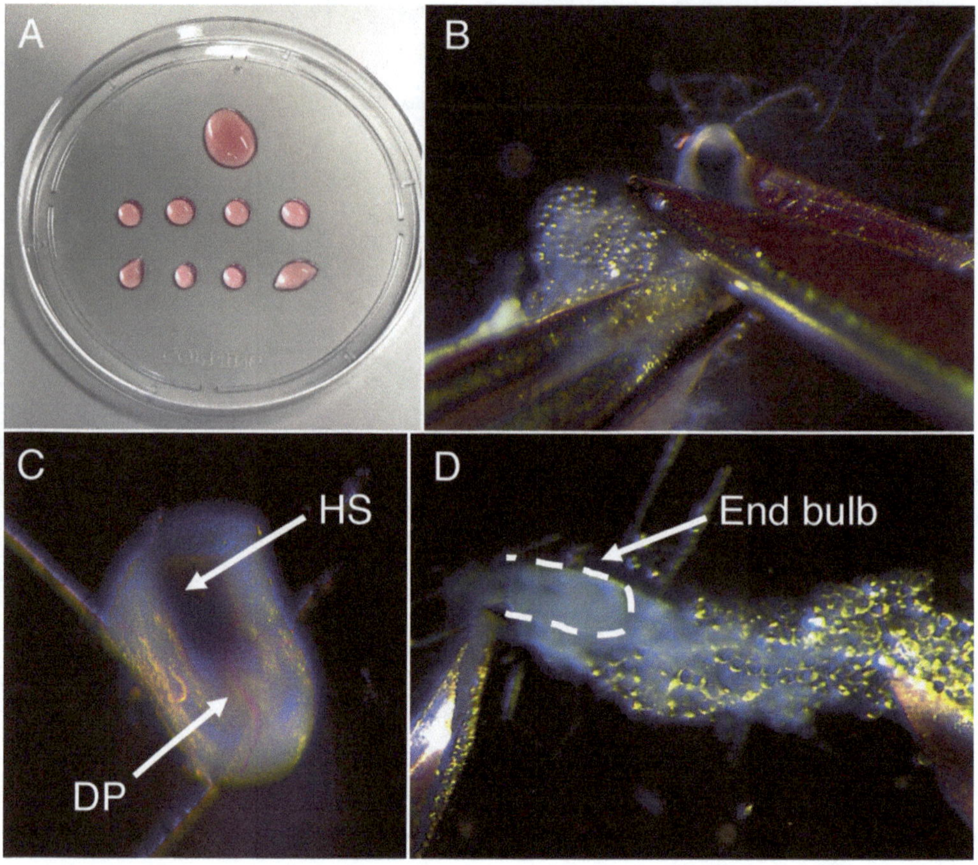

Fig. 3 Transection of end bulbs. (**a**) Make eight small droplets and one large droplet of 1 × DMEM which will be used to transfer end bulbs. (**b**) Hold the hair follicle with forceps and use a pair of scissors to transect the follicle, removing an end bulb. (**c**) Both the base of the hair shaft (HS), and the DP are located within the transected end bulb. (**d**) Excess fat which is present on end bulbs should be removed

6. Set up plates for dissection; on the lid of a 100 mm petri dish, use a Pasteur pipette to make eight small droplets and one big droplet of dissection media (Fig. 3a). Small droplets will be used for dissection while the large droplet is for collecting inverted samples.
7. Transect end bulbs; hold one follicle in place using forceps held in your left hand. Use scissors, held in your right hand, to cut the end bulb of the follicle containing the DP (*see* **Notes 1** and **2**, Fig. 3b, c).
8. After cutting, the end bulb will tend to stay attached to the scissors. Use these to transfer the end bulb into an empty small droplet on the plate prepared in **step 6**. Do not let the end bulb float away in the media. If it does, use the forceps to gently grab

onto the end bulb and transfer it into an empty droplet. Repeat with seven more hair follicles, so each droplet contains one end bulb (*see* **Note 3**).

9. Place any remaining follicles at 4 °C, in dissection media, until use.

3.1.2 Inversion of End Bulbs

Imagine a ball at the base of a sock. To remove the ball either the sock can be cut open, or it can be turned inside out. The sock is essentially the end bulb in our scenario, while the ball is the DP. Here, we describe an inversion method to isolate the DP; however, the cutting open method can also be applied [17]. Two sterile needles placed in 1 mL syringes as holders are required to invert the end bulb and expose the DP (*see* **Note 4**) .

1. If there is any remaining fat on the end bulbs, hold the cut edge of the end bulb using a needle in your left hand and drag fat away from the end bulb using a needle in your right hand (*see* **Note 5**, Fig. 3d).
2. To invert the end bulb, first use a needle in your left hand to pin down at the cut edge (*see* **Note 5**, Fig. 4a). Use a needle in your right hand to gently push the DP from the bottom of the end bulb (*see* **Note 6**, Fig. 4b, c). After the DP is out of the end bulb, use a needle in your left hand to push the end bulb cup back over the tip of the right-hand needle (Fig. 4d).
3. Use the needle in your left hand to gently prise the inverted end bulb off the tip of the needle in your right hand. The inverted end bulb should now be floating in the droplet.
4. Pin the inverted cup of the end bulb with the needle in your left hand, then use the needle in your right hand to clean the DP of any hair matrix which remains attached.
5. Use the needles to pick up and transfer the inverted end bulb into the big droplet.
6. Repeat **steps 1–7** with rest of the end bulbs and gather all eight of the inverted DPs together (Fig. 4e) before proceeding to the next stage of the dissection procedure.

3.1.3 Separating the DP from the Inverted End Bulb

1. Use new, clean, needles for this stage. To separate the DP, first use the needle in your left hand to pin the inverted end bulb cup. Then use the needle in your right hand like a knife edge and cleave through the stalk which connects the DP with the inverted cup (Fig. 5a, b).
2. Repeat **step 1** with the rest of the end bulbs.
3. Move all the DPs to a clean area of the larger drop, so they are separate from the inverted cups. The DPs should be clustered together (Fig. 5c).

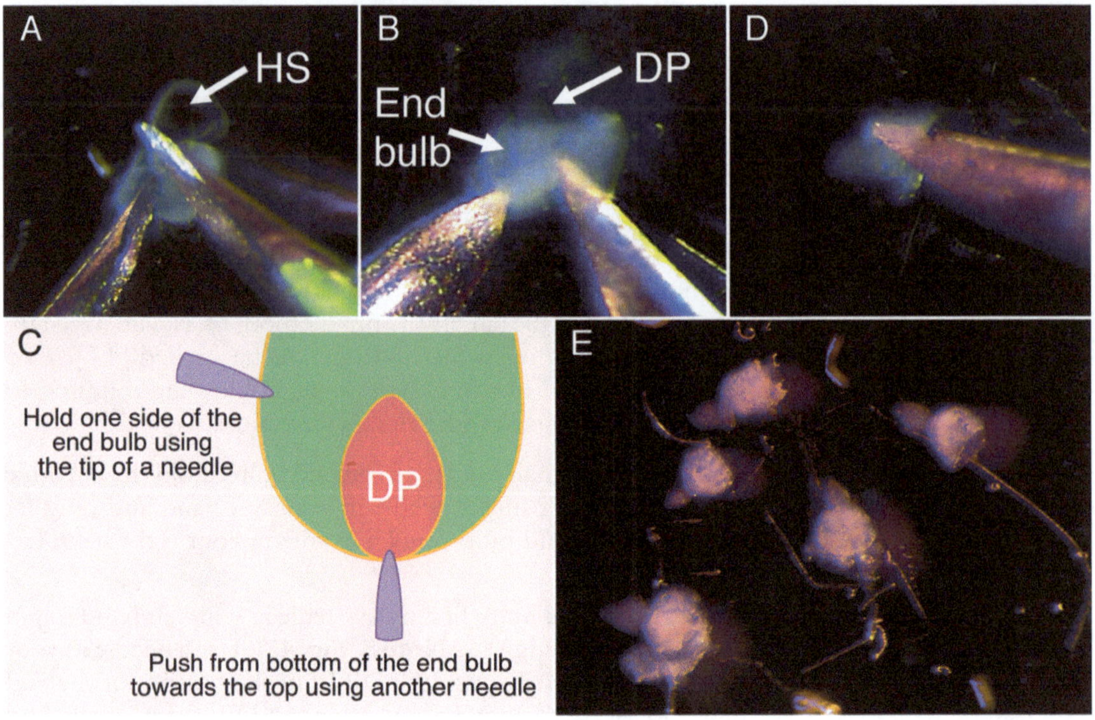

Fig. 4 Inverting the end bulb. (**a**) Hold the end bulb on the left side, then use the right hand needle to invert the bulb. The hair shaft will usually fall out at this stage. (**b**) To complete the inversion, push the needle in your right hand from the bottom toward the top of the end bulb. (**c**) Illustration to show the placement of needles around the end bulb during the process of inversion. (**d**) After inverting, the end bulb will be on the tip of the right hand needle. Clean, then transfer the end bulb to the big droplet. (**e**) Gathered inverted end bulbs together in the big droplet

3.2 Adherence of the DP

There are two variations on adhering the DP to the cell culture plate; attachment using needles and enzymatic digestion. Both alternatives are described below.

3.2.1 Attachment Using Needles

1. Place a 35 mm dish into a 100 mm dish. Add 2.5–3 mL of growth media into the 35 mm dish in a cell culture hood (*see* **Note** 7, Fig. 6a). Close both lids.
2. Carefully transfer the 100 mm dish (containing the 35 mm dish) so it is located next to the stereo microscope.
3. Transfer the DPs from the big droplet into the 35 mm dish (there are two methods for transferring the DP—*see* **Notes 8 (a)** and **(b)**).
4. Once DPs have been transferred, check the 35 mm dish under the stereo microscope. If any DP cells are floating, use a needle tip to push them below the surface. Once they are beneath the surface, they will sink to the bottom of the 35 mm dish.

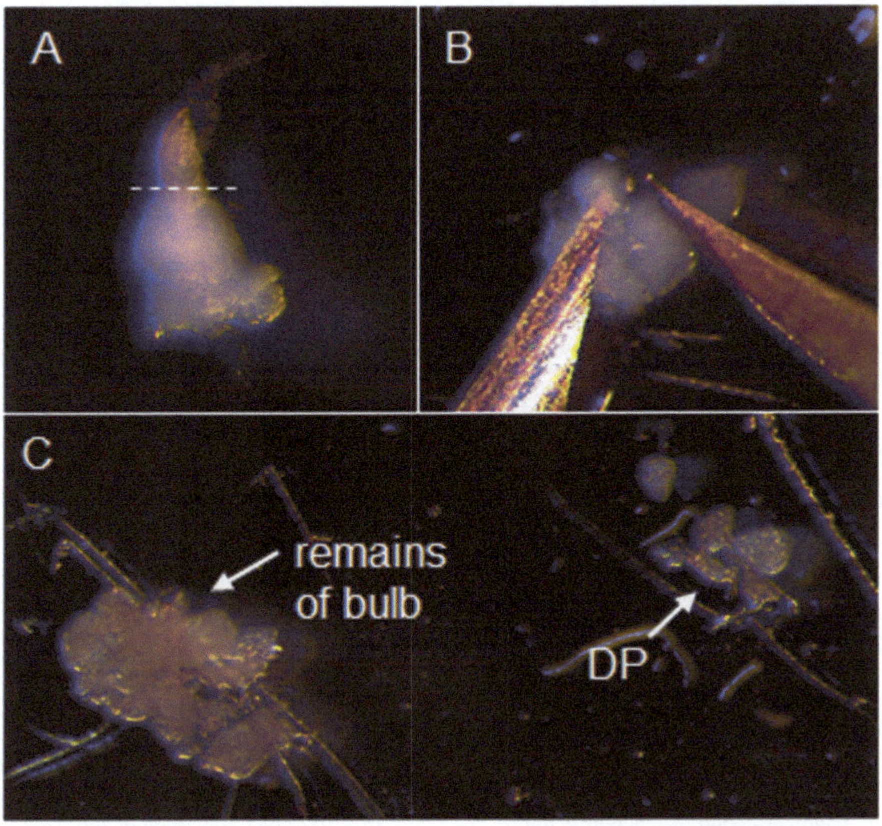

Fig. 5 Separating the DP from the inverted end bulb. (**a**) The line shows where the cut should be made, in the neck of the DP. (**b**) The DP is cut along the stalk to separate it from the inverted end bulb. Note the orientation of the needle tip. Using this orientation makes it easier to cut the DP from the inverted bulb. The motion should be like a cleaver, rather than a slice. (**c**) All the DPs are collected together in one place away from inverted end bulbs

5. To attach the DP to the plate, first use a needle in your left hand to keep the DP in place. Turn a needle in your right hand so the bevelled edge is facing down. Then, pin the DP to the plate by pushing the needle in your right hand into it. Turn the needle 90° clockwise so the bevelled edge is now facing to the left. Release pressure on the needle and drag the needle away from the DP, from left to right, keeping in contact with the plastic. This process will attach the DP to the plate while simultaneously cutting the DP; this allows cells to migrate out of the intact DP structure in a starburst formation (Fig. 6b, c).
6. Repeat **step 5** until all eight DP cells are adhered to the dish.
7. Close the lid of the 35 mm dish, and the outer 100 mm dish. Carefully transfer the dishes to an incubator.
8. Incubate at 37 °C, 5% CO_2 (*see* **Note 9**).

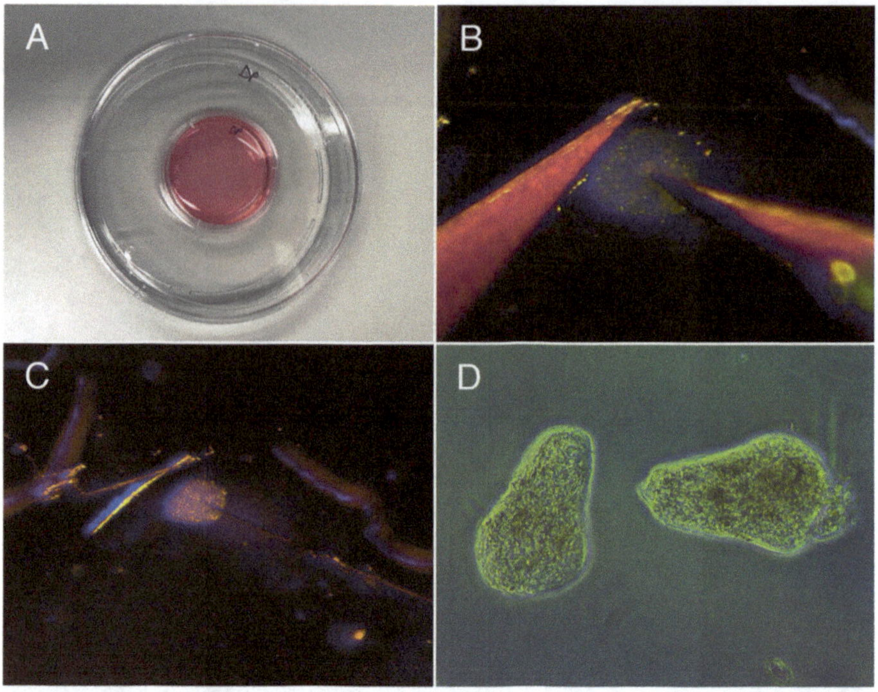

Fig. 6 Attaching the DP to the plate. (**a**) Place a 35 mm dish inside a 100 mm dish and add 2.5–3 mL of growth media in a cell culture hood. Take this dish to the stereo microscope. (**b**) "Pinning" method: to attach the DP to the plate, one needle is used to hold it in place, then the other needle is used to cut DP and pin it down. (**c**) After DP has been scratched, it stays attached to the plate. (**d**) After collagenase digestion, DPs are not fully digested and do not attach to the plate straight away

9. Do not move the dish for 10 days, to ensure strong attachment of the DP. After this time, change growth media every 2–3 days until confluency.

3.2.2 Enzyme Digestion

1. Place a 35 mm dish into a 100 mm dish inside a cell culture hood. Add 100 μL of collagenase 1 and DNase 1 mixture into the centre of the 35 mm dish.
2. Carefully move the dishes so they are positioned next to the stereo microscope.
3. Transfer DPs into the collagenase 1 and DNase 1 droplet (*see* **Note 8(a)** and **(b)**).
4. Place dish in to an incubator and leave for 30 min at 37 °C, 5% CO_2. The DP will not digest into single cells; they will remain intact (Fig. 6d). However, the edges will become sticky, so they will adhere to the base of the 35 mm dish.
5. After incubation, move the dishes to a cell culture hood. Use a Pasteur pipette to slowly add 2.5–3 mL of growth media into the 35 mm dish. The DPs do not attach to the bottom of plate during enzyme digestion; they will be free floating in the media at this stage.

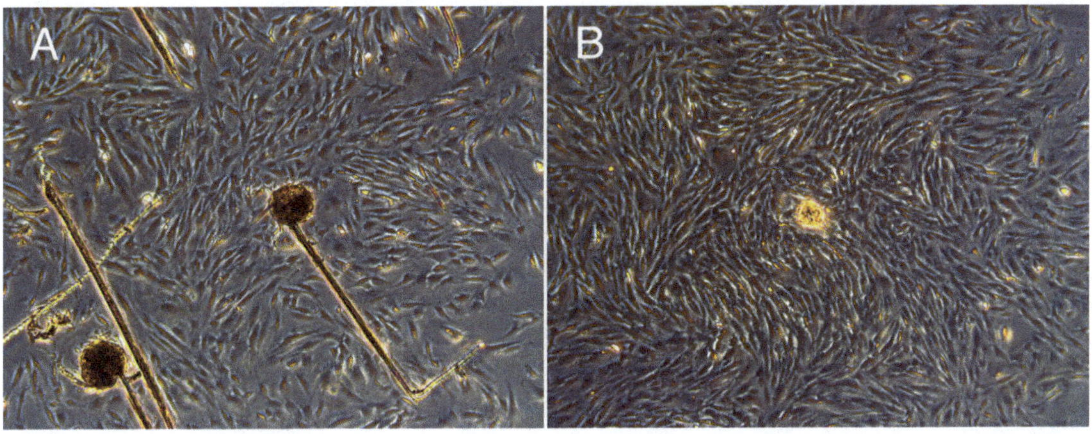

Fig. 7 Growth of DP cells 10 days after isolation. (**a**) DP attached using needle. (**b**) DP treated with collagenase type 1

6. Move dishes to the incubator, and incubate at 37 °C, 5% CO_2 (*see* **Note 9**).
7. Check for growth after 10 days (*see* **Note 10**). After this growth media should be changed every 2–3 days until confluency.

After 10 days, the DP structures start to collapse, and cells can be seen migrating out of the DPs (Fig. 7a, b). With DP cells that were attached using the "pinning" method, you will first observe cells migrating out of the "cut" which was made in the DP when it was pinned. Contrastingly, in the collagenase method, cells migrate from all edges of the DP. The number of cells observed migrating can be quite variable; we sometimes have DP with only 5–6 cells after 10 days, while surrounding others there will be 50–60 cells.

4 Notes

1. Cut the end bulb as close to the DP as possible. This makes it easier to invert.
2. If the hair follicle has fat attached, remove using scissors before cutting the end bulb. Do not pull at the fat using forceps, as it can disturb the DP and pull it away from the hair shaft, which can make it difficult to see the DP later on. If all of the fat cannot be removed, cut end bulb with some of the fat.
3. It is advisable to dissect and to do the whole process one plate at a time. Do not collect lots of end bulbs in multiple plates.
4. Always keep the needles in the petri dishes rather than the sides of the bench or the microscope. This ensures the needles are clean and also reduces the possibility of accidently grabbing onto the needles rather than the syringe.

5. Holding down the end bulb with a needle can dislodge the hair shaft out of the end bulb. If the hair shaft does not come out, then remove during the DP cleaning step.
6. Ensure that the needle in your right hand is pushing the DP toward the cut edge of the end bulb, not to the side. Be gentle; if too much pressure is applied, the needle could push through the DP and end bulb.
7. It is important to use 35 mm dish inside a 100 mm dish as moving the 35 mm dish by itself to an incubator could result in contamination.
8. There are two ways to transfer DPs into the 35 mm dish.
   (a) To transfer DPs using needles, turn the needle in your right hand to face the lumen of the bevel upward, so it is like a spoon. Use the needle in your left hand to gently push DP so they are sitting in the bevel. Move the needle slowly from the droplet and submerge beneath the media in the 35 mm dish. Tap the needle on the bottom of the 35 mm dish to dislodge the DP from the bevel. Check the needle in your right hand under the microscope to ensure no DP remain within it. Using this method, 1–2 DP can be transferred a time.
   (b) The DPs can also be transferred using a 200 μL pipette and tips. First, coat the interior of the tip with media containing FBS by taking up and pushing out media 3–4 times. Next, use the pipette to draw up DPs from the droplet, then transfer and expel into the 35 mm dish containing growth medium. If all of the DPs cannot be taken up in one go, transfer the DPs that have been already taken up and then repeat.
9. Place in an incubator that is not used frequently as the temperature changes during opening and closing of the incubator door can affect the growth of DP cells.
10. After 10 days, the enzyme digested DPs attach to the plate themselves and they seem to attach mostly on the edges of the dish. Thus, check the edges of the dish for growth.

Acknowledgments

This work was supported by an EPSRC iCASE award, sponsored by HairClone, to C.A.H.

References

1. Elliott K, Stephenson TJ, Messenger AG (1999) Differences in hair follicle dermal papilla volume are due to extracellular matrix volume and cell number: implications for the control of hair follicle size and androgen responses. J Invest Dermatol 113(6):873–877
2. Westgate GE, Messenger AG, Watson LP et al (1991) Distribution of proteoglycans during the hair growth cycle in human skin. J Invest Dermatol 96(2):191–195
3. Messenger AG, Elliott K, Westgate GE et al (1991) Distribution of extracellular matrix molecules in human hair follicles. Ann N Y Acad Sci 642:253–262
4. Yang H, Adam RC, Ge Y et al (2017) Epithelial-mesenchymal micro-niches govern stem cell lineage choices. Cell 169 (3):483–496.e13
5. Oliver RF (1967) The experimental induction of whisker growth in the hooded rat by implantation of dermal papillae. J Embryol Exp Morphol 18(1):43–51
6. Jahoda CAB, Horne KA, Oliver RF (1984) Induction of hair growth by implantation of cultured dermal papilla cells. Nature 311:560–562
7. Reynolds AJ, Jahoda CAB (1992) Cultured dermal papilla cells induce follicle formation and hair growth by transdifferentiation of an adult epidermis. Development 115:587–593
8. Jahoda CAB, Reynolds AJ, Oliver RF (1993) Induction of hair growth in ear wounds by cultured dermal papilla cells. J Investig Dermatol 101(4):584–590
9. Mistriotis P, Andreadis ST (2013) Hair follicle: a novel source of multipotent stem cells for tissue engineering and regenerative medicine. Tissue Eng Part B Rev 19(4):265–278
10. Logan NJ, Camman M, Williams G et al (2018) Demethylation of ITGAV accelerates osteogenic differentiation in a blast-induced heterotopic ossification in vitro cell culture model. Bone 117:149–160
11. Jahoda C, Oliver RF (1981) The growth of vibrissa dermal papilla cells in vitro. Br J Dermatol 105(6):623–627
12. Messenger AG (1984) The culture of dermal papilla cells from human hair follicles. Br J Dermatol 110(6):685–689
13. Wu JJ, Liu RQ, Lu YG et al (2005) Enzyme digestion to isolate and culture human scalp dermal papilla cells: a more efficient method. Arch Dermatol Res 297(2):60–67
14. Warren R, Chestnut MH, Wong TK et al (1992) Improved method for the isolation and cultivation of human scalp dermal papilla cells. J Investig Dermatol 98(5):693–699
15. Rendl M, Lewis L, Fuchs E (2005) Molecular dissection of mesenchymal-epithelial interactions in the hair follicle. PLoS Biol 3(11):e331
16. Topouzi H, Logan NJ, Williams G et al (2017) Methods for the isolation and 3D culture of dermal papilla cells from human hair follicles. Exp Dermatol 26(6):491–496
17. Magerl M, Kauser S, Paus R et al (2002) Simple and rapid method to isolate and culture follicular papillae from human scalp hair follicles. Exp Dermatol 11(4):381–385

Chapter 9

Methods to Study Human Hair Follicle Growth Ex Vivo: Human Microdissected Hair Follicle and Human Full Thickness Skin Organ Culture

Janin Edelkamp, Jennifer Gherardini, and Marta Bertolini

Abstract

The culture of microdissected hair follicles (HFs) and scalp skin enriched in terminal HFs are the best currently available preclinical assays for studying hair and skin biology/pathology in the human system. While microdissected HF organ culture only allows the testing of compounds added into the culture medium, mimicking a systemic application, the scalp skin organ culture also is suitable to test topical and intradermal applications. Here, we describe different methods for isolation of human scalp HFs, the procedures for culturing the scalp skin and microdissected HFs and we also outline different delivery techniques (e.g., topical, systemic) to test active and control substances.

Key words Human hair follicle, Human scalp skin, Organ culture, Hair growth, Compound testing

1 Introduction

Philpott et al. were the pioneers of hair follicle organ culture (HFOC). They demonstrated that microdissected rat HFs could be maintained for up to 7 days ex vivo. They also were first who developed the ex vivo model for culturing freshly isolated the so-called amputated human scalp HFs, that are cut below the bulge, therefore lacking the epithelial and melanocyte stem cell populations under serum-free conditions [1, 2]. In this model, Philpott and colleagues reported that the length of anagen VI HFs increases during their culture ex vivo, achieving hair shaft production with a similar rate (about 0.3 mm per day) to what can be seen in vivo [3, 4]. Since its original description, this model has been extensively used by several working groups investigating hair follicle biology, with some slight modification at times from the original protocol [5–8]. In this system, anagen VI HFs can spontaneously transition to catagen [1, 2, 5, 6], the regression phase of the hair cycle. In catagen, HF matrix cells cease dividing and hair

Natalia V. Botchkareva and Gillian E. Westgate (eds.), *Molecular Dermatology: Methods and Protocols*,
Methods in Molecular Biology, vol. 2154, https://doi.org/10.1007/978-1-0716-0648-3_9,
© Springer Science+Business Media, LLC, part of Springer Nature 2020

shaft formation is replaced by formation of a club hair [9]. This transition in HFs ex vivo can happen within different time in culture, which reflects inter-individual variability and can be accelerated by eliminating Insulin/IGF1 from the culture medium [10]. Therefore, the human amputated HFOC has been used to understand the key signaling pathways involved in the control of anagen–catagen transition and to its modulation ex vivo [2, 5–7, 11–14]. Additional techniques have recently been implemented for isolation of not only amputated but also full-length anagen HFs, containing the bulge area, which is bearing the stem cell populations [15], and HFs at different hair cycle stages, that is, telogen [16].

Using the same serum-free culture conditions, Lu et al. achieved the maintaining of viable scalp skin and HFs producing the hair shafts for up to 17 days [17], although epidermal thinning and HF keratinocyte apoptosis were already detected after 5 days of culture. For this reason, current methods restrict the culture of scalp skin using serum free-medium condition for up to 7 days [18, 19]. This assay is the closest ex vivo model that recapitulates in vivo situation and offers to study not only HFs activities, but also interactions of HFs with their surrounding environment. In this assay, test substances can be administered not only into the medium, mimicking systemic application, but also injected intradermally, or applied topically using viscous formulations [18, 19].

The original methods [1, 2, 17] have been substantially extended for a wide range of applications permitting a variety of instructive functional and mechanistic studies, and to mimic pathological conditions of the HF (e.g., HF immune privilege collapse [20–23], and epithelial–mesenchymal transition in the HF stem cells [15]). It is now also possible to silence the expression of the defined genes in the individual HFs [11, 24–28] and skin organ culture [26].

The possibility to maintain viable human HFs and scalp skin for several days in culture offers insights into the pathogenesis of hair growth disorders, including androgenetic alopecia, alopecia areata, scarring alopecia, and hirsutism.

In this chapter, we provide an overview of the methods to (1) microdissect amputated and full-length anagen HFs from scalp skin or follicular unit extractions (FUEs), (2) the procedure to culture human microdissected HFs and scalp skin, and (3) pitfalls and methodological tricks to be applied in order to perform a meaningful study. In microdissected amputated HFOC the major stem cell component (bulge) is cut off but HFs produce the hair shaft at a similar rate as in vivo. In contrast, microdissected full-length HFOC can be employed for research focused on the stem cell compartment or the infundibulum, or where emergence of a hair shaft from skin surface is a key measure.

2 Materials

Prepare all solutions under sterile conditions. Store all reagents according to manufacturer's instructions; the culture medium has to be stored at +4 °C. Diligently follow all waste disposal regulations, especially with regard to human tissue left overs.

2.1 Instruments and Tools

1. Long, curved tip forceps.
2. Short, curved tip forceps.
3. Short tip straight forceps (max 0.5 mm diameter tip).
4. Scalpel.
5. 4- or 6-mm biopsy punch.
6. Insulin syringe for intradermal injection.
7. 100/20 mm sterile petri dish.
8. 35/10 mm sterile petri dish.
9. 48-Well plate.
10. Binocular and/or digital bright field microscope.
11. CO_2 Incubator.
12. Laminar flow hood.

2.2 William's E Culture Medium (WCM)

William's E medium (WEM) supplemented with 2 mM L-glutamine, 10 ng/ml hydrocortisone, 10 μg/ml insulin, and 1% penicillin–streptomycin mix (*see* **Note 1**).

2.3 Human Samples

"Clinically" healthy human scalp skin or FUEs from donors undergoing routine face-lifts or hair transplantations obtained after ethics approval and signed informed consent by the donor. Skin samples and FUEs should arrive to the laboratory within 24 h after the surgical procedure and stored in William's E medium at 4 °C.

3 Methods

Before deciding which microdissection technique to use it is important to clarify the questions addressed in the experiment. This chapter describes methods for dissection of amputated HFs (without bulge region), from both facelift and FUE donor tissue (Fig. 1a), and microdissection of full-length pilosebaceous units (Fig. 1b), each having advantages and disadvantages for the question being addressed.

All procedures should be carried out at room temperature, unless indicated otherwise. Make sure that the pH of the culture medium in which human skin and freshly microdissected HFs are

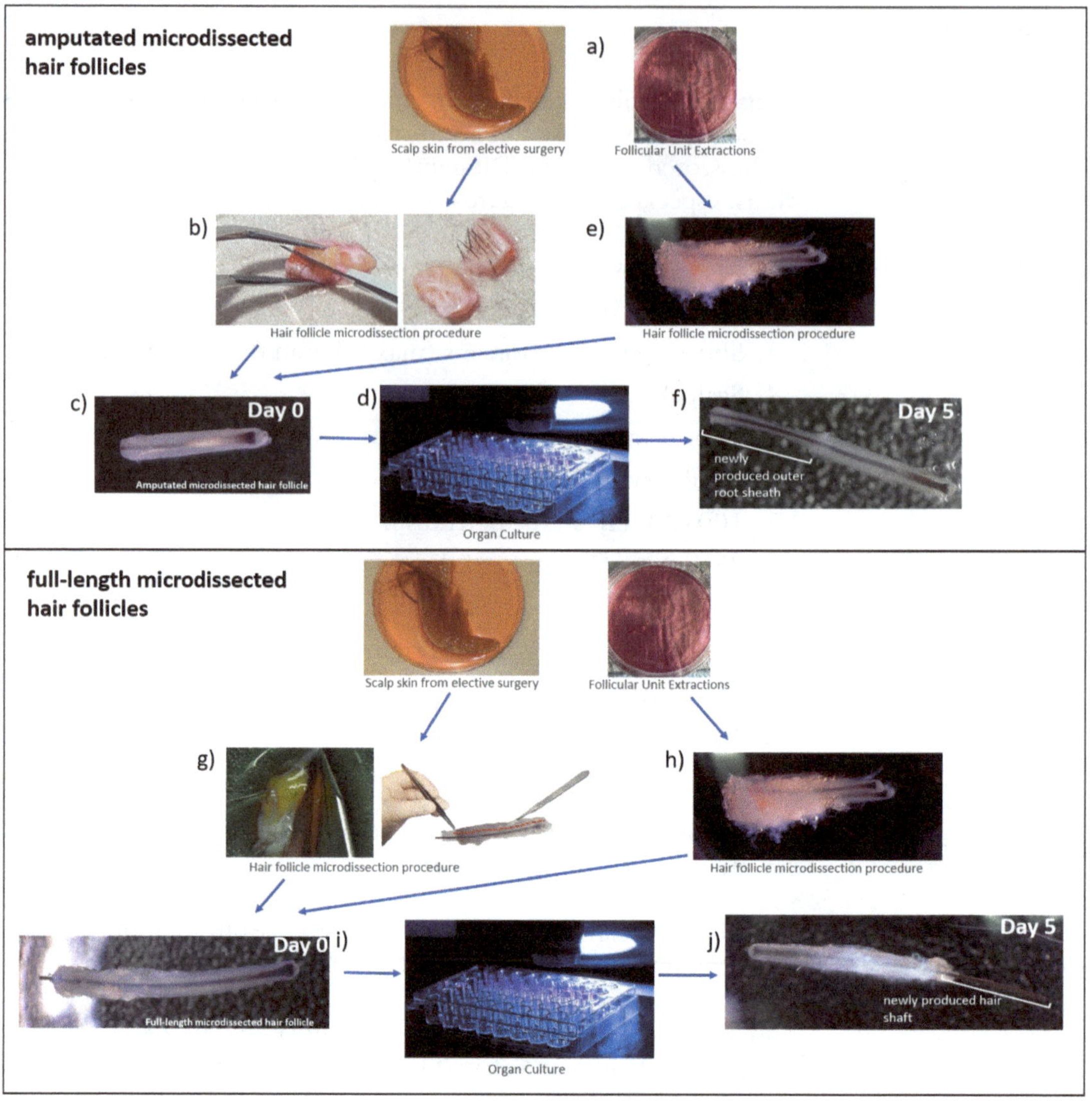

Fig. 1 Microdissection and culture of amputated (upper panel) or full-length (lower panel) hair follicles. The skin or FUEs are placed in a new petri dish upon arrival (**a**). Skin and FUEs are visually inspected and gently shaved/trimmed if necessary. The skin is cut into smaller pieces. For isolating amputated HFs, this is followed by an incision at the dermal–subcutaneous fat junction (**b**). If cut correctly, a "lattice structure" becomes visible. HFs are then isolated with forceps, cleaned if necessary, and placed singly in a 48-well plate containing supplemented William's E culture medium (**c**, **d**). FUEs are cut below the bulge (**e**). Afterward, follicles are cleaned from remaining tissue, and also placed singly in a 48-well plate containing supplemented William's E culture medium (**d**). For isolating full-length HFs, HFs are separated by performing a longitudinal cut using a scalpel between the HFs without damaging the connective tissue sheath of the neighboring HF (**g**). Afterward, HFs are cleaned from the surrounding tissue, by performing small cuts with a scalpel. The same procedure can be applied to clean HFs from FUEs (**h**). During the culture newly produced outer root sheath (amputated HFs), or hair shaft (full-length HFs) can be observed during the period of culture (e.g., from day 0 to day 5) (**c**–**f**, **i**, **j**)

placed remains red (a color shift to pink indicates alkaline pH, and will cause damage to the HFs). Do not dry the skin samples, FUEs, microdissected HFs, or skin fragments/punches.

3.1 Scalp Skin Preparation for HF Microdissection

1. Visually inspect the scalp skin sample in order to identify growth direction of the HFs and ensure that the hair bulbs are located deep within the subcutaneous fat, protecting the bulbs.
2. Place the skin in a 100 mm petri dish on a cooling pad with enough William's E-media to prevent dryness. Carefully shorten the hair shafts up to 2 mm length with a round-edge scissor.
3. Cut off a small skin fragments (approximately 0.5 × 0.5 cm) from the big skin sample, taking into account the growth direction of the hair, so as not to cut off the hair bulbs, and transfer skin fragments into a 100 mm petri dish lid (Figs. 2a and 3a).

3.2 Microdissection of Amputated HFs from Scalp Skin

1. Hold the skin with long, curved tip forceps (Fig. 2b) and make an incision with a scalpel at the level of the dermal–subcutaneous junction (*see* **Note 2**) (Fig. 2b–g).
2. Transfer and place the lid of the 100 mm petri dish under a binocular.
3. Use short, curved tip forceps to gently squeeze the skin piece containing the hair bulbs (subcutis) so that HFs pop out independently (Fig. 2h). With short tip straight forceps (max 0.5 mm diameter tip) gently pull out the HFs by grabbing the HFs from the top (Fig. 2i, j). Do not touch the bulb.
4. Transfer the microdissected HFs (Fig. 2k) with short tip straight forceps to a 35 mm dish containing fresh WCM for max 20–30 min (*see* **Notes 3** and **4**).

3.3 Microdissection of Full-Length HFs from Scalp Skin

1. Hold the skin with short tip straight forceps (max 0.5 mm diameter tip) at the level of the HF infundibulum or tissue surrounding the lower connective tissue (Fig. 3a, b).
2. Separate the HFs by performing a longitudinal cut using a scalpel between the HFs without damaging the connective tissue sheath of the neighboring HF (Fig. 3c–g).
3. Clean the HFs from the surrounding tissue by performing small cuts with a scalpel (Fig. 3i–k).
4. Transfer the microdissected HFs (Fig. 3l) with a short tip straight forceps in a 35 mm dish containing fresh WCM for max 20–30 min.

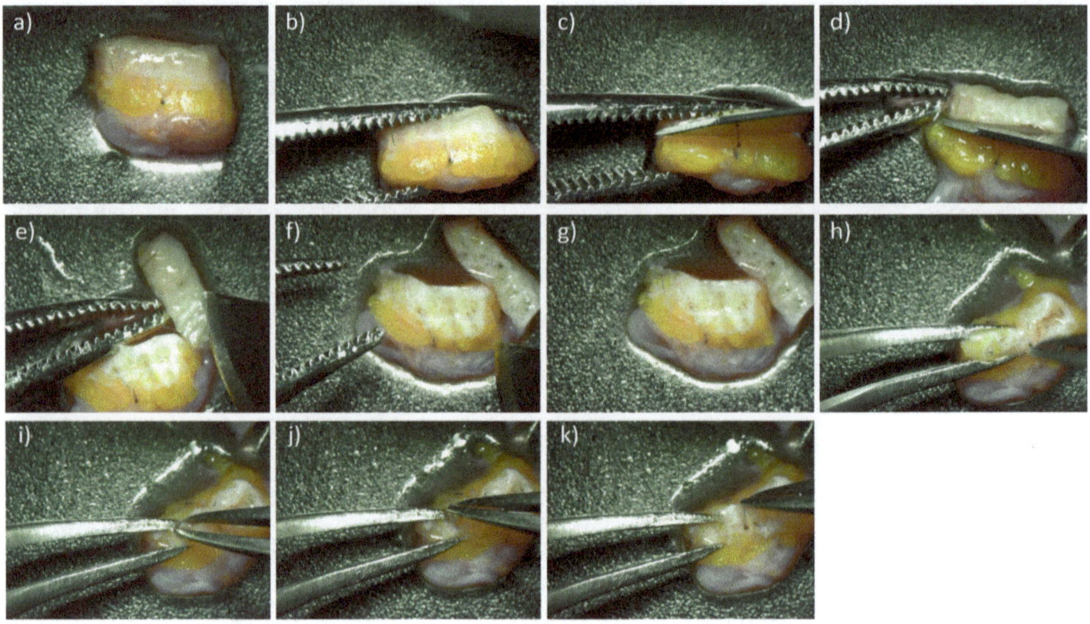

Fig. 2 Detailed procedure of isolating amputated microdissected hair follicles from full-thickness scalp skin (**a**–**k**)

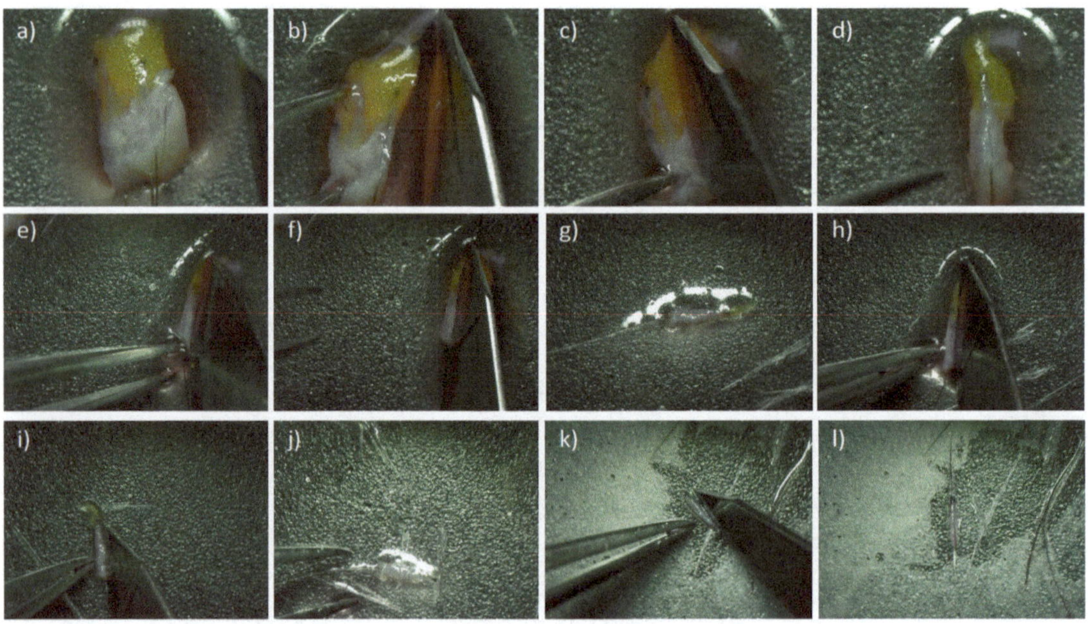

Fig. 3 Detailed procedure of isolating full-length microdissected hair follicles from full-thickness scalp skin (**a**–**l**)

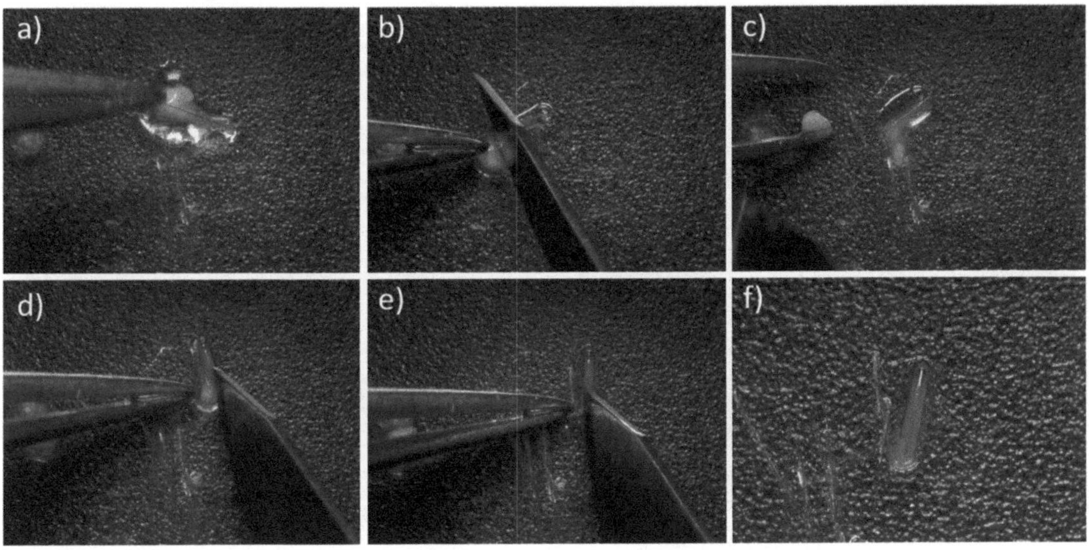

Fig. 4 Detailed procedure of isolating amputated microdissected hair follicles from FUEs (**a–f**)

3.4 Microdissection of Amputated HFs Ex Vivo from FUEs

1. Upon arrival, transfer FUEs into a 100 mm petri dish on a cooling pack with enough William's E-media to prevent dryness, and inspect for integrity of the connective tissue and outer root sheath, as well as for intact bulbs using a binocular microscope (*see* **Note 5**).
2. Transfer a single FUE into the lid of a 100 mm petri dish with some medium to prevent dryness.
3. Place the lid under a binocular microscope. Hold the FUE with short tip straight forceps (max 0.5 mm diameter tip) at the level of the HF infundibulum or tissue surrounding the lower connective tissue.
4. Make a sharp cut for each HF below the bulge region, that is, the one which corresponds approximately to the dermal subcutaneous junction (*see* Fig. 4). By doing that, usually, the 3–4 terminal HFs within FUEs are separated.
5. Clean the HFs by using a scalpel to scrape off the remaining surrounding dermal and subcutaneous tissue, and eccrine glands (*see* Fig. 4). Be careful to not cut the connective tissue sheath.
6. Transfer microdissected HFs with short tip straight forceps to a 35 mm dish containing fresh WCM.

3.5 Microdissection of Full-Length HFs Ex Vivo from FUEs

1. To prepare full-length HFs, transfer a single FUE into the lid of a 100 mm petri dish with some culture medium to prevent dryness.
2. Place the lid under a binocular microscope. Hold the FUE with a short tip straight forceps (max 0.5 mm diameter tip) at the

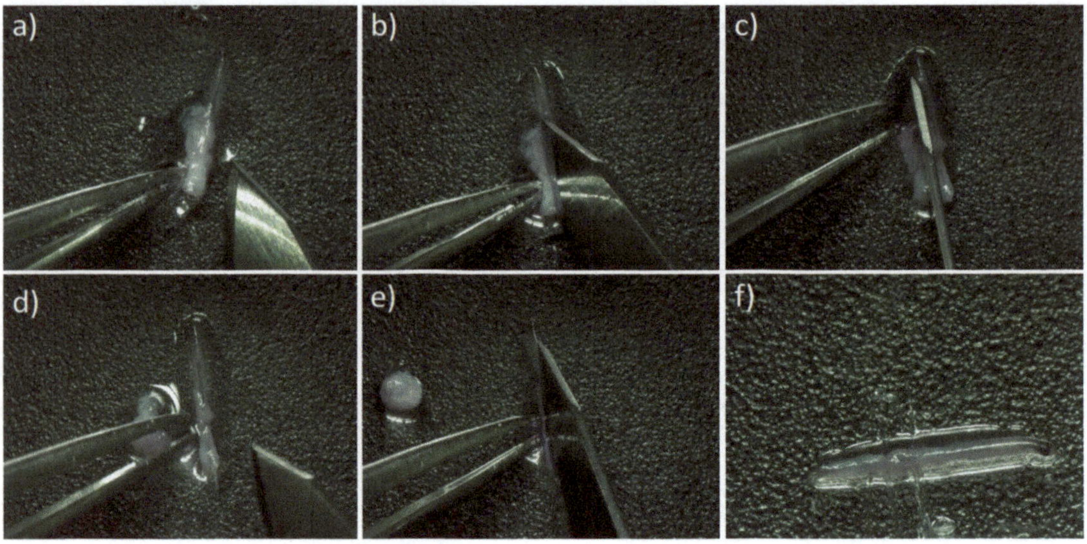

Fig. 5 Detailed procedure of isolating full-length microdissected hair follicles from FUEs (**a–f**)

level of the HF infundibulum or tissue surrounding the lower connective tissue.

3. Separate the HFs within a FUE by performing a longitudinal cut between the HFs without damaging the connective tissue sheath of the neighboring HFs.
4. Afterward, clean the HFs from the surrounding tissue, by performing small cuts with a scalpel (*see* Fig. 5).
5. Transfer microdissected HFs with short tip straight forceps to a 35 mm dish containing fresh WCM.

3.6 Plating of Amputated or Full-Length Microdissected HFs

1. Take a 48 well plate and add 250 μl of WCM per well.
2. Select anagen VI HFs under the binocular microscope and distribute one HF per well. This should be done by holding HF very gently with short tip straight forceps (max 0.5 mm diameter tip) from the distal connective tissue sheath (*see* **Note 6**).
3. Store the plate in the incubator at 37 °C with 5% CO_2 until the initiation of quality control (*see* **Note 7**).
4. Ensure that the microdissected HFs are not damaged and free from adnexal structures, such as sebaceous or eccrine glands, by examining them under a brightfield microscope (magnification approximately 50×).
5. Examine the microdissected HFs under the microscope to confirm that all HFs are in anagen VI; high magnification approximately 200× is recommended (*see* **Notes 6** and 7).

6. Measure HF length. To calculate HF length for amputated microdissected HFs, measure the distance from the lower point of the bulb connective tissue sheath to the upper point of the longer side of the outer root sheath. To calculate HF length for full-length microdissected HFs, measure the distance from the lower point of the bulb connective tissue sheath to the upper point of the hair shaft. This is the day 0 baseline measurement.

3.7 Initiation of the Culture of Amputated or Full-Length Microdissected HFs Ex Vivo

1. Culture the microdissected anagen VI scalp HFs at 37 °C with 5% CO_2 in serum-free William's E culture medium (WCM) for 24 h.
2. Inspect HFs under the microscope to confirm that all HFs are remained in anagen VI after the overnight incubation (high magnification approximately 200× is recommended). Discard HFs that are damaged or show signs that they have transitioned into catagen (*see* **Note 4**).
3. Measure HF length using a brightfield microscope. This is the day 1 measurement, which is important as a baseline measure for analyzing the effect of a test agent on HF growth. Only HFs that grew a certain percentage from day 0 to day 1 can be used in the experiment (*see* **Note 6**).

3.8 Continuation and Termination of the Culture of Amputated or Full-Length Microdissected HFs Ex Vivo

1. Divide the remaining anagen HFs into the corresponding experimental groups (*see* **Notes** 7 and **8**).
2. Replace the medium with fresh WCM or WCM + substance/ compound to test. One group should always be treated as vehicle control, that is, containing the solvent in which the test substance is dissolved at an appropriate concentration.
3. In a standard organ-culture, media is changed every other day and HFs are cultured for a maximum of 5–6 days (*see* Table 1) (*see* **Note 9**).

3.9 Full-Thickness Human Scalp Skin Organ Culture with Terminal Hair Follicles Ex Vivo

1. Visually inspect the skin in order to identify growth direction of the HFs, and ensure that the hair bulbs are located deep within the subcutaneous fat, so that the bulbs are protected.
2. Prepare square skin fragments using a scalpel or punch biopsies with a biopsy puncher (2, 4 or 6 mm diameter) along the growth direction of the HFs (*see* **Note 10**).
3. Place the skin samples either at air liquid interface within 500 μl/well of WCM in a 24 well plate (for topical application), or floating in 5000 μl/well of WCM in a 6-well plate (for systemic treatment). Incubate in the CO_2 incubator at 37 °C in a humidified atmosphere of 5% CO_2.

Table 1
Experimental procedure for HFOC

Day 0	Day 1	Day 2	Day 3	Day 4	Day 5	Day 6
Microdissected anagen VI HFs from scalp skin. Select anagen VI HFs using a binocular or a digital brightfield microscope, and plate them in 48well plate, one HFs/well, 250 µl medium/well. Measure HF length Incubate at 37 °C and 95% air–5% CO_2	Select anagen VI HFs for further culture, according to quality control guide lines, using a binocular or a digital brightfield microscope. Measure HF length. Change the culture medium. Apply the vehicle or test substance to the corresponding experimental group. Incubate at 37 °C and 95% air–5% CO_2	Rest	Measure HF length, and track macroscopic hair cycle changes. Change the culture medium. Apply the vehicle or test substance to the corresponding experimental group. Incubate at 37 °C and 95% air–5% CO_2	Rest	Measure HF length, and track macroscopic hair cycle changes. Change the culture medium. Apply the vehicle or test substance to the corresponding experimental group. Incubate at 37 °C and 95% air–5% CO_2	Terminate the culture

4. After 1 day of culture for equilibration, biopsies can be treated systemically, topically, or intradermally. For systemic treatment, replace the medium with fresh WCM containing the substance/compound, or the corresponding vehicle. For topical treatment, apply 2–5 µl of the formulation containing the test agent (*see* **Note 11**). For intradermal injection, inject the test agent into the dermis with max 20 µl of solution, using an insulin syringe (*see* Fig. 6).
5. Treatment schedule depends on the experimental design, but usually treatment/change of media is performed every other day, and skin biopsies are cultured for up to 7 days (*see* Table 2) (*see* **Note 9**).

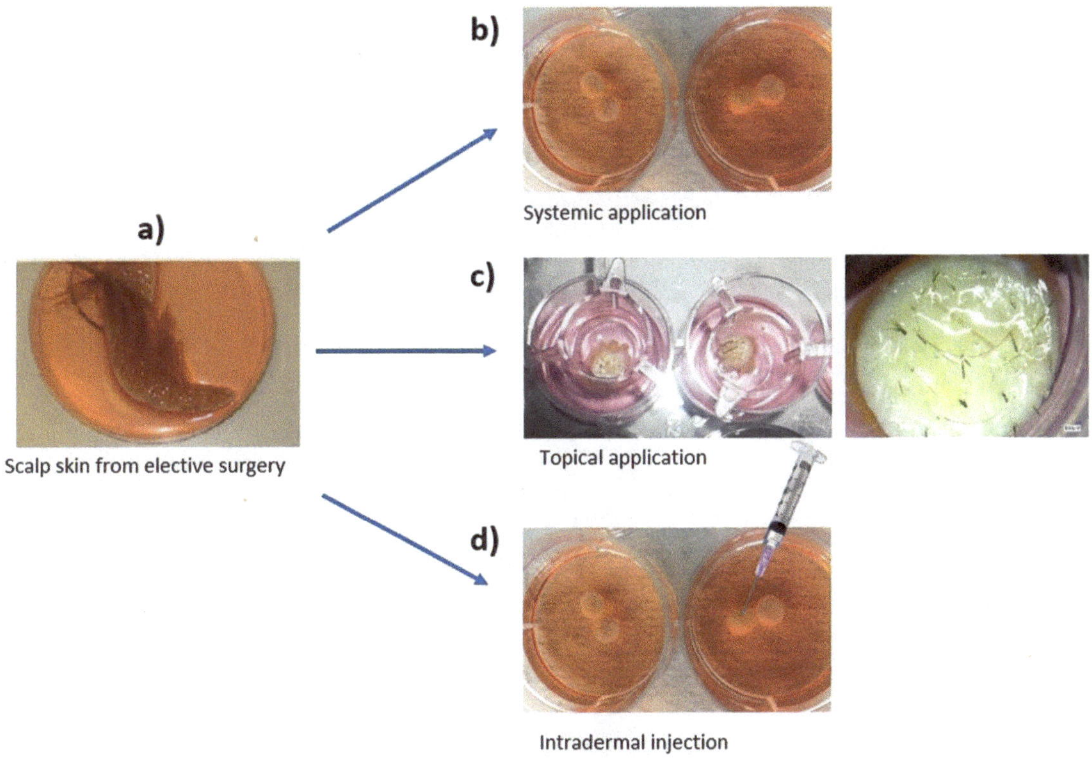

Fig. 6 Human full-thickness scalp skin organ culture—substance application. The scalp skin is placed in a new petri dish upon arrival (**a**). Skin is visually inspected and gently shaved/trimmed. 2–6 mm punch biopsies, or skin fragments are prepared for treatment. After 1 day of rest, substances/compounds can be either applied systemically, within the supplemented William's E culture medium (**b**), topically, by applying 2–5 μl of viscous formulation containing PEG6000 in the middle of the punch (**c**), or injected intradermally (**d**)

4 Notes

1. William's E culture medium can be reused if stored at +4 °C. It is important to carefully check the color of the medium because it can be reused as long as the color remains red.
2. The level of the incision is very important. If the cut is too superficial or too deep, it is not possible to isolate HFs with connective tissue sheath, as the latter would remain inside the skin. The incision is correct when a "lattice" structure (subcutis) is visible (Fig. 2g).
3. When the culture medium is left outside for too long, pH will shift to the alkaline range, which is damaging for the hair follicles.
4. We suggest not discarding catagen HFs. They can be used for isolation of distinct HF cell populations, or for gene expression studies, including immunohistology, Western blotting, and qRT-PCR. They can also be used for training and mastering new techniques.

Table 2
Experimental procedure for full-thickness human scalp skin organ culture with terminal hair follicles

Day 0	Day 1	Day 2	Day 3	Day 4	Day 5	Day 6
Prepare 2–6 mm skin punch biopsies. Place the skin samples either at air liquid interface within 500 µl/well of WCM in a 24 well plate, or floating in 5000 µl/well of WCM in a 6-well plate. Incubate at 37 °C and 95% air–5% CO_2	Assign the skin punches to the treatment groups. Change the culture medium. Apply the vehicle or test substance to the corresponding experimental group. Incubate at 37 °C and 95% air–5% CO_2	Rest	Change the culture medium. Apply the vehicle or test substance to the corresponding experimental group. Incubate at 37 °C and 95% air–5% CO_2	Rest	Change the culture medium. Apply the vehicle or test substance to the corresponding experimental group. Incubate at 37 °C and 95% air–5% CO_2	Terminate the culture

5. Sometimes, FUEs are damaged when they are extracted during a hair transplantation surgery, which can be seen as longitudinal cuts in the outer root sheath, or squeezed bulbs. Only intact FUEs can be used for the successful HF growth ex vivo, damaged ones may have uses as described in **Note 4**.
6. When studying hair growth ex vivo, it is important to keep in mind that different factors influence the outcome of the experiment, for example, individual rate of growth variability, intrinsic hair cycle of the individual HFs, and other factors [6, 29]. To ensure robust and reproducible results, the quality control of the isolated HFs is very important. Therefore, it is crucial to select only anagen VI HFs that have grown between day 0 and day 1 for starting the treatment. It is also very important to make sure that HFs remained in anagen VI after the resting overnight incubation, because HF keratinocyte apoptosis causing transition to catagen can induce a pseudoelongation of the HF, as a result of the pushing hair shaft outward due to the "apoptotic force" during the regression of the HF epithelium [30].
7. The HF is considered as an independent mini-organ [6, 31]; individual HFs can exhibit different responses to treatments, even if anagen VI HFs were obtained from the same donor at the same time, and from the same skin location. Therefore, it is

very essential to have a minimum of five anagen VI HFs (preferably more than ten) per experimental group to obtain meaningful results. Lastly, it is important to understand that the model also reflects interindividual differences, as well as sex-dependent differences in HF biology. Therefore, it is highly important to select the most representative cohort (e.g., sex, age), and to perform a minimum of $n = 3$ independent experiments to confirm the results.

8. In order to study hair growth ex vivo, it is advisable to include positive and/or negative controls in the HFOC or HSOC. Minoxidil is not an optimal "positive control" for anagen prolongation in HFOC [6, 31], while thyrotropin-releasing hormone [32] and cyclosporin A [12] significantly prolong anagen ex vivo. There are many hair growth-inhibitory, catagen-promoting agents proven to work in human HFOC or HSOC; however, interferon gamma (IFNγ) [33], testosterone [8], and nicotinamide [18] are the most potent.
9. At the termination of the culture, HFs or skin samples should be processed for the analysis. HFs or skin samples can be immediately snap-frozen in liquid nitrogen, stored in RNAlater or lysis buffer and then snap frozen in liquid nitrogen, placed into a droplet of OCT (maximum three HFs per droplet) and then snap frozen in liquid nitrogen, or fixed in 4% PFA and embedded in paraffin. For the analysis of basic hair growth markers in situ, HFs or skin samples should be either embedded within a droplet of OCT (maximum three HFs per droplet) and snap frozen in liquid nitrogen, or fixed in paraffin, followed by their sectioning at 6 μm-thickness. The following analyses should be performed: % of HF elongation (as indicator of hair shaft production); hair cycle-dependent proliferation and apoptosis of hair matrix keratinocytes, using Ki-67/TUNEL immunofluorescence detection; hair cycle dependent pigmentation, using Masson Fontana or Warthin Starry histochemistry; intrafollicular expression of anagen or catagen associated growth factors (e.g., IGF-1 and TGFß2). Hair cycle associated HF gross morphology, as a preliminary investigation, can also be performed, but it should be confirmed by histomorphological analysis [1, 2, 5–8, 11–14, 18–20, 26, 28, 29, 33].
10. Carefully select the size of the skin punches/fragments (max 6 mm) for the experiment that enables medium perfusion into the center of the tissue biopsy and that has the required density of the HFs.
11. Carefully select a vehicle for the topical application that enhances substance/compound penetration and prevents compound spillover from the skin surface into the medium [18, 19].

References

1. Philpott MP, Green MR, Kealey T (1989) Studies on the biochemistry and morphology of freshly isolated and maintained rat hair follicles. J Cell Sci 93:409–418
2. Philpott MP, Green MR, Kealey T (1990) Human hair growth in vitro. J Cell Sci 97:463–471
3. Trotter M (1924) The life cycles of hair in selected regions of the body. Am J Phys Anthropol VII:427–437
4. Myers RJ, Hamilton JB (1951) Regeneration and rate of growth of hairs in man. Ann N Y Acad Sci 53:562–568
5. Kloepper JE, Sugawara K, Al-Nuaimi Y et al (2010) Methods in hair research: how to objectively distinguish between anagen and catagen in human hair follicle organ culture. Exp Dermatol 19:305–312
6. Langan E, Philpott MP, Kloepper JE et al (2015) Human hair follicle organ culture: theory, application and perspectives. Exp Dermatol 24(12):903–911
7. Buscone S, Mardaryev AN, Raafs B, Bikker JW, Sticht C, Gretz N, Farjo N, Uzunbajakava NE, Botchkareva NV (2017) A new path in defining light parameters for hair growth: discovery and modulation of photoreceptors in human hair follicle. Lasers Surg Med 49(7):705–718
8. Fischer TW, Hipler UC, Elsner P (2007) Effect of caffeine and testosterone on the proliferation of human hair follicles in vitro. Int J Dermatol 46(1):27–35
9. Oh JW, Kloepper J, Langan EA, Kim Y, Yeo J, Kim MJ, Hsi TC, Rose C, Yoon GS, Lee SJ, Seykora J, Kim JC, Sung YK, Kim M, Paus R, Plikus MV (2016) A guide to studying human hair follicle cycling in vivo. J Invest Dermatol 136(1):34–44
10. Philpott MP, Sanders DA, Kealey T (1994) Effects of insulin and insulin-like growth factors on cultured human hair follicles: IGF-I at physiologic concentrations is an important regulator of hair follicle growth in vitro. J Invest Dermatol 102(6):857–861
11. Chéret J, Bertolini M, Ponce L, Lehmann J, Tsai T, Alam M, Hatt H, Paus R (2018) Olfactory receptor OR2AT4 regulates human hair growth. Nat Commun 9(1):3624
12. Hawkshaw NJ, Hardman JA, Haslam IS, Shahmalak A, Gilhar A, Lim X, Paus R (2018) Identifying novel strategies for treating human hair loss disorders: Cyclosporine A suppresses the Wnt inhibitor, SFRP1, in the dermal papilla of human scalp hair follicles. PLoS Biol 16(5): e2003705
13. Oláh A, Gherardini J, Bertolini M et al (2016) The thyroid hormone analogue KB2115 (eprotirome) prolongs human hair growth (anagen) ex vivo. J Invest Dermatol 36 (8):1711–1714
14. Harel S, Higgins CA, Cerise JE, Dai Z, Chen JC, Clynes R, Christiano AM (2015) Pharmacologic inhibition of JAK-STAT signaling promotes hair growth. Sci Adv 1(9):e1500973
15. Imanishi H, Ansell DM, Chéret J et al (2018) Epithelial-to-mesenchymal stem cell transition in a human organ: lessons from lichen Planopilaris. J Invest Dermatol 38(3):511–519
16. Alam A, Hernandez I, Paus R et al (2018) Image gallery: intravital visualization of the dynamic changes in human hair follicle cycling. Br J Dermatol 178(6):e396
17. Lu Z, Hasse S, Bodo E et al (2007) Towards the development of a simplified long-term organ culture method for human scalp skin and its appendages under serum-free conditions. Exp Dermatol 16(1):37–44
18. Haslam I, Hardman JA, Paus R (2018) Topically applied nicotinamide inhibits human hair follicle growth ex vivo. J Invest Dermatol 138 (6):1420–1422
19. Gherardini J, Wegner J, Chéret J et al (2019) Transepidermal UV radiation of scalp skin ex vivo induces hair follicle damage that is alleviated by the topical treatment with caffeine. Int J Cosmet Sci 41(2):164–182
20. Ito T, Ito N, Bettermann A, Tokura Y, Takigawa M, Paus R (2004) Collapse and restoration of MHC class-I-dependent immune privilege: exploiting the human hair follicle as a model. Am J Pathol 164(2):623–634
21. Meyer KC, Bodó E, Brzoska T, Abels C, Paus R (2009) Immunomodulatory effects of the alpha-melanocyte-stimulating hormone-related tripeptide K(D)PT on human scalp hair follicles under proinflammatory conditions. Br J Dermatol 161(6):1400–1403
22. Kinori M, Bertolini M, Funk W, Samuelov L, Meyer KC, Emelianov VU, Hasse S, Paus R (2012) Calcitonin gene-related peptide (CGRP) may award relative protection from interferon-γ-induced collapse of human hair follicle immune privilege. Exp Dermatol 21 (3):223–226
23. Bertolini M, Pretzlaff M, Sulk M, Bahr M, Gherardini J, Uchida Y et al (2016) Vasoactive intestinal peptide, whose receptor-mediated signalling may be defective in alopecia areata, provides protection from hair follicle immune

privilege collapse. Br J Dermatol 175 (3):531–541

24. Hardman JA, Tobin DJ, Haslam IS, Farjo N, Farjo B, Al-Nuaimi Y, Grimaldi B, Paus R (2015) The peripheral clock regulates human pigmentation. J Invest Dermatol 135 (4):1053–1064
25. Al-Nuaimi Y, Hardman JA, Bíró T, Haslam IS, Philpott MP, Tóth BI, Farjo N, Farjo B, Baier G, Watson REB, Grimaldi B, Kloepper JE, Paus R (2014) A meeting of two chronobiological systems: circadian proteins Period1 and BMAL1 modulate the human hair cycle clock. J Invest Dermatol 134(3):610–619
26. Samuelov L, Sprecher E, Sugawara K, Singh SK, Tobin DJ, Tsuruta D, Bíró T, Kloepper JE, Paus R (2013) Topobiology of human pigmentation: P-cadherin selectively stimulates hair follicle melanogenesis. J Invest Dermatol 133(6):1591–1600
27. Samuelov L, Sprecher E, Tsuruta D, Bíró T, Kloepper JE, Paus R (2012) P-cadherin regulates human hair growth and cycling via canonical Wnt signaling and transforming growth factor-β2. J Invest Dermatol 132 (10):2332–2341
28. Paus R, Cotsarelis G (1999) The biology of hair follicles. N Engl J Med 341(7):491–497
29. Stenn KS, Paus R (2001) Controls of hair follicle cycling. Physiol Rev 81(1):449–494
30. Schneider MR, Schmidt-Ullrich R, Paus R (2009) The hair follicle as a dynamic miniorgan. Curr Biol 19(3):R132–R142
31. Magerl M, Paus R, Farjo N, Müller-Röver S, Peters EM, Foitzik K, Tobin DJ (2004) Limitations of human occipital scalp hair follicle organ culture for studying the effects of minoxidil as a hair growth enhancer. Exp Dermatol 13(10):635–642
32. Gáspár E, Hardenbicker C, Bodó E, Wenzel B, Ramot Y, Funk W, Kromminga A, Paus R (2010) Thyrotropin releasing hormone (TRH): a new player in human hair-growth control. FASEB J 24(2):393–403
33. Ito T, Ito N, Saathoff M, Bettermann A, Takigawa M, Paus R (2005) Interferon-gamma is a potent inducer of catagen-like changes in cultured human anagen hair follicles. Br J Dermatol 152(4):623–631

Chapter 10

Nonsurgical Induction of Alopecia Areata in C3H/HeJ Mice via Adoptive Transfer of Cultured Lymphoid Cells

Eddy H. C. Wang and Kevin J. McElwee

Abstract

Surgical induction of alopecia areata (AA) via full-thickness grafting of spontaneous AA-affected C3H/HeJ mouse skin to naïve recipients has been a primary method of transferring the AA disease model phenotype. However, this method is associated with the need to perform an invasive procedure that could negatively impact animal wellbeing. Therefore, a rodent model that rapidly develops AA at a predictable rate and without the need to perform invasive surgical procedures on the mice is essential for studying the pathogenesis of AA. Here we describe a cell injection technique using cultured skin-draining lymph node cells (LNCs) injected intradermally into naïve recipients to induce rapid AA development. The cultured LNCs can reach ~ten fold expansion after 6 days with specific cytokine stimulation. The LNCs derived from a single AA affected mouse donor can induce AA development in more than 80 naïve mice within 2–18 weeks. For comparative control studies, mice receiving cultured LNCs from normal donors remain normally haired. The method enables the production of large numbers of AA mice for use in research and treatment development studies while avoiding the use of surgical procedures. We anticipate that the protocol can also be adapted for use in other mouse autoimmune disease models.

Key words Alopecia areata, C3H/HeJ, Mouse model, Adoptive transfer, T cells

1 Introduction

Alopecia areata (AA) is believed to be an inflammatory, nonscarring, cell-mediated autoimmune hair loss disease. There is a 1.7% lifetime risk for AA in humans; it can affect both genders in all age groups [1]. Patients affected with AA typically experience patchy hair loss on the scalp, but it can progress to complete loss of scalp or even body hair. AA has an unpredictable nature and there are currently no effective treatments. Because of this, AA can have a severe psychological burden, especially in women and children [1].

There are several different species of animals that spontaneously develop AA-like symptoms similar to humans. However, the rate of spontaneous AA development in most of these species is extremely low making them difficult to use as practical models to study AA

Natalia V. Botchkareva and Gillian E. Westgate (eds.), *Molecular Dermatology: Methods and Protocols*, Methods in Molecular Biology, vol. 2154, https://doi.org/10.1007/978-1-0716-0648-3_10,
© Springer Science+Business Media, LLC, part of Springer Nature 2020

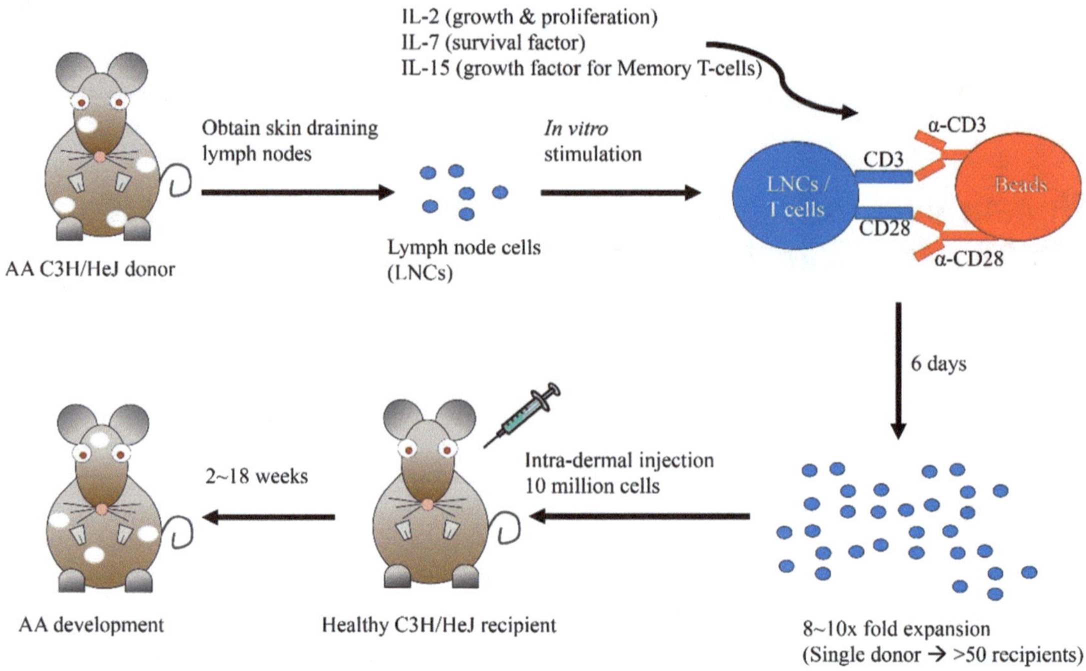

Fig. 1 Simplified cartoon flowchart of LNC isolation, culture, and injection into C3H/HeJ mice. Skin draining lymph nodes were obtained from an AA affected donor and were processed via single cell suspension to obtain LNCs. The LNCs were cultured for 6 days with the supplementation of IL-2, IL-7, and IL-15 as well as magnetic beads coated with anti-CD3 and anti-CD28 cross-linking antibodies. Ten million cells were injected intradermally to naïve recipients and AA onset first developed 2–18 weeks post injection

[2, 3]. Therefore, an easily accessible, small animal model that can be consistently induced to develop AA, while sharing a high biological resemblance to human AA, is required.

The C3H/HeJ mouse model is one of the most popular and well-defined rodent models for AA research [4]. It is an inbred strain where individual mice spontaneously develop AA-like hair loss at a low frequency. Up to 20% of C3H/HeJ mice aged 18 months and older will spontaneously develop AA [5]. Previously, a surgical method was developed to transfer the AA phenotype from AA affected mice to healthy histocompatible recipients as a way to control disease onset and increase numbers [6]. This was achieved by grafting full-thickness skin taken from spontaneous AA affected C3H/HeJ donors, to naïve recipients of the same strain or related histocompatible strains [6, 7]. The skin-graft AA model has proven useful with its similarity to human AA as well as being relatively easy to generate. The model has been used to test several experimental treatments for human AA [8, 9].

However, the major draw-backs are the need to perform a relatively long, invasive surgery on the mice, and the use of controlled drugs for anesthesia and recovery during surgery. Additionally, the time it takes for the mice to recover from the skin-graft surgery is relatively long and there is a risk of infection and graft rejection [6]. Most importantly, only 10–20 recipient mice can be grafted with skin from a single AA donor and it takes about 10 weeks before AA starts to develop [6].

We developed a culturing technique to significantly increase the skin-draining lymph node cell (LNC) population isolated from spontaneously affected AA mice and adoptively transfer the cells into naïve C3H/HeJ mice to induce AA development (Fig. 1). To facilitate the activation and expansion of LNCs, we stimulated the LNCs with magnetic beads coated with anti-CD3 and anti-CD28 cross-linking antibodies optimized for nonspecific expansion of T-cells. In tandem, interleukin-2 (IL-2), an important T-cell growth factor, was added to the culture medium. To ensure the survival of important T-cell clones and the growth of memory T-cells that may be important for triggering the development of AA, interleukin-7 (IL-7) and interleukin-15 (IL-15) were also added to the culture medium [10–12].

By adoptively transferring the cultured LNCs from an AA-affected donor to naïve recipients, we were able to induce AA development with a success rate over 90% [13]. The AA mice generated using this method displayed similar pathological features as observed in AA mice induced via fresh (noncultured) LNC injection, as well as AA induced by skin grafting. Comparative control littermates that received cultured LNCs derived from normal haired mice had a very low frequency of AA development—consistent with rates of spontaneous AA in standard C3H/HeJ colonies. This alternate AA-induction method has made in vivo AA research much more feasible, because of its simplicity and consistency, and the technique has been adopted by several investigators as well as pharmaceutical companies [14, 15].

2 Materials

2.1 Reagents and Animals

1. Naïve female C3H/HeJ mice as cell recipients (at least 10 weeks old, no AA phenotype; The Jackson Laboratory, Bar Harbor, ME) (*see* **Note 1**).
2. AA affected female C3H/HeJ mice as cell donors (at least 70% hair loss. The Jackson Laboratory).
3. Dynabeads Mouse T-Activator CD3/CD28 (Gibco, Burlington, ON).
4. Human Recombinant IL-2 (Roche Life Science, Laval, QC).
5. Mouse Recombinant IL-7 (R&D Systems, Minneapolis, MN).
6. Mouse Recombinant IL-15 (R&D Systems, Minneapolis, MN).
7. Sterile PBS, pH = 7.4.

8. Sterile DPBS.
9. EDTA (0.5 M), pH 8.0.
10. Advanced RPMI 1640 (Gibco, Burlington, ON).
11. GlutaMAX (Gibco, Burlington, ON).
12. Fetal Bovine Serum.
13. Penicillin-Streptomycin.
14. Trypan blue.

2.2 Equipment

1. Centrifuge tubes.
2. Petri dishes, 35 × 10 mm.
3. Centrifuge capable of up to 600 × *g*.
4. Hemocytometer.
5. Inverted microscope.
6. 37 °C Incubator with 5% CO_2.
7. Surgical scissors.
8. Forceps.
9. Electric hair shaver.
10. Nontreated tissue culture plates—24 well.
11. Suspension cell TC flasks T25/T75 vented.
12. Sterile round bottom falcon tubes, 12 × 75 mm.
13. 70 μM cell strainer.
14. 10 mL Syringes.
15. EasySep™ Magnet (StemCell Technologies, Vancouver, BC).
16. 1 cc insulin syringes.
17. Heat pad or heat lamp.
18. Anesthesia and euthanasia apparatus.

2.3 Reagent Setup

1. Dynabead wash buffer: Sterile PBS with 0.1% FBS and 2 mM EDTA, pH 7.4.
2. 1× PBS and DPBS: Dilute with sterile distilled H_2O.
3. Complete medium AR10: Advanced RPMI 1640 with 10% FBS, 2 mM GlutaMAX, and 100 U/mL Penicillin–Streptomycin.
4. Mouse Recombinant IL-7: Reconstitute to 25 ng/μL in sterile PBS with 0.1% FBS. Aliquot and store in −20 °C.
5. Mouse Recombinant IL-15: Reconstitute to 50 ng/μL in sterile PBS with 0.1% FBS. Aliquot and store in −20 °C.
6. Complete medium AR10 supplemented with cytokines: AR10 supplemented with 30 U/mL human recombinant IL-2,

25 ng/mL mouse recombinant IL-7 and 50 ng/mL mouse recombinant IL-15.

7. Dynabead mixture: Follow the manufacturer's protocol to resuspend the Dynabeads via vortex for 30 s. Twenty-five microliters of Dynabeads is required for each one million LNCs. Pipet Dynabeads into a sterile round bottom centrifuge tube and add 1 mL of Dynabead wash buffer, pipet gently 30 times without generating air bubbles. Place the centrifuge tube into an EasySep magnet for 1 min. Decant/pipet the wash buffer into a waste container while keeping the tube inside the magnet; the Dynabeads will adhere to the wall of the tube. Remove the tube from the magnet and add AR10 supplemented with cytokines at 500 μL/25 μL Dynabeads. Make sure all of the Dynabeads on the wall of the tube are resuspended into the medium.

All reagents should be made fresh prior to use.

3 Methods

3.1 Isolation of Skin-Draining Lymph Nodes

This procedure will take approximately 30 min.

1. Euthanize the donor mice using institution approved procedures. Clean mouse skin with 70% ethanol.
2. Use surgical scissors and forceps to dissect the mice from the ventral side.
3. Excise the inguinal, auxiliary and cervical skin-draining lymph nodes with forceps or scissors and trim away the extra fat and connective tissues.
4. Place the trimmed skin-draining lymph nodes into complete AR10 medium and keep on ice (*see* **Note 2**).

3.2 Separation of Lymph Node Cells (LNCs) into Single Cell Suspension

This step will take approximately 1 h.

1. Work in a biosafety cabinet, with aseptic techniques, fill a small petri dish with 1 mL of 1× DPBS.
2. Remove the lymph nodes from AR10 and transfer into DPBS to wash by brief rinsing.
3. In another clean petri dish, add 1 mL of fresh AR10 and place a 70 μM cell strainer in the middle of the dish.
4. Pick up the lymph nodes with a pair of forceps and transfer into the middle of the cell strainer.
5. Use the plunger end of a syringe to gently mash the lymph nodes against the cell strainer for about 5 min.

6. The culture medium will turn cloudy as the lymph nodes disintegrate, leaving a small amount of white connective tissue in the strainer (*see* **Note 3**).
7. Pipet 1 mL of fresh AR10 and rinse the bottom of the cell strainer to wash the extra LNCs into the petri dish. There should be 2 mL of LNC suspension at this step.
8. Remove the cell strainer and transfer all of the LNC suspension into a clean 15 mL centrifuge tube.
9. Wash the petri dish with another 1 mL of fresh AR10 and transfer into the centrifuge tube.
10. Top up centrifuge tube to 10 mL with another 7 mL of fresh AR10 and pipet thoroughly but gently to wash the LNC suspension.
11. Take out 10 μL and count the cell number with a hemocytometer and trypan blue or equivalent method to count the cells (*see* **Note 4**).
12. Prepare complete medium AR10 with cytokines. Make enough AR10 with cytokines to resuspend the cell pellet into two million cells per mL.
13. Centrifuge the LNC suspension for 5 min at 350 × *g* at room temperature.
14. Remove the supernatant carefully leaving the cell pellet.
15. Resuspend the cell pellet to two million cells per mL with AR10 supplemented with cytokines.
16. Pipet 1 mL of LNC suspension into a 24-well, non-tissue culture treated plate until desired number of wells are filled. Each single well should now have two million LNCs and each is expected to expand eight- to tenfold after 6 days (*see* **Note 5**).
17. Place the 24-well plates in the 37 °C incubator with 5% CO_2.
18. Prepare Dynabead mixture at this point, resuspend well.
19. Take out the 24-well plates from the incubator and add 500 μL of Dynabead mixture into each well of the plate. Each well should have two million cells in 1.5 mL of medium supplemented with cytokine and Dynabeads.
20. Transfer the plate back into 37 °C incubator with 5% CO_2.

3.3 Activation and Expansion of LNCs

This procedure will take 6 days.

1. Check the culture daily but do not disturb the LNC suspension (*see* **Note 6**).
2. The LNCs will expand and cover the entire bottom of the wells; after 48 h, the color of culture medium will turn light yellow.

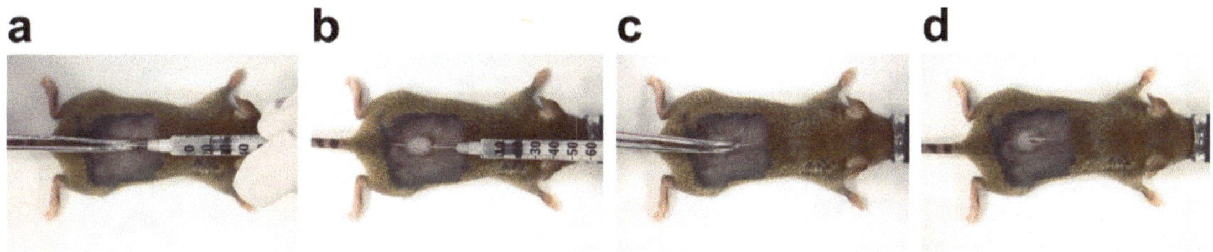

Fig. 2 Injection of cultured LNCs intradermally to the back skin of C3H/HeJ mice. A small area on the back of the recipient mice is shaved and cleaned with 70% ethanol before injection. The needle needs to enter the skin almost parallel to the pinched skin to ensure the cells will not go into the subcutaneous layer (**a**). LNC suspension (ten million cells in 100 μL PBS) is injected slowly into the dermis layer and a small bulge forms (**b**). A pair of blunt forceps is used to clamp the needle wound opening from the injection briefly to close the wound (**c**). The injected LNCs and fluid volume in the bulge slowly dissipate into the skin (**d**)

3. Split each well to two by gently resuspending the culture and pipetting 750 μL to an adjacent well. Add 750 μL of AR10 supplemented with cytokines so each well contains 1.5 mL.
4. After another 24 h, combine 2× wells into 1× T25 flask and add 3 mL of AR10 supplemented with cytokines.
5. After another 24 h, combine 2× T25 into 1× T75 flask and add 8 mL of AR10 supplemented with cytokines (*see* **Note 7**).

3.4 Preparation of Expanded LNCs for Injection

This step will take approximately 1.5 h.

1. Combine the LNC suspension in T75 flasks into 50 mL centrifuge tube(s).
2. Place a clean round bottom centrifuge tube into an EasySep magnet or similar.
3. Transfer LNCs suspension into the centrifuge tube to a level at the same height as the magnet and let it set for 1 min.
4. Gently pipet the LNC suspension from the centrifuge tube to a clean 50 mL centrifuge tube.
5. Repeat **steps 3–4** until removal of all Dynabeads from all LNC suspension has been achieved (*see* **Note 8**).
6. Take 10 μL and count the number of cells with a hemocytometer (*see* **Note 9**).
7. Centrifuge the LNC suspension for 5 min at 350 × *g* at room temperature (*see* **Note 10**).
8. Remove the supernatant from the LNC pellet and resuspend into ten million cells per 100 μL with sterile PBS.
9. Draw the resuspended LNC suspension into insulin syringe (s) or similar and keep on ice (*see* **Note 11**).

This procedure will take approximately 5–10 min per mouse.

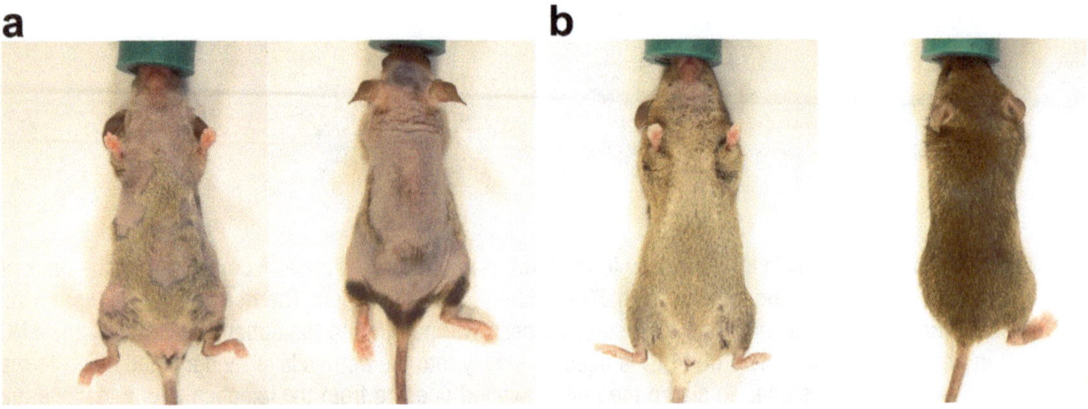

Fig. 3 Hair loss in mice receiving cultured AA LNCs usually initially develops on the ventral skin away from the injection site. The mice that receive cultured AA LNCs usually first develop patchy AA on their ventral skin and eventually the alopecia progresses to the dorsal skin (**a**). The mice that receive cultured control LNCs obtained from healthy mice without AA remain fully haired and have a very low chance of developing spontaneous hair loss (**b**)

3.5 Adoptive Transfer of LNCs into Naïve C3H/HeJ Mice by Intradermal Injection

1. Anaesthetize the recipient mice that are at least 10 weeks old using isoflurane or similar following the standardized protocols as provided by the local institution (*see* **Note 1**).
2. Shave a small area on the lower back of the mice to expose an area of skin for intradermal injection.
3. Gently pinch up a section of the skin with a pair of blunt forceps and insert the syringe needle almost parallel to the plane of the skin pinched up by the forceps, into the dermis layer (*see* **Note 12** and Fig. 2).
4. Inject 100 μL of LNC suspension slowly into the skin. A small bulge will appear at the site of injection.
5. Retract the syringe, but use the forceps to hold the skin for a few seconds to allow the injected LNCs to dissipate a little bit as well as to allow the injection site opening to close.
6. Transfer the injected mice back to their cage but take care not to press on the injection site.
7. The mice should recover from isoflurane anesthesia within 10 min.

3.6 Development of Alopecia Areata

Initial development of hair loss will take approximately 8–10 weeks.

1. Continue to feed the mice with a low-fat diet. High-fat/oil diets reduce success rates of AA transfer.
2. Hair at the site of injection will start to grow back after about 1 week due to injury induced hair growth.
3. Monitor the ventral side of the mice every few days as hair loss can start to develop as early as 2 weeks post injection and

typically initially occurs away from the injection site (*see* **Notes 13** and **14**).

3.7 Anticipated Results

3.7.1 Single Cell Suspension and LNC Expansion

The number of fresh LNCs from each donor AA mouse is variable; we can typically obtain around 50–90 million cells from one donor. The rate of cell expansion can also be variable, usually at 48 h changes in cell morphology can be observed. By 72 h, the number of cells should cover over 90% of the bottom of the 24-well plate. The rate of cell expansion will start to slow down after 6 days; therefore, it is not advisable to culture cells beyond that point.

3.7.2 The Progression of Alopecia Areata in Cell-Injected Mice

We have observed the mice start to lose hair as early as 2 weeks and as late as 18 weeks after injection with AA cells; between 7 to 10 weeks seems to be the peak where most mice will first develop AA hair loss. The hair loss does not begin at the site of injection, rather, it usually progresses from patches first observed on the ventral side to the dorsal skin; a pattern similar to that observed with skin grafted mice (Fig. 3). It should be anticipated that not all AA mice receiving cultured AA LNCs will develop AA within 20 weeks; some mice are apparently resistant to AA induction while any AA development after 20 weeks is unlikely to be a result of cell injection [16]. Conversely, it is also possible for a small number of control mice that received cultured control LNCs taken from normal haired donors to develop AA simply because the C3H/HeJ strain is known to develop spontaneous AA at a low rate.

4 Notes

1. Mice younger than 10 weeks old are relatively resistant to AA induction.
2. Process the lymph nodes as soon as possible, prevent drying of tissues.
3. While grinding the lymph nodes, change the angle and direction often to ensure complete breakdown of the tissues to recover the maximum amount of LNCs.
4. Once the total number of cells is calculated, determine how many cells will be used for expansion. Ten million cultured cells are required to inject into a single recipient; the number of cells will expand around eight- to tenfold during the 6-day culture. Therefore, we initiate the culture with two million cells for each mouse we expect to inject to ensure we will have enough cells for the target number of cell recipients.
5. Keeping the LNCs at a high density throughout the culturing process allows better cell-to-cell contact as well as interaction

with the antibody coated magnetic beads. In pilot experiments, initiating the cell culture in a T25 flask with a larger volume of media did not yield as high cell number compared to initiating cultures in a 24-well culture plate.

6. Clumping of LNCs with the Dynabeads can be observed after 24 h. Do not attempt to break up the clumps as they will disappear after 48 h.
7. The passaging protocol here serves as a general guideline only. The purpose is to keep the LNCs at very high density, around 1.5–2 million cells/mL.
8. Remember to keep the centrifuge tube inside the magnet while transferring LNC suspension to a new 50 mL falcon tube. If pipetting, do not scratch the wall of the tube as the Dynabeads will come off back into the LNC suspension.
9. If cell recovery is low even though expansion was noted (total cell number less than eight-fold of initial number), wash the plate or flask again with AR10. Even though LNCs are suspension cells, some of them still tend to adhere to the bottom of plates or flasks.
10. If Dynabeads were not completely removed, red colored precipitates can be seen within the cell pellets after centrifuge.
11. It is preferable to avoid using normal syringes as there will be a small volume inside the syringe that is very difficult to eject, insulin syringes bypass this problem because syringe and needle are sealed as one unit. Inject the LNCs as soon as possible. Resuspend the cell pellet with 1–2 mL of PBS in the round bottom centrifuge tubes and place into the magnet for another 2 min; repeat the wash step.
12. Be careful not to go too deep or the LNCs will be injected into the subcutaneous layer. Preliminary investigations revealed significantly lower AA induction success rates with subcutaneous injection.
13. Mice can be reinjected with another ten million cultured LNCs a week after the initial injection to increase AA development rate.
14. If the cell injected mice do not develop hair loss within 20 weeks, it could be because female C3H/HeJ used have been too young. Ensure they are 10 weeks or older before starting cell culture. Ensure they are on the right type of low-fat diet. High-fat diets prevent AA development [17]. Inject the LNCs as soon as they are removed from culture, preferably within 1 h. During injection, make sure to inject slowly into the dermis, not into the subcutaneous layer. Do not inject male-derived LNCs into female recipients; the injected cells may be rejected (female-derived LNCs injected to male recipients should be accepted).

References

1. Strazzulla LC, Wang EHC, Avila L et al (2018) Alopecia areata: disease characteristics, clinical evaluation, and new perspectives on pathogenesis. J Am Acad Dermatol 78:1–12
2. McElwee KJ, Yu M, Park SW et al (2005) What can we learn from animal models of alopecia areata? Dermatology 211:47–53
3. Gilhar A, Schrum AG, Etzioni A et al (2016) Alopecia areata: animal models illuminate autoimmune pathogenesis and novel immunotherapeutic strategies. Autoimmun Rev 15:726–735
4. McElwee KJ, Gilhar A, Tobin DJ et al (2013) What causes alopecia areata? Exp Dermatol 22:609–626
5. Sundberg JP, Cordy WR, King LE Jr (1994) Alopecia areata in aging C3H/HeJ mice. J Invest Dermatol 102:847–856
6. McElwee KJ, Boggess D, King LE Jr et al (1998) Experimental induction of alopecia areata-like hair loss in C3H/HeJ mice using full-thickness skin grafts. J Invest Dermatol 111:797–803
7. McElwee KJ, Boggess D, Miller J et al (1999) Spontaneous alopecia areata-like hair loss in one congenic and seven inbred laboratory mouse strains. J Investig Dermatol Symp Proc 4:202–206
8. Tang L, Cao L, Sundberg JP et al (2004) Restoration of hair growth in mice with an alopecia areata-like disease using topical anthralin. Exp Dermatol 13:5–10
9. Sun J, Silva KA, McElwee KJ et al (2008) The C3H/HeJ mouse and DEBR rat models for alopecia areata: review of preclinical drug screening approaches and results. Exp Dermatol 17:793–805
10. Harnaha J, Machen J, Wright M et al (2006) Interleukin-7 is a survival factor for CD4+ CD25+ T-cells and is expressed by diabetes-suppressive dendritic cells. Diabetes 55:158–170
11. Picker LJ, Reed-Inderbitzin EF, Hagen SI et al (2006) IL-15 induces CD4 effector memory T cell production and tissue emigration in non-human primates. J Clin Invest 116:1514–1524
12. Bere A, Denny L, Hanekom W et al (2010) Comparison of polyclonal expansion methods to improve the recovery of cervical cytobrush-derived T cells from the female genital tract of HIV-infected women. J Immunol Methods 354:68–79
13. Wang EH, Khosravi-Maharlooei M, Jalili RB et al (2015) Transfer of alopecia areata to C3H/HeJ mice using cultured lymph node-derived cells. J Invest Dermatol 135 (10):2530–2532
14. Jalili RB, Kilani RT, Li Y et al (2018) Fibroblast cell-based therapy prevents induction of alopecia areata in an experimental model. Cell Transplant 27:994–1004
15. Suzuki T, Tokura Y, Ito T (2016) Similarities of dermoscopic findings in alopecia areata between human and C3H/HeJ mouse. J Dermatol Sci 83:154–157
16. McElwee KJ, Hoffmann R, Freyschmidt-Paul P et al (2002) Resistance to alopecia areata in C3H/HeJ mice is associated with increased expression of regulatory cytokines and a failure to recruit CD4+ and CD8+ cells. J Invest Dermatol 119:1426–1433
17. McElwee KJ, Niiyama S, Freyschmidt-Paul P et al (2003) Dietary soy oil content and soy-derived phytoestrogen genistein increase resistance to alopecia areata onset in C3H/HeJ mice. Exp Dermatol 12:30–36

Chapter 11

Collapse and Restoration of Hair Follicle Immune Privilege Ex Vivo: A Model for Alopecia Areata

Markus Fehrholz and Marta Bertolini

Abstract

Alopecia areata (AA) is a hair follicle (HF) autoimmune disease leading to hair loss. Interferon gamma (IFNγ) is regarded as the key cytokine involved in AA development, which has been shown to play a role in the collapse of HF immune privilege (IP), $CD8^+$ T-cell-driven inflammation toward the HF bulb, and premature catagen development. We present here the procedure to induce IP collapse in human healthy HFs ex vivo. This assay can be suitable not only to investigate mechanisms involved in the HF IP collapse but also to test substances for either preventing and/or reversing the HF IP collapse for the management of AA or other autoimmune diseases sharing similar pathogenesis.

Key words Alopecia areata, Hair follicle, Hair follicle organ culture, IFNγ, Immune privilege, Prevention, Restoration

1 Introduction

Alopecia areata (AA) is an organ-restricted autoimmune hair loss in which hair follicles (HFs) that lose their immune privilege (IP) are attacked by a $CD8^+$ T-cell-driven immune response toward unknown autoantigens and are rapidly transformed into catagen [1–3]. Interferon-γ (IFNγ) is considered one of the key cytokines involved in AA development. In a spontaneous mouse model for AA, the knocking-out of IFNγ prevents the development of the disease [4, 5]. In addition, low doses of IFNγ induce bulb IP collapse (75 IU/mL) [6] and premature catagen development (100 IU/mL) [7] in human healthy HFs treated ex vivo. Recently, the new concept that AA-like hair loss is a stereotypic response of the HFs to any insults, including an autoimmune response that increases local exposition of the HFs to local stressors, including IFNγ, is currently taking hold in the field [8]. Therefore, effective treatments for the management of AA should not only aim to

Natalia V. Botchkareva and Gillian E. Westgate (eds.), *Molecular Dermatology: Methods and Protocols*, Methods in Molecular Biology, vol. 2154, https://doi.org/10.1007/978-1-0716-0648-3_11,
© Springer Science+Business Media, LLC, part of Springer Nature 2020

decrease inflammation but also to prevent/reverse HFs from IP collapse.

The model we present here is based on the use a low dose of IFNγ for inducing the hair bulb IP collapse without induction of premature catagen development in healthy HFs ex vivo. This method has been extensively used for the identification of the so-called "HF-IP guardians", endogenous or exogenous immunosuppressants capable of preventing HFs from IP collapse and/or restoring the IP in HFs once collapsed [6, 9–11]. This assay is based on the standard HF ex vivo organ culture [12–14], and consists in treating microdissected anagen VI scalp HFs with 75 IU/mL IFNγ for at least 3 days [6]. The basic experimental set up has been further developed to test and identify agents capable of preventing IP collapse, in a so-called "Prevention HF-IP assay," and/or rescuing HFs once the IP is collapsed, named as "Restoration HF-IP assay" [9–11].

These assays can not only be used for further investigation of mechanisms involved in the HF IP collapse but also for the screening of novel agents that are able to prevent and/or reverse IP collapse for the management of AA or other autoimmune diseases sharing similar pathogenesis.

2 Materials

Prepare all solutions under sterile conditions. Store all reagents according to manufacturer's instructions.

2.1 Instruments and Tools

1. Short tip straight forceps (max 0.5 mm diameter tip).
2. 48-Well plate.
3. Binocular and/or digital brightfield microscope.
4. CO_2 incubator.
5. Laminar flow hood.

2.2 Tissue Preparation

1. William's E culture medium (WCM): William's E medium (WEM) supplemented with 2 mM L-glutamine, 10 ng/mL hydrocortisone, 10 μg/mL insulin, and 1% penicillin–streptomycin mix. Store the medium at +4 °C and reequilibrate at RT before use (*see* **Note 1**).
2. Human microdissected amputated anagen VI scalp hair follicles: Microdissected amputated anagen VI HFs from "clinically" healthy human scalp skin or follicular unit extractions (FUEs) obtained from healthy donors undergoing routine face-lift surgery after ethical approval and signed informed consent of the donor [12–14].

3. IFNγ WCM: Dissolve human IFNγ in Dulbecco's phosphate-buffered saline (DPBS) supplemented with 0.1% bovine serum albumin (BSA) (*see* **Note 2**). Dilute the stock solution of IFNγ in WCM accordingly to obtain 75 IU/mL concentrations.
4. Vehicle WCM: WCM + the final concentration of DPBS supplemented with 0.1% BSA in IFNγ WCM.
5. Test substance WCM: Use vehicle WCM to dissolve the test substance of interest.
6. Test substance/IFNγ WCM: Use IFNγ WCM to dissolve the test substance of interest.

3 Methods

3.1 Experimental Induction of the Collapse of Immune Privilege in Human Scalp Hair Follicles Ex Vivo

1. Microdissect anagen VI HFs from scalp skin as described [12].
2. Incubate HFs for 24 h at 37 °C and 95% air–5% CO_2.
3. Divide the HFs into corresponding experimental groups: (1) vehicle (WCM + DPBS 0.1% BSA) and (2) IFNγ (WCM + 75 IU/mL IFNγ) (*see* **Note 3**).
4. Replace the medium with vehicle WCM or IFNγ WCM of the corresponding experimental group (*see* **Note 4**).
5. Incubate HFs for 24 h at 37 °C and 95% air–5% CO_2.
6. Replace the medium with vehicle WCM or IFNγ WCM of the corresponding experimental group (*see* **Note 4**).
7. Incubate HFs for 24 h at 37 °C and 95% air–5% CO_2.
8. Replace the medium with vehicle WCM or IFNγ WCM of the corresponding experimental group (*see* **Note 4**).
9. Terminate the culture at day 4 (e.g., freeze the HFs in Optimal Cutting Temperature [OCT]) (*see* **Note 5**).

Table 1 summarizes the experimental procedure.

3.2 Assay for Evaluation of Prevention of Hair Follicle Immune Privilege Collapse

1. Microdissect anagen VI HFs from scalp skin as described [12].
2. Incubate HFs for 24 h at 37 °C and 95% air–5% CO_2.
3. Divide the HFs into the corresponding experimental groups: (1) vehicle, (2) vehicle and preventive test substance, (3) IFNγ, and (4) a preventive test substance + IFNγ (*see* **Notes 3**, **4**, and **6**).
4. Replace the medium with vehicle WCM or preventive test substance WCM of the corresponding experimental group. Do not use IFNγ in this step. Vehicle WCM is also used for the IFNγ group and test substance WCM is also used for the test substance + IFNγ group.
5. Incubate HFs for 48 h at 37 °C and 95% air–5% CO_2.

Table 1
Experimental induction of HF-IP collapse in human microdissected anagen VI scalp hair follicles ex vivo

Day 0	Day 1	Day 2	Day 3	Day 4
Microdissect anagen VI HFs from scalp skin or FUEs. Select anagen VI HFs using a binocular or a digital brightfield microscope, and plate them in a 48-well plate, one HF/well, 250 μL medium/well. Measure HF length. Incubate at 37 °C and 95% air–5% CO_2	Select anagen VI HFs for further culture, according to quality control guidelines, using a binocular or a digital brightfield microscope. Measure HF length. Change the culture medium. Apply vehicle WCM or IFNγ WCM to the corresponding experimental group. Incubate at 37 °C and 95% air–5% CO_2	Change the culture medium. Apply vehicle WCM or IFNγ WCM to the corresponding experimental group. Incubate at 37 °C and 95% air–5% CO_2	Change the culture medium. Apply vehicle WCM or IFNγ WCM to the corresponding experimental group. Incubate at 37 °C and 95% air–5% CO_2	Terminate the culture

6. Replace the medium with vehicle WCM, IFNγ WCM, test substance WCM, or preventive test substance + IFNγ WCM of the corresponding experimental group (*see* **Note 4**).
7. Incubate for 48 h at 37 °C and 95% air–5% CO_2.
8. Replace the medium with vehicle WCM, IFNγ WCM, test substance WCM, or preventive test substance + IFNγ WCM of the corresponding experimental group (*see* **Note 4**).
9. Incubate for 48 h at 37 °C and 95% air–5% CO_2.
10. Terminate the culture at day 7 (e.g., freeze the HFs in OCT) (*see* **Note 5**).

Table 2 summarizes the experimental procedure.

3.3 Assay for Evaluation of Restoration of Hair Follicle Immune Privilege Collapse

1. Microdissect the anagen VI HFs from scalp skin as described [12].
2. Incubate HFs for 24 h at 37 °C and 95% air–5% CO_2.
3. Divide the HFs into the corresponding experimental groups: (1) vehicle, (2) restoration test substance, (3) IFNγ, and (4) restoration test substance + IFNγ (*see* **Notes 3, 4** and **6**).

Table 2
Prevention HF-IP assay

Day 0	Day 1	Day 2	Day 3	Day 4	Day 5	Day 6	Day 7
Microdissect amputated anagen VI HFs from scalp skin or FUEs. Select anagen VI HFs using a binocular or a digital brightfield microscope, and plate them in a 48-well plate, one HF/well, 250 μL medium/well. Measure HF length. Incubate at 37 °C and 95% air–5% CO_2	Select anagen VI HFs for further culture, according to quality control guidelines, using a binocular or a digital brightfield microscope. Measure HF length. Change the culture medium. Apply vehicle WCM, or test substance WCM to the corresponding experimental group. Incubate at 37 °C and 95% air–5% CO_2	Rest	Change the culture medium. Apply vehicle WCM, test substance + vehicle WCM, IFNγ + vehicle WCM, or test substance + IFNγ WCM to the corresponding experimental group. Incubate at 37 °C and 95% air–5% CO_2	Rest	Change the culture medium. Apply vehicle WCM, test substance + vehicle WCM, IFNγ + vehicle WCM, or test substance + IFNγ WCM to the corresponding experimental group. Incubate at 37 °C and 95% air–5% CO_2	Rest	Terminate the culture

Table 3
Restoration HF-IP assay

Day 0	Day 1	Day 2	Day 3	Day 4	Day 5
Microdissect amputated anagen VI HFs from scalp skin or FUEs. Select anagen VI HFs using a binocular or a digital brightfield microscope, and plate them in a 48-well plate, one HF/well, 250 μL medium/well. Measure HF length. Incubate at 37 °C and 95% air–5% CO_2	Select anagen VI HFs for further culture, according to quality control guidelines, using a binocular or a digital brightfield microscope. Measure HF length. Change the culture medium. Apply vehicle WCM, or IFNγ WCM to the corresponding experimental group. Incubate at 37 °C and 95% air–5% CO_2	Rest	Change the culture medium. Apply vehicle WCM, test substance + vehicle WCM, IFNγ + vehicle WCM, or test substance + IFNγ WCM to the corresponding experimental group. Incubate at 37 °C and 95% air–5% CO_2	Rest	Terminate the culture

4. Replace the medium with vehicle WCM or IFNγ WCM of the corresponding experimental group. Do not use the restoration test substance in this step. Vehicle WCM is also used for the restoration test substance group and IFNγ WCM is also used for the restoration test substance + IFNγ group (*see* **Note 4**).
5. Incubate for 48 h at 37 °C and 95% air–5% CO_2.
6. Replace the medium with vehicle WCM, IFNγ WCM, restoration test substance WCM, or restoration test substance + IFNγ WCM of the corresponding experimental group (*see* **Note 4**).
7. Incubate for 48 h at 37 °C and 95% air–5% CO_2.
8. Terminate the culture at day 5 (e.g., freeze the HFs in OCT) (*see* **Note 5**).

Table 3 summarizes the experimental procedure.

4 Notes

1. William's E culture medium can be reused if stored at +4 °C. It is important to carefully check the color of the medium because it can be reused until the color remains red.
2. We use recombinant IFNγ (MW 16879.29 g/mol) expressed in E. coli without any modifications for all our experiments. Prepare 2000 and 20 IU/μL stock solutions of IFNγ in DPBS 0.1% BSA and store at −20 °C upon use. Avoid repeated freeze–thaw cycles. Dilute 20 IU/μL stock solutions of IFNγ in WCM accordingly to obtain 75 IU/mL concentrations.

3. It is well established that each human HF is best considered as an independent mini-organ [12], with no two individual HFs being or responding identically to each other, even in a group of anagen VI HFs harvested at the same time and at the same skin location from one individual. Therefore, it is very important to have a minimum of five anagen VI HFs (preferably more than ten) per experimental group to have meaningful results. Lastly, it is important to understand that the model also reflects interindividual differences, as well as gender-dependent differences in HF biology. Hence, it is advisable to select the most representative cohort possible and at least perform a minimum of $n = 3$ independent experiments to confirm the results.
4. Culture medium can be collected for further analysis (e.g., LDH assay).
5. At the termination of the culture, HFs should be processed for analysis. HFs can be immediately snap frozen in liquid nitrogen, stored in RNA later or lysis buffer and snap frozen in liquid nitrogen, placed into a droplet of OCT (maximum three HFs per droplet) and snap frozen in liquid nitrogen, or fixed in 4% paraformaldehyde (PFA) and embedded in paraffin or placed into a droplet of OCT (maximum three HFs per droplet) and snap frozen in liquid nitrogen. For the analysis of standard HF-IP markers in situ [6, 9–11], the following immunostainings/histochemical should be performed: MHC class I, MHC class II, β2-microglobulin, TGFβ2, α-MSH. If the induction of HF-IP was successful, the expression of MHC class I, MHC class II, and β2-microglobulin (molecules presenting antigens) in the bulb should be significantly increased compared to vehicle HFs. On the contrary, the expression of the "HF-IP guardians" TGFβ2, and α-MSH, should be significantly decreased compared to vehicle HFs [6, 9–11] (*see* Fig. 1).
6. It is also possible to test different concentrations of test substances or combinations of test substances. It is advisable to include positive control substances (e.g., IGF-1 or FK506) [6, 9–11].

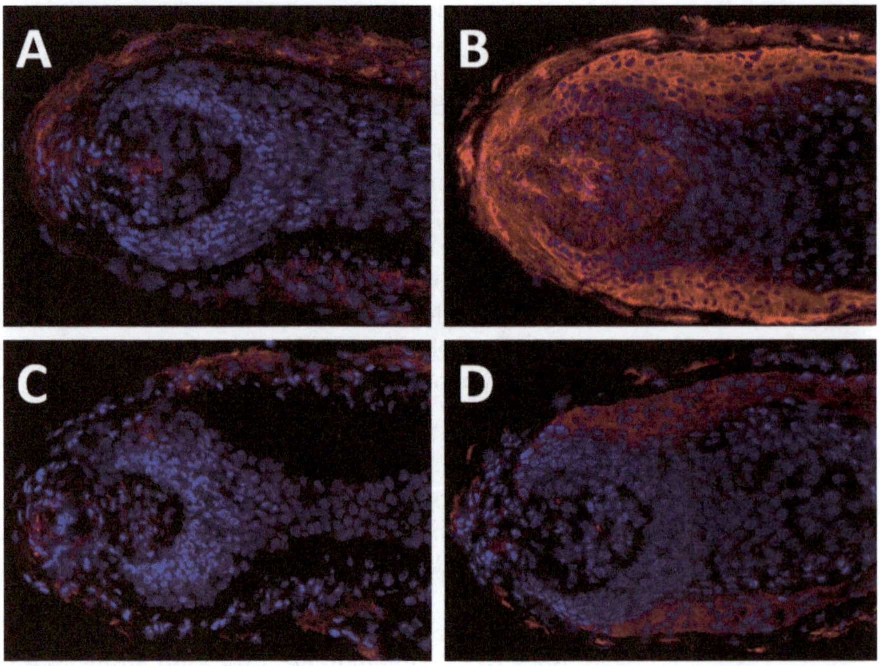

Fig. 1 Induction of immune privilege collapse in hair follicle organ culture ex vivo. HFs were cultured as described in Subheading 3.1. Immunofluorescent staining of MHC class I and II was performed on 6 μm sections as described. While very few cells in the HF epithelium and mesenchyme express MHC class I (**a**) or MHC class II (**c**) in HFs treated with vehicle WCM, HFs treated with IFNγ WCM are characterized by strong expression of MHC class I (**b**) and MHC class II (**d**), indicating HF IP collapse

References

1. Gilhar A, Etzioni A, Paus R (2012) Alopecia areata. N Engl J Med 366:1515–1525
2. McElwee KJ, Gilhar A, Tobin DJ et al (2013) What causes alopecia areata? Exp Dermatol 22:609–626
3. Darwin E, Hirt PA, Fertig R et al (2018) Alopecia areata: review of epidemiology, clinical features, pathogenesis, and new treatment options. Int J Trichology 10:51–60
4. Freyschmidt-Paul P, Zoller M, McElwee KJ et al (2005) The functional relevance of the type 1 cytokines IFN-gamma and IL-2 in alopecia areata of C3H/HeJ mice. J Investig Dermatol Symp Proc 10:282–283
5. Freyschmidt-Paul P, McElwee KJ, Hoffmann R et al (2006) Interferon-gamma-deficient mice are resistant to the development of alopecia areata. Br J Dermatol 155:515–521
6. Ito T, Ito N, Bettermann A et al (2004) Collapse and restoration of MHC class-I-dependent immune privilege: exploiting the human hair follicle as a model. Am J Pathol 164:623–634
7. Ito T, Ito N, Saathoff M et al (2005) Interferon-gamma is a potent inducer of catagen-like changes in cultured human anagen hair follicles. Br J Dermatol 152:623–631
8. Paus R, Bulfone-Paus S, Bertolini M (2018) Hair follicle immune privilege revisited: the key to alopecia areata management. J Investig Dermatol Symp Proc 19:S12–S17
9. Meyer KC, Bodó E, Brzoska T et al (2009) Immunomodulatory effects of the alpha-melanocyte-stimulating hormone-related tripeptide K(D)PT on human scalp hair follicles under proinflammatory conditions. Br J Dermatol 161:1400–1403
10. Kinori M, Bertolini M, Funk W et al (2012) Calcitonin gene-related peptide (CGRP) may award relative protection from interferon-γ-induced collapse of human hair follicle immune privilege. Exp Dermatol 21:223–226
11. Bertolini M, Pretzlaff M, Sulk M et al (2016) Vasoactive intestinal peptide, whose receptor-mediated signalling may be defective in alopecia areata, provides protection from hair follicle immune privilege collapse. Br J Dermatol 175:531–541

12. Langan EA, Philpott MP, Kloepper JE et al (2015) Human hair follicle organ culture: theory, application and perspectives. Exp Dermatol 24:903–911
13. Philpott MP, Green MR, Kealey T (1990) Human hair growth in vitro. J Cell Sci 97 (Pt 3):463–471
14. Kloepper JE, Sugawara K, Al-Nuaimi Y et al (2010) Methods in hair research: how to objectively distinguish between anagen and catagen in human hair follicle organ culture. Exp Dermatol 19:305–312

Chapter 12

Experimentally Induced Epithelial–Mesenchymal Transition of Human Hair Follicle Stem Cells as a Model of Scarring Alopecia Ex Vivo

Ilaria Piccini and Marta Bertolini

Abstract

Primary cicatricial alopecia is characterized by a permanent "scarring" alopecia. This condition is characterized by the irreversible loss of hair follicles (HF) as a result of apoptosis and epithelial–mesenchymal transition (EMT) of epithelial stem cells localized in the HF bulge.

We here report the procedure for experimentally induced EMT in healthy human epidermal stem cells (eSCs) using full-length HF organ culture ex vivo. The present model can be used to recapitulate the complex processes observed in scarring alopecia patient tissues, to further investigate the mechanisms involved in EMT transformation of HFeSCs, and to test substances that could prevent and/or rescue HFeSCs from EMT for the management of scarring alopecias.

Key words Epithelial to mesenchymal transition (EMT), Hair follicle, Ex vivo organ culture, Hair follicle bulge, Hair follicle stem cells, Scarring alopecia

1 Introduction

Epithelial–mesenchymal transition (EMT) is a biological process in which epithelial cells undergo a functional transition into a mesenchymal cell phenotype. EMT is a central process during embryonic development (Type1 EMT), but also occurs in chronically inflamed tissues (Type 2 EMT) and in neoplasia (Type 3 EMT) [1]. We have shown that pathological EMT is also occurring in hair follicle (HF) epithelial stem cells (eSCs) located in the bulge [2–4], after loss of immune privilege and immune attack [5] in patients suffering from lichen planopilaris. Lichen planopilaris is a type of primary cicatricial alopecias, a group of hair loss disorders, which includes also frontal fibrosis alopecia, and central centrifugal cicatricial alopecia. These conditions are characterized by permanent destruction of the HFs and formation of scar tissue leading to irreversible hair loss [6]. Therefore, effective treatments for the management of

Natalia V. Botchkareva and Gillian E. Westgate (eds.), *Molecular Dermatology: Methods and Protocols*,
Methods in Molecular Biology, vol. 2154, https://doi.org/10.1007/978-1-0716-0648-3_12,
© Springer Science+Business Media, LLC, part of Springer Nature 2020

lichen planopilaris and possibly other primary cicatricial alopecias should also be able to prevent/reverse EMT in HFeSCs.

A cocktail of EMT-promoting agents has been established to experimentally induce EMT in healthy human eSCs in full-length HF organ culture ex vivo [2]. The assay, based on the standard HF ex vivo organ culture [7–9], consists of treating microdissected full-length anagen VI scalp HFs with transforming growth factor beta 1 (TGFβ 1), epidermal growth factor (EGF), interferon-γ (INFγ), and a selective E-cadherin inhibiting peptide SWELYYPLRANL (H-SWELYYPLRANL-NH2) (peptide A) for at least 3 days [2]. The basic experimental set up has been further developed to test and identify agents capable of preventing HFeSCs from experimentally induced EMT, in a so-called "Prevention EMT assay," and/or rescuing HFeSCs from experimentally induced EMT, named as "Restoration EMT assay" [2]. In the "Prevention EMT assay," HFs are treated with the test substance for 24 h followed by the cotreatment with the test substance and the cocktail of EMT-promoting agents for 4 days [2]. In the "Restoration EMT assay," HFs are first treated with the cocktail of EMT-promoting agents for 48 h and then with the test agent in the presence of the cocktail of EMT-promoting agents for 3 days [2]. These assays have been successfully used to reveal the beneficial effect of PPARγ modulators [2], whose signaling is abnormal in lichen planopilaris and possibly frontal fibrosis alopecia [10, 11], in rescuing and/or protection of HFeSCs from experimentally induced EMT.

Therefore, the presented model can be used to recapitulate the complex processes observed in vivo in scarring alopecia patient tissues and further investigate the mechanisms involved in the EMT transformation of HFeSCs. Furthermore, the experimentally induced EMT model can be useful to test substances that could prevent and/or restore HFeSCs from EMT transformation for the management of lichen planopilaris and possibly other scarring alopecias.

2 Materials

Prepare all solutions under sterile conditions. Store all reagents according to the manufacturer's instructions.

2.1 Instruments and Tools

1. Long, curved tip forceps.
2. Short, curved tip forceps.
3. Short tip straight forceps (max 0.5 mm diameter tip).
4. Scalpel.
5. 100/20 mm sterile petri dish.
6. 35/10 mm sterile petri dish.
7. 48-Well plate.

8. Binocular and/or Digital brightfield microscope.
9. CO_2 Incubator.
10. Laminar flow hood.

2.2 Human Microdissected Full-Length Anagen VI Scalp Hair Follicles

Microdissected full-length anagen VI HFs from clinically "healthy" human scalp skin or follicular unit extractions (FUEs) obtained from healthy donors undergoing routine facelift/hair transplant surgery after ethics approval and signed informed consent of the donor [7].

2.3 Reagents

1. *William's E culture medium (WCM)*: William's E medium (WEM) supplemented with 2 mM L-glutamine, 10 ng/ml hydrocortisone, 10 μg/ml insulin, and 1% penicillin/streptomycin. Store the medium at +4 °C and reequilibrate the medium at RT before using it (*see* **Note 1**).
2. *EMT cocktail medium*: 3 ng/ml TGFβ1, 10 ng/ml EGF, 500 IU/ml INFγ, and 500 nM Peptide A, dissolved in WCM under sterile conditions (to be prepared fresh before use).
3. *Test substance medium (for EMT prevention or restoration)*: Test materials are dissolved in appropriate solvents according to their physiochemical properties. For the prevention/restoration assays, test materials are then dissolved in WCM at the appropriate concentration(s) by preparing intermediate concentration solutions when necessary. To be prepared fresh before use.
4. *Test substance + EMT cocktail medium (for EMT prevention or restoration)*: Test materials are then dissolved in EMT cocktail WCM medium at the appropriate concentration(s) by preparing intermediate concentration solutions when necessary. To be prepared fresh before use.
5. *Vehicle WCM medium*: A vehicle control where solvent(s) alone in WCM (without test agent and/or EMT cocktail but at the same final concentration as in the test substance WCM and/or in the EMT Cocktail WCM samples) is used is always essential. To be prepared fresh before use.

3 Methods

3.1 Experimental EMT-Induction in Human Microdissected Full-Length Anagen VI Scalp Hair Follicles Ex Vivo

Day 0

1. Microdissect full-length HFs as described in [7].
2. Incubate overnight in WCM culture medium at 37 °C and 95% air–5% CO_2.

Table 1
Experimental EMT-induction in human microdissected full-length anagen VI scalp hair follicles ex vivo

Ex vivo culture day						
0 **HF microdissection**	**1** **EMT induction**	**2** **Rest**	**3** **EMT induction**	**4** **Rest**	**5** **EMT induction**	**6** **Culture termination**
Anagen VI HF selection	Anagen VI HFs selection/ Experimental groups assignment					
HF plating: 1 HF/well 250 μl medium/ well 48-well plate	Culture medium change (accordingly to experimental groups)		Culture medium change (accordingly to experimental groups)		Culture medium change (accordingly to experimental groups)	
Incubation 37 °C and 95% air–5% CO_2	Incubation 37 °C and 95% air–5% CO_2		Incubation 37 °C and 95% air–5% CO_2		Incubation 37 °C and 95% air–5% CO_2	

Note: Culture can be terminated at days 3, 5 and 6

Day 1

3. Select anagen VI HFs for further culture, according to quality control guide lines [7], using a binocular or a digital brightfield microscope.
4. Divide microdissected HFs into the corresponding experimental groups: (a) vehicle control and (b) EMT cocktail (*see* **Notes 1** and **2**).
5. Refresh the medium.
   - For vehicle HFs group (a), use Vehicle WCM medium.
   - For EMT cocktail HFs group (b) use EMT Cocktail medium.
6. Culture HFs at 37 °C and 95% air–5% CO_2.

Days 3–5

7. Change culture medium accordingly to the two different groups every other day.
8. Terminate the culture at day 3 or at latest day 6 (*see* **Notes 3** and **4**).
9. Table 1 summarizes the experimental procedure.

3.2 Prevention EMT Assay

Day 0

1. Microdissect full-length HFs as described in [7].
2. Incubate overnight in WCM culture medium at 37 °C and 95% air–5% CO_2.

Day 1

3. Select anagen VI HFs for further culture, according to quality control guide lines [7], using a binocular or a digital brightfield microscope.
4. Divide microdissected HFs into the corresponding experimental groups: (a) vehicle, (b) EMT cocktail, and (c) prevention test substance + EMT cocktail (*see* **Notes 1** and **5**).
5. Refresh the medium.
   - For vehicle HFs group (a), use Vehicle WCM medium.
   - For EMT cocktail HFs group (b) use Vehicle WCM medium.
   - For prevention test substance + EMT Cocktail HFs group (c) use Test substance medium (*see* **Note 2**).
6. Incubate for 24 h at 37 °C and 95% air–5% CO_2.

Day 2

7. Replace the medium.
   - For vehicle HFs group (a), use vehicle WCM medium.
   - For EMT cocktail HFs group (b), EMT Cocktail medium.
   - For prevention test agent + EMT cocktail HFs group (c), use Test substance + EMT cocktail medium (*see* **Note 2**).
8. Incubate for 48 h at 37 °C and 95% air–5% CO_2.

Day 4

9. Replace the medium.
   - For vehicle HFs group (a), use Vehicle WCM medium.
   - For EMT cocktail HFs group (b), EMT Cocktail medium.
   - For prevention test agent + EMT cocktail HFs group (c), use Test substance + EMT cocktail medium (*see* **Note 2**).
10. Incubate for 48 h at 37 °C and 95% air/5% CO_2.

Day 6

11. Terminate the culture (*see* **Note 4**).
12. Table 2 summarizes the experimental procedure.

Table 2
EMT prevention assay

Ex vivo culture day						
0 **HF microdissection**	**1** **EMT prevention**	**2** **EMT induction**	**3** **Rest**	**4** **EMT induction**	**5** **Rest**	**6** **Culture termination**
Anagen VI HF selection	Anagen VI HFs selection/ Experimental groups assignment					
HF plating: 1 HF/well 250 μl medium/ well 48-well plate	Culture medium change (accordingly to experimental groups)	Culture medium change (accordingly to experimental groups)		Culture medium change (accordingly to experimental groups)		
Incubation 37 °C and 95% air–5% CO_2	Incubation 37 °C and 95% air–5% CO_2	Incubation 37 °C and 95% air–5% CO_2		Incubation 37 °C and 95% air–5% CO_2		

3.3 Restoration EMT Assay

Day 0

1. Microdissect full-length HFs as described in [7].
2. Incubate overnight in WCM culture medium at 37 °C and 95% air–5% CO_2.

Day 1

3. Select anagen VI HFs for further culture, according to quality control guide lines [7], using a binocular or a digital brightfield microscope.
4. Divide microdissected HFs into the corresponding experimental groups: (a) vehicle, (b) EMT cocktail, and (c) prevention test substance + EMT cocktail (*see* **Notes 1** and **5**).
5. Refresh the medium.
   - For vehicle HFs group (a), use Vehicle WCM medium.
   - For EMT cocktail HFs group (b), use EMT Cocktail medium.
   - For restoration test substance + EMT Cocktail HFs group (c) use EMT Cocktail medium (*see* **Note 2**).
6. Incubate for 48 h at 37 °C and 95% air–5% CO_2.

Day 3

7. Replace the medium.
   - For vehicle HFs group (a), use Vehicle WCM medium.
   - For EMT cocktail HFs group (b), EMT Cocktail medium.

Table 3
EMT restoration assay

Ex vivo culture day						
0 **HF microdissection**	**1** **EMT induction**	**2** **Rest**	**3** **EMT restoration**	**4** **Rest**	**5** **EMT restoration**	**6** **Culture termination**
Anagen VI HF selection	Anagen VI HFs selection/ Experimental groups assignment					
HF plating: 1 HF/well 250 μl medium/ well 48-well plate	Culture medium change (accordingly to experimental groups)		Culture medium change (accordingly to experimental groups)		Culture medium change (accordingly to experimental groups)	
Incubation 37 °C and 95% air–5% CO_2	Incubation 37 °C and 95% air–5% CO_2		Incubation 37 °C and 95% air–5% CO_2		Incubation 37 °C and 95% air–5% CO_2	

- For restoration test agent + EMT cocktail HFs group (c), use Test substance + EMT cocktail medium (*see* **Note 2**).

8. Incubate for 48 h at 37 °C and 95% air–5% CO_2.

Day 5

9. Replace the medium.
   - For vehicle HFs group (a), use vehicle WCM medium.
   - For EMT cocktail HFs group (b), EMT Cocktail medium.
   - For restoration test agent + EMT cocktail HFs group (c), use Test substance + EMT cocktail medium (*see* **Note 2**).
10. Incubate for 48 h at 37 °C and 95% air–5% CO_2.

Day 6

11. Terminate the culture (*see* **Note 4**).
12. Table 3 summarizes the experimental procedure.

4 Notes

1. It is well established that each human HF is considered as an independent mini-organ [12], with no two individual HFs being identical to each other, even among of anagen VI HFs harvested from the same donor, the same time and the same skin location. Therefore, it is very important to have a

minimum of five anagen VI HFs (preferably more than ten) per experimental group to obtain meaningful results. It is also important to understand that the model also reflects inter-individual differences, as well as gender-dependent differences in HF biology. Therefore, it is advisable to select the most homogeneous cohort possible and at to perform a minimum of $n = 3$ independent experiments to confirm the results.

2. Culture medium can be collected for further analyses (e.g., LDH assay).
3. EMT transition in the bulge HFeSCs is induced already after 3 days of culture (treatment for 2 days), and longer culture period does not make the EMT phenotype more severe [2].
4. At the termination of the culture, HFs should be collected for the desired downstream assays (e.g., immunohistochemistry, Western blot, or RNA extraction). For immunohistochemistry, fresh HFs should be placed into a droplet of OCT (Optimal Cutting Temperature) compound, maximum three HFs per droplet, and snap frozen in liquid nitrogen, or fixed in 4% PFA (paraformaldehyde) and embedded in paraffin or placed into a droplet of OCT (maximum three HFs per droplet) and snap-frozen in liquid nitrogen. For the analysis of standard EMT markers in situ [2], HFs should be placed into a droplet of OCT (maximum three HFs per droplet) and snap frozen in liquid nitrogen, and tissue sections of 6 μm should be cut.
5. To validate the EMT model the following immunostainings should be performed: α6-integrin/E-cadherin, α6-Integrin/vimentin, α6-Integrin/SLUG, keratin 15/vimentin (Fig. 1).

   α6-integrin is used to mark the bulge area. If the EMT induction was successful, E-cadherin (an epithelial marker) expression in the bulge should be significantly reduced compared to vehicle HFs. On the contrary, the number of vimentin or SLUG (mesenchymal markers) positive cells should be increased after treatment with EMT-inducing cocktail. Importantly, the number of keratin 15+ HFeSCs in the bulge should be decreased after EMT induction and remaining keratin 15+ cells should express vimentin (a mesenchymal marker) [2].
6. It is possible to test also different concentrations of the test substance or combination of test substances, depending on the number of HFs remaining in anagen VI at day 1 [2]. It is advisable to include positive control substances that are known to prevent or restore EMT (e.g., PPARγ agonist Pioglitazone [30 μM] and PPARγ modulator N-Acetyl-GED [AGED 0.1–1 mM]) [2].

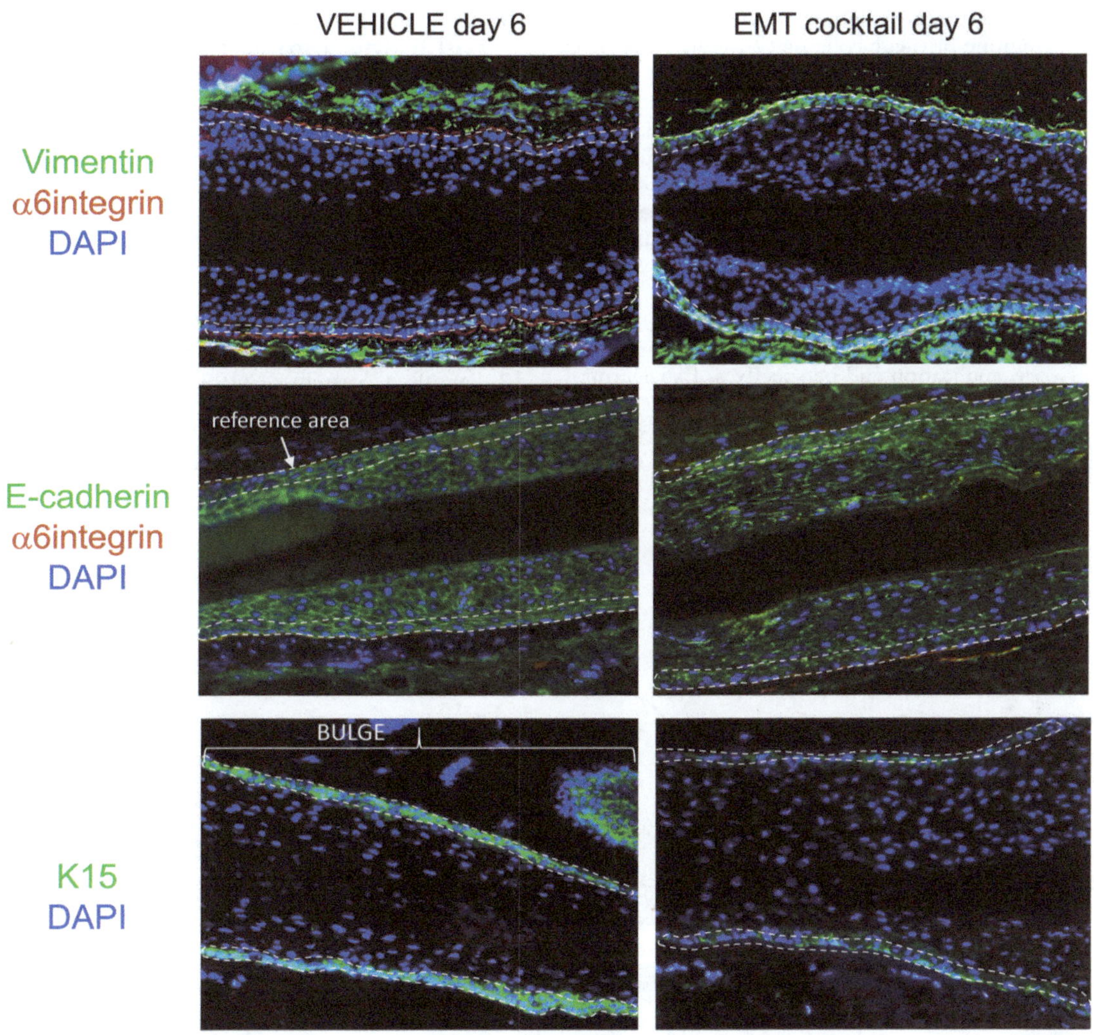

Fig. 1 Representative images of vimentin, E-cadherin, and K15 immunostainings of the bulge region of healthy HFs cultured ex vivo with VEHICLE medium and with EMT cocktail medium. Organ-cultured human full-lengths anagen VI hair follicles treated with the EMT cocktail showed increased vimentin+ cell number, decreased E-cadherin expression, and K15+ HFeSC number in the bulge region compared to v ehicle treated hair follicles

References

1. Kalluri R, Weinberg RA (2009) The basics of epithelial-mesenchymal transition. J Clin Invest 119(6):1420–1428
2. Imanishi H, Ansell DM, Chéret J, Harries M, Bertolini M, Sepp N, Bíró T, Poblet E, Jimenez F, Hardman J, Panicker SP, Ward CM, Paus R (2018) Epithelial-to-mesenchymal stem cell transition in a human organ: lessons from lichen planopilaris. J Invest Dermatol 138(3):511–519
3. Harries MJ, Jimenez F, Izeta A, Hardman J, Panicker SP, Poblet E, Paus R (2018) Lichen planopilaris and frontal fibrosing alopecia as model epithelial stem cell diseases. Trends Mol Med 24(5):435–448
4. Harries M, Imanishi H, Paus R (2019) Fibrosis and stem cell epithelial-mesenchymal transition in primary cicatricial alopecias. J Am Acad Dermatol 80(6):e165–e166
5. Harries MJ, Meyer K, Chaudhry I, E Kloepper J, Poblet E, Griffiths CE, Paus R

(2013) Lichen planopilaris is characterized by immune privilege collapse of the hair follicle's epithelial stem cell niche. J Pathol 231 (2):236–247

6. Bolduc C, Sperling LC, Shapiro J (2016) Primary cicatricial alopecia: lymphocytic primary cicatricial alopecias, including chronic cutaneous lupus erythematosus, lichen planopilaris, frontal fibrosing alopecia, and Graham-Little syndrome. J Am Acad Dermatol 75 (6):1081–1099
7. Langan EA, Philpott MP, Kloepper JE, Paus R (2015) Human hair follicle organ culture: theory, application and perspectives. Exp Dermatol 24(12):903–911
8. Philpott MP, Green MR, Kealey T (1990) Human hair growth in vitro. J Cell Sci 97:463–471
9. Kloepper JE, Sugawara K, Al-Nuaimi Y et al (2010) Methods in hair research: how to objectively distinguish between anagen and catagen in human hair follicle organ culture. Exp Dermatol 19:305–312
10. Karnik P, Tekeste Z, McCormick TS, Gilliam AC, Price VH, Cooper KD, Mirmirani P (2009) Hair follicle stem cell-specific PPAR-gamma deletion causes scarring alopecia. J Invest Dermatol 129(5):1243–1257
11. Harries MJ, Paus R (2009) Scarring alopecia and the PPAR-gamma connection. J Invest Dermatol 129(5):1066–1070
12. Paus R, Cotsarelis G (1999) The biology of hair follicles. N Engl J Med 341(7):491–497

Chapter 13

Generation of Hair Follicle Germs In Vitro Using Human Postnatal Skin Cells

Ekaterina P. Kalabusheva and Ekaterina A. Vorotelyak

Abstract

Modeling organoids with hair follicle germ-like properties provides an opportunity for developing strategies for alopecia drug discovery and replacement therapy, as well as investigating the molecular mechanisms underlying human hair follicle regeneration in vitro. Hair follicle germ reconstruction in vitro is based on dermal papilla hair-inducing abilities and the plasticity of skin epidermal keratinocytes. The current protocol describes a highly efficient approach suitable for adult human skin cell applications. This method allows to obtain hair follicle germs using tissues from one donor. Isolated and cultured for 2 weeks, adult hair follicle dermal papilla cells and skin epidermal keratinocytes self-organize in hanging drop cultures generating organoids that exhibit the features of folliculogenesis onset.

Key words Hair follicle, Dermal papilla, Hair germ, Organoid, Keratinocytes, Cell culture

1 Introduction

The pioneering experiment by Jahoda and colleagues with cultured hair follicle cells [1] gave rise to the specialized segment of a modern tissue engineering and cell therapy: hair follicle reconstruction. They demonstrated the unique abilities of the hair follicle mesenchymal compartment—the dermal papilla—to induce new hair follicles development after subepidermal transplantation. One of the key points of this study is the fact that dermal papilla cells preserve these abilities in vitro which provides an opportunity for increasing the number of cells essential for hair follicle neogenesis. In 1991, Ihara et al. [2] summarized data on the self-assembly of cells isolated from mammal embryonic and fetal tissues. He performed a series of experiments focusing on integument reconstruction in culture with promising results: rat fetal dissociated skin cell suspension self-organized in structures bearing hair follicles. Taken together, these studies defined the potential of this new field

Natalia V. Botchkareva and Gillian E. Westgate (eds.), *Molecular Dermatology: Methods and Protocols*,
Methods in Molecular Biology, vol. 2154, https://doi.org/10.1007/978-1-0716-0648-3_13,
© Springer Science+Business Media, LLC, part of Springer Nature 2020

and future perspectives of future in vitro hair follicle reconstruction studies.

Nevertheless, extensive studies focusing on the generation of functional hair follicle equivalents highlights several restrictions of the in vitro approaches. Investigators identified the harmful influence of long-term culturing on dermal papilla cells leading to the loss of their ability to induce hair follicle neogenesis [3–5]. The withdrawal of dermal papilla cells from the hair bulb in hair follicles seemed to be crucial for culture maintenance. Many studies were undertaken to modify the culture conditions by adding growth factors, involved in hair cycle regulation [6–8], extracellular matrix components [9, 10], or their analogues [11, 12]. Significant progress in dermal papilla intrinsic capacity restoration was achieved in 2010 by the Higgins group [13]: 3D spheroid cultures of dermal papilla cells not only maintained but upregulated the hair-inducing abilities of dermal papilla cells [14]. This influenced the development of approaches to human hair follicle reconstruction. Obtaining the living skin equivalent with mouse dermal papilla cells led to the initiation of hair follicle development, while human dermal papilla cells did not have the same effect [15, 16]. The application of novel 3D printing technologies and spheroid cultures in combination with genetic modification activating the WNT pathway allowed scientists to produce fully functional mature hair follicles in skin equivalents in 2018 [17].

The current protocol proposes the generation of hair follicle germ-like organoids from easily accessible cell sources—human adult skin biopsies and employs the hanging drop culture system, which does not require highly specialized techniques. Dermal papilla cells and epidermal keratinocytes are self-organized into aggregates, which is associated with the expression of folliculogenesis onset markers [18]. Epidermal keratinocytes assemble into trabecula-like structures and acquire a follicle-like phenotype: switch E cadherin expression to P cadherin and produce keratin 6 and AE 13. Dermal papilla cells aggregate in spheroid-like structures that enhance their hair-inducing abilities. We also identified that 0.5 mg/ml hyaluronic acid supplementation promotes the organoids' organization and development. One of the key points of the protocol is the possibility to apply the cells from a single patient for organoid generation that fits the requirements of disease modeling: it is possible to obtain the organoids from epidermal keratinocytes and dermal papilla cells from patients with rare or widespread diseases or genetic mutations. Organoids with hair follicle germ features will become a suitable system for personalized drug development, replacement therapy optimization and a model for hair follicle biology investigation.

2 Materials

2.1 Human Tissue Samples

1. Human scalp biopsies, average size 6–12 cm^2 (*see* **Note 1**).
2. Human skin biopsies, average size 1–10 cm^2.

2.2 Equipment

Biological safety cabinet, class II:

1. 37 °C/5% CO_2 humidified tissue culture incubator.
2. 37 °C thermostat.
3. Centrifuge equipped with a rotor for 15 and 50 ml conical tubes.
4. Inverted microscope.

2.3 Human Dermal Papilla Isolation and Culturing

1. Hank's Balanced Salt Solution (HBSS) supplemented with 50 U/ml penicillin/streptomycin.
2. Small sharp straight or curved scissors, scalpel, toothed tissue forceps, sterile.
3. 100 mm petri dishes, 15 and 50 ml conical tubes, T-25 culture flasks, 5 and 10 ml serological pipettes.
4. 0.5% dispase I solution in DMEM medium (*see* **Note 2**).
5. 0.2% collagenase I solution in DMEM medium (*see* **Note 3**).
6. AmnioMAX™-II medium.
7. Versene solution.
8. 0.05% trypsin–EDTA solution.

2.4 Human Primary Keratinocyte Isolation and Culturing

1. HBSS containing 50 U/ml penicillin/streptomycin.
2. Scalpel, curved tip forceps, forceps, with wide flat straight jaws.
3. 100 mm petri dishes, 15 and 50 ml conical tubes, 5 and 10 ml serological pipettes.
4. 0.2% dispase solution in DMEM medium (*see* **Note 2**).
5. Dulbecco's phosphate-buffered saline (DPBS) without Ca^{2+} and Mg^{2+}.
6. Trypsin–DPBS solution: add 5 ml 0.05% trypsin–EDTA solution and 5 ml DPBS without Ca^{2+} and Mg^{2+} into 15 ml conical tube and incubate for 15–20 min at 37 °C.
7. Fetal bovine serum (FBS).
8. Collagen I–treated flasks: add 5 ml of 30 μg/ml collagen I solution in 0.1% glacial acetic acid in a T-25 flask, incubate for 15 min at 37 °C. Aspirate collagen solution and rinse three times with HBSS. Add 3 ml of primary keratinocyte medium to prevent it from drying before plating the cells.

9. Primary keratinocyte medium: DMEM/F12 medium, containing 4 mM glutamine, 10% FBS, 10 ng/ml EGF, 5 mg/ml insulin, and 0.25 mg/ml isoproterenol.
10. Epidermal keratinocyte medium: CNT-07 medium.

2.5 Reagents for Organoid Modeling

1. 100 mm petri dishes, 15 ml conical tubes, 5 and 10 ml serological pipettes.
2. Versene solution.
3. 0.05% and 0.25% trypsin–EDTA solution.
4. DMEM medium, containing 4 mM glutamine and 10% FBS.
5. Organoid medium: DMEM medium, containing 4 mM glutamine, 10% FBS, and 0.5 mg/ml hyaluronic acid (MW 1–2 kDa) (*see* **Note 4**).
6. HBSS containing 50 U/ml penicillin/streptomycin.

3 Methods

All procedures should be performed by aseptic technique using a biological safety cabinet. All the solutions should have room temperature unless otherwise specified.

3.1 Human Dermal Papilla Isolation and Culturing

The protocol of dermal papilla cells isolation is based on a Wu method [19] with Chermnykh modifications [20].

1. Wash human scalp skin samples 7–10 times with 100 ml HBSS containing 50 U/ml penicillin/streptomycin (*see* **Note 5**).
2. Place skin samples on a 100 mm petri dish and cut the tissue into 1 cm wide strips using a sterile scalpel.
3. Transfer the tissue strips using forceps into 15–30 ml of 0.5% dispase I solution in a 50 ml conical tube. Incubate skin samples at 4 °C overnight.
4. Rinse samples with 30 ml HBSS containing 50 U/ml penicillin/streptomycin.
5. Using forceps with toothed tips and scissors separate subcutaneous fat with hair bulbs and collect it into new 100 mm petri dish. Mince the fat using the scissors (*see* **Note 6**).
6. Transfer minced fat with hair bulbs using a 5 ml serological pipette into a 15 ml conical tube with 10 ml of 0.2% collagenase I solution.
7. Incubate at 37 °C till complete adipose tissue dissociation and fat drops appearance in the tube for 90 min.

8. Homogenize the cell suspension by pipetting and centrifuge for 10 min at 200 × *g*. Remove supernatant with fat tissue by inverting the tube.
9. Add 10 ml 0.2% collagenase I solution to the pellet, resuspend and incubate for 4 h at 37 °C.
10. Centrifuge for 7 min at 50 × *g*. Remove supernatant by inverting the tube.
11. Perform a series of sequential centrifuging at low speed to remove single cells. Add 5 ml HBSS, resuspend the pellet and centrifuge for 5 min at 20 × *g*. Remove supernatant by inverting the tube. Repeat this step three times.
12. Add 5 ml DMEM medium, pipette and centrifuge for 5 min at 20 × *g*. Repeat this step at least three times (*see* **Note 7**).
13. Resuspend the pellet in 3 ml AmnioMAX™-II medium for each T-25 flask to be inoculated. Transfer the inoculated flasks into a tissue culture incubator (*see* **Note 8**).
14. After 3 days hair bulbs and dermal papillae should attach. Single fibroblasts migrated from papillae should be visible under the microscope. Aspirate the medium and add 3 ml of fresh AmnioMAX™-II. Change the medium every 3 days.
15. Monitor the explant outgrowth after inoculation. After the size of clones reaches a 1–1.5 cm diameter, split the culture:
    - Aspirate the culture medium.
    - Wash twice with 5 ml Versene solution.
    - Add 1 ml 0.05% trypsin–EDTA.
    - Incubate the flasks for 3–7 min at 37 °C till cell detachment.
    - Add 4 ml AmnioMAX™-II medium, resuspend and transfer into a new T-25 flask.
    - After 24 h make medium change with 3 ml of fresh AmnioMAX™-II medium.
16. Change the medium every 2–3 days till the culture reaches 90% confluency. After this, split the culture 1:3 using the technique described in p.15 (*see* **Note 9**).
17. For obtaining aggregates, use passage 2–4 of dermal papilla cells (*see* **Note 10**).

3.2 Human Skin Keratinocyte Isolation and Culturing

1. Wash human skin samples 7–10 times with 100 ml HBSS containing 50 U/ml penicillin/streptomycin.
2. Place the samples onto a 100 mm Petri dish. Using curved tip forceps and scalpel remove the dermis till the thickness of the skin reaches 0.5 mm or less, then cut the epidermal sheets into 2–5 mm strips using scalpel.

3. Transfer the strips into a 15 ml conical tube with 0.2% dispase solution and incubate at 4 °C overnight.
4. Prepare trypsin/DPBS solution.
5. Aspirate the dispase I solution, add 10 ml DPBS without Ca^{2+} and Mg^{2+}. Transfer skin strips with DPBS into a 10 cm petri dish.
6. Add 10 ml DPBS without Ca^{2+} and Mg^{2+} to the new petri dish. Separate the epidermis from dermis with curved tip forceps and flat tip forceps.
7. Transfer the epidermis to the petri dish with fresh DPBS.
8. When all epidermal strips are collected, transfer them into prewarmed trypsin/DPBS solution using forceps.
9. Incubate for 30 min at 37 °C. Shake the conical tube with the skin strips; the solution should become cloudy due to keratinocyte detachment from the basement membrane, otherwise incubate further for 15 min.
10. Add 500 μl FBS into a 50 ml conical tube.
11. Using a 10 ml serological pipette, release the cells by pipetting up and down.
12. Transfer the cell suspension without skin strips to the 50 ml conical tube with FBS.
13. Add 10 ml fresh DPBS without Ca^{2+} and Mg^{2+} to the 15 ml conical tube with skin strips and continue the pipetting. Repeat this step till the DPBS remains clear.
14. Centrifuge the 50 ml conical tube for 15 min at $200 \times g$.
15. Prepare collagen I treated flasks.
16. Resuspend the pellet into 10–20 ml primary keratinocyte medium. Count the cells and dilute the suspension to 1×10^6 cells per 1 ml.
17. Plate 5 ml of suspension per one T-25 flask. Transfer the flasks into a tissue culture incubator.
18. After 3 days, change the medium: half of the flasks keep the primary keratinocyte medium, the other half switch to the CNT-07 medium. Use 5 ml of medium per flask (*see* **Note 11**).
19. Change the medium every 3–5 days. Primary keratinocyte medium stimulates keratinocytes to produce a layered sheet, while CNT-07 maintains only basal keratinocyte proliferation without stratification.
20. Culture the keratinocytes until confluent.

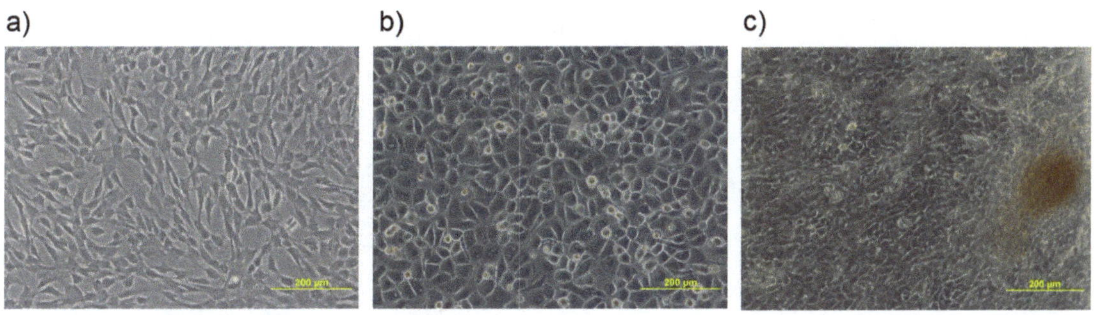

Fig. 1 The cultures suitable for organoids modeling. Phase-contrast microscopy. Scale bar: 200 μm. (**a**) Dermal papilla cell culture (passage 2); (**b**) keratinocyte culture in CNT-07 medium; and (**c**) keratinocyte culture in primary keratinocyte medium

3.3 Organoid Modeling

3.3.1 Cell Suspension Preparation

Dermal Papilla Cell Suspension

1. Passage 2–4 80–90% confluent dermal papilla culture (Fig. 1a), rinse twice with 5 ml of Versene solution, and add 2 ml of 0.05% trypsin–EDTA.
2. Incubate for 3–7 min at 37 °C until complete cell dissociation.
3. Add 5 ml of DMEM medium containing 10%FBS and 4 mM Glutamine; resuspend and calculate cell concentration.

Keratinocyte Cell Suspension

CNT-07

1. Human primary keratinocytes cultured in CNT-07 medium reach the confluency after 5–7 days, over the next 3–10 days any differentiated keratinocyte cells retained after isolation are discarded. Days 10–15 of culturing are the optimal for obtaining aggregates (Fig. 1b).
2. Rinse flasks twice with 5 ml of Versene solution; add 2 ml of 0.25% trypsin–EDTA.
3. Incubate for 5–15 min at 37 °C till complete cell dissociation.
4. Add 5 ml of DMEM medium containing 10% FBS and 4 mM glutamine; resuspend and calculate cell concentration.

Primary Keratinocyte Medium

1. Cells cultured in primary keratinocyte medium reach the confluency after 7–10 days, producing differentiated spinous layer keratinocytes during the same time. Days 7–10 of culturing are optimal for obtaining aggregates (Fig. 1c).
2. Rinse flasks twice with 5 ml of Versene solution; add 5 ml of 0.25% trypsin–EDTA.
3. Incubate for 2 min at 37 °C and aspirate the trypsin solution. Add 5 ml of fresh 0.25% trypsin–EDTA.
4. Incubate for 5–10 min at 37 °C till complete cell dissociation.
5. Add 10 ml of DMEM medium containing 10% FBS and 4 mM glutamine; resuspend and calculate cell concentration.

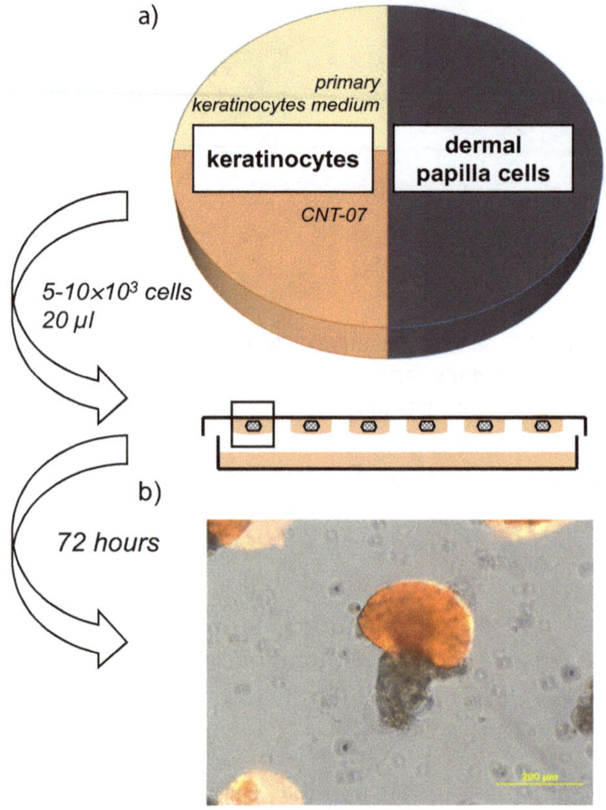

Fig. 2 Organoid generation. (**a**) The ratio of each culture type in one aggregate, (**b**) organoids after 3 days in hanging drop culture. Dermal papilla cells were preliminarily labeled with RFP. Fluorescent and phase-contrast microscopy. Scale bar: 200 μm

3.3.2 Obtaining Organoids

1. Calculate the necessary number of cells based on the following instructions (Fig. 2a):
   - Use 5–10 $\times$ 10^3 cells per organoid (*see* **Note 12**).
   - The ratio of dermal papilla cells and skin keratinocytes should be 1:1.
   - The ratio of keratinocytes cultured in primary keratinocyte medium and CNT-07 medium should be 1:1 (*see* **Note 11**).

   For example, if you want to use 10 $\times$ 10^3 cells per organoid, you need 5 $\times$ 10^3 dermal papilla cells, 2.5 $\times$ 10^3 keratinocytes cultured in primary keratinocyte medium, and 2.5 $\times$ 10^3 keratinocytes cultured in CNT-07 medium. If you need 500 organoids you should collect 2.5 $\times$ 10^6 dermal papilla cells and 1.25 $\times$ 10^6 each type of keratinocytes.

2. Transfer the necessary amount of each cell suspension into a 15 ml conical tube. Centrifuge for 10 min at 200 $\times$ *g*.

3. Aspirate supernatant; resuspend the pellet into appropriate volume of organoid medium. Use 20 μl medium per organoid (*see* **Note 13**).
4. Prepare the 100 mm petri dishes for hanging drop culture (*see* **Note 14**): add 5–7 ml of HBSS containing 50 U/ml penicillin/streptomycin to prevent drying out and contamination.
5. Invert the lid and place the 20 μl drops of cell suspension at a 3–5 mm distance from each other.
6. Invert the lid with drops and close the petri dish. Transfer the dishes into tissue culture incubator.
7. During the next 48 h spontaneous cell sorting occurs initializing the folliculogenesis events. Dermal papilla cells produce spheroid-like structures enveloped by keratinocyte trabeculae (Fig. 2b). After 3 days, early hair follicle development markers, such as P cadherin, Lefl, keratin 6, and AE13, could be detected by immunohistochemical staining.
8. Culture the organoids for 3–5 days and collect for the following experiments.

4 Notes

1. Human scalp skin contains hair follicles producing terminal hairs, whose hair bulbs are located in subcutaneous fat that is important for efficient dermal papillae isolation. Use a microdissection approach for obtaining dermal papillae from vellus hair follicles.
2. Prepare and aliquot 2% dispase stock solution in DMEM medium or DPBS and store at −20 °C. Use these aliquots for preparing 0.5% and 0.2% dispase solutions in DMEM medium right before the procedures.
3. Prepare and aliquot 1% collagenase I stock solution in DMEM medium or DPBS and store at −20 °C. Use these aliquots for preparing 0.2% collagenase I solution in DMEM medium immediately before the procedures.
4. Do not use hyaluronic acid with a molecular mass less than 500 kDa due to effects on proliferation.
5. Trim the hair shafts using scissors or scalpel before the washing procedures to prevent contamination.
6. Separate only subcutaneous fat and hair bulbs avoiding collagen-rich dermal tissue.
7. The pellet could be invisible at this stage.
8. Use 1–2 T-25 culture flasks for rapid dermal papilla and hair bulb attachment.

9. During passage 0 and 1 several hair bulb keratinocyte colonies could be observed. Application of 0.05% trypsin–EDTA, AmnioMAX™-II medium, and new flasks during the passaging suppresses keratinocyte growth.
10. Dermal papilla cells can be genetically transfected at this stage, for example with fluorescent protein vectors.
11. It is not obligatory to use both types of keratinocyte cultures. We recommend the application of CNT-07 and primary keratinocyte medium for aggregates modeling for maintaining the balance between progenitors and differentiating keratinocytes. Keratinocytes cultured in CNT-07 may not incorporate into organoids while cells in primary keratinocyte medium could be over-differentiated to support hair follicle germ formation.
12. The number of cells per organoid depends on the purpose of the experiment.
13. If the final volume of cell suspension is more than 5 ml aliquot the suspension in several conical tubes. Otherwise cells will settle down during an excessively long procedure with one tube that will affect cell concentration per organoid.
14. We do not recommend an application of specialized hanging drop plates, as dermal papilla cells attach to the walls of the wells.

Acknowledgments

The work was funded by the Russian Science Foundation (Project no. 16-14-00204).

References

1. Jahoda CA, Horne KA, Oliver RF (1984) Induction of hair growth by implantation of cultured dermal papilla cells. Nature 311 (5986):560–562
2. Ihara S, Watanabe M, Nagao E, Shioya N (1991) Formation of hair follicles from a single-cell suspension of embryonic rat skin by a two-step procedure in vitro. Cell Tissue Res 266(1):65–73
3. Horne KA, Jahoda CA, Oliver RF (1986) Whisker growth induced by implantation of cultured vibrissa dermal papilla cells in the adult rat. J Embryol Exp Morphol 97:111–124
4. Lichti U, Weinberg WC, Goodman L, Ledbetter S, Dooley T, Morgan D, Yuspa SH (1993) In vivo regulation of murine hair growth: insights from grafting defined cell populations onto nude mice. J Invest Dermatol 101(1 Suppl):124S–129S
5. Weinberg WC, Goodman LV, George C, Morgan DL, Ledbetter S, Yuspa SH, Lichti U (1993) Reconstitution of hair follicle development in vivo: determination of follicle formation, hair growth, and hair quality by dermal cells. J Invest Dermatol 100(3):229–236
6. Kishimoto J, Burgeson RE, Morgan BA (2000) Wnt signaling maintains the hair-inducing activity of the dermal papilla. Genes Dev 14(10):1181–1185
7. Aoi N, Inoue K, Chikanishi T, Fujiki R, Yamamoto H, Kato H, Eto H, Doi K, Itami S, Kato S, Yoshimura K (2012) 1α,25-dihydroxyvitamin D3 modulates the hair-inductive capacity of dermal papilla cells:

therapeutic potential for hair regeneration. Stem Cells Transl Med 1(8):615–626

8. Rendl M, Polak L, Fuchs E (2008) BMP signaling in dermal papilla cells is required for their hair follicle-inductive properties. Genes Dev 22(4):543–557
9. Bratka-Robia CB, Mitteregger G, Aichinger A, Egerbacher M, Helmreich M, Bamberg E (2002) Primary cell culture and morphological characterization of canine dermal papilla cells and dermal fibroblasts. Vet Dermatol 13 (1):1–6
10. Young TH, Tu HR, Chan CC, Huang YC, Yen MH, Cheng NC, Chiu HC, Lin SJ (2009) The enhancement of dermal papilla cell aggregation by extracellular matrix proteins through effects on cell-substratum adhesivity and cell motility. Biomaterials 30(28):5031–5040
11. Ohyama M, Kobayashi T, Sasaki T, Shimizu A, Amagai M (2012) Restoration of the intrinsic properties of human dermal papilla in vitro. J Cell Sci 125(Pt 17):4114–4125
12. Lee SH, Yoon J, Shin SH, Zahoor M, Kim HJ, Park PJ, Park WS, Min DS, Kim HY, Choi KY (2012) Valproic acid induces hair regeneration in murine model and activates alkaline phosphatase activity in human dermal papilla cells. PLoS One 7(4):e34152
13. Higgins CA, Richardson GD, Ferdinando D, Westgate GE, Jahoda CA (2010) Modelling the hair follicle dermal papilla using spheroid cell cultures. Exp Dermatol 19(6):546–548
14. Higgins CA, Itoh M, Inoue K, Richardson GD, Jahoda CA, Christiano AM (2012) Reprogramming of human hair follicle dermal papilla cells into induced pluripotent stem cells. J Invest Dermatol 132(6):1725–1727
15. Sriwiriyanont P, Lynch KA, Maier EA, Hahn JM, Supp DM, Boyce ST (2012) Morphogenesis of chimeric hair follicles in engineered skin substitutes with human keratinocytes and murine dermal papilla cells. Exp Dermatol 21 (10):783–785
16. Sriwiriyanont P, Lynch KA, McFarland KL, Supp DM, Boyce ST (2013) Characterization of hair follicle development in engineered skin substitutes. PLoS One 8(6):e65664
17. Abaci HE, Coffman A, Doucet Y, Chen J, Jacków J, Wang E, Guo Z, Shin JU, Jahoda CA, Christiano AM (2018) Tissue engineering of human hair follicles using a biomimetic developmental approach. Nat Commun 9 (1):5301
18. Kalabusheva E, Terskikh V, Vorotelyak E (2017) Hair germ model in vitro via human postnatal keratinocyte-dermal papilla interactions: impact of hyaluronic acid. Stem Cells Int 2017:9271869
19. Wu JJ, Liu RQ, Lu YG, Zhu TY, Cheng B, Men X (2005) Enzyme digestion to isolate and culture human scalp dermal papilla cells: a more efficient method. Arch Dermatol Res 297 (2):60–67
20. Chermnykh ES, Vorotelyak EA, Gnedeva KY, Moldaver MV, Yegorov YE, Vasiliev AV, Terskikh VV (2010) Dermal papilla cells induce keratinocyte tubulogenesis in culture. Histochem Cell Biol 133(5):567–576

Chapter 14

Method for Human Eccrine Sweat Gland Isolation from the Scalp by Means of the Micropunch Technique and Vital Dyes

Francisco Jiménez

Abstract

The isolation of eccrine sweat glands from human skin has always been a difficult task. The human scalp contains thousands of eccrine glands. Recently, the close anatomic relationship of the eccrine gland with the scalp hair follicle has been described. Taking advantage of this anatomic relationship as well as of the availability of follicular units (FUs) obtained in hair transplant procedures, we describe here a simple and efficient method to isolate eccrine sweat glands from the human scalp. This method is identical to the micropunch hair graft harvesting method known as follicular unit excision (FUE), used in modern hair transplantation. Once the FU has been extracted, it needs to be stained with methylene blue or neutral red in order to make the sweat gland visible for stereoscopic microdissection. Only the secretory (coiled) portion of the sweat gland can be obtained with this method. The efficiency of this isolation method should encourage further research into human eccrine sweat glands and opens possibilities for new translational applications.

Key words Eccrine glands, Sweat glands, Hair follicle, Hair transplantation, Follicular unit, Neutral red, Methylene blue

1 Introduction

Humans have the highest density of eccrine glands (EGs) of any mammal. It is thought that there are approximately 2–4 million EGs dispersed throughout the human skin [1, 2]. Morphologically, the EG consists of a coiled tubular structure (secretory portion) which connects to the skin surface via an intradermal portion (duct) and an intraepidermal segment (acrosyringium) [3].

Primary sweat secretion is produced in the coiled portion of the EGs and mainly consists of a sodium chloride rich solution with some potassium that is isotonic to plasma. The duct also acts to

Electronic supplementary material The online version of this chapter (https://doi.org/10.1007/978-1-0716-0648-3_14) contains supplementary material, which is available to authorized users.

Natalia V. Botchkareva and Gillian E. Westgate (eds.), *Molecular Dermatology: Methods and Protocols*, Methods in Molecular Biology, vol. 2154, https://doi.org/10.1007/978-1-0716-0648-3_14,
© Springer Science+Business Media, LLC, part of Springer Nature 2020

reabsorb sodium and chloride from the primary sweat, minimizing excessive salt loss [4].

The most important function of EGs is body temperature regulation through evaporative heat loss. The EGs also play a less understood yet important role in cutaneous wound healing [5] and in secreting antimicrobial peptides such as dermcidin [6].

The isolation of human EGs ex vivo that maintain their morphological integrity and function can be an invaluable source for the cosmetic industry interested in studying the sweat response to pharmacological products. A further fascinating challenge lies in the possibility of transplanting exclusively EGs in special clinical situations, for example to extensive deep-burn survivors whose scarring skin covered with skin grafts is devoid of sweat glands and suffers from heat intolerance.

1.1 Anatomy and Distribution of Eccrine Glands in the Scalp

Recent publications have described the close anatomic relationship between EGs (specifically the secretory coil portion) and scalp hair follicles [7–9]. This finding has led to the original idea described in this paper of harvesting EGs from the scalp in the same way as hair grafts are harvested from the donor area in the follicular unit excision (FUE) method of hair transplant surgery, a procedure for treatment of androgenetic alopecia [10, 11].

The follicular unit (FU) is the prime autograft used in hair transplant procedures. The FU is a histological structure composed of 1–4 terminal hair follicles, 1–2 vellus follicles and their associated sebaceous glands and arrector pili muscles. Although the EGs were not considered in the original description by Headington [12] to be a component of the FU, in light of current knowledge it seems that the majority of scalp FUs do in fact contain an associated eccrine gland [8]. In particular, the secretory portion of the EG (the coil) is located in a very close spatial relationship with the lower half of anagen terminal follicles, normally at a depth of 2–3 mm below the epidermal surface of human scalp and always below the sebaceous gland and arrector pili muscle (Fig. 1).

1.2 Other Methods for Eccrine Gland Isolation

Research into human eccrine sweat gland function has been greatly hindered by problems of gland isolation. Obtaining full intact EGs by stereoscopic microdissection of human skin is a very difficult task. Sato and Sato [13] described how to isolate single sweat glands under a stereomicroscope by gently teasing away periglandular collagen fibers with sharp forceps. However, this method is very time-consuming and has a low rate of efficiency. Lee et al. [14] and Brayden and Fitzpatrick [15] described a better way to isolate human sweat glands by chopping skin samples with sharp scissors. The principle was that sweat glands "pop out" from surrounding collagen and fat due to shearing forces created by the chopping.

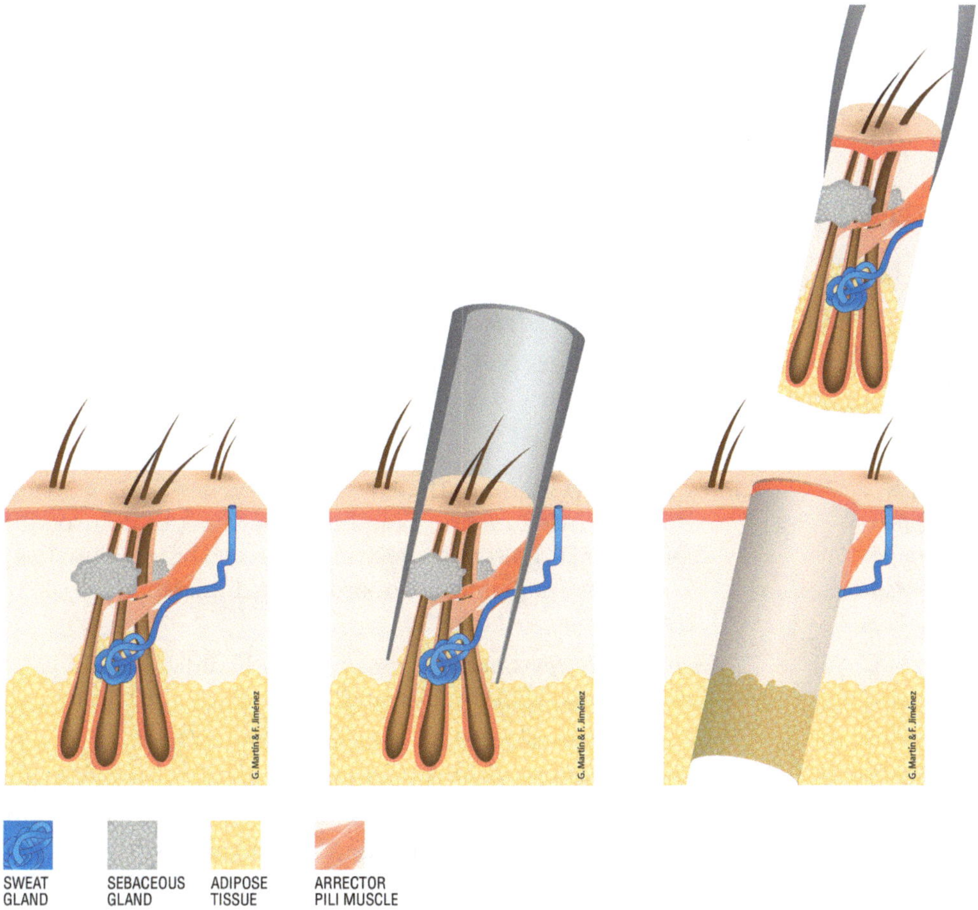

Fig. 1 This figure illustrates the technical procedure of follicular unit (FU) harvesting using a micropunch (FUE technique). The left drawing depicts the anatomy of a scalp FU, showing the location of the eccrine coil (in blue) in relation to the hair follicles, sebaceous glands, and arrector pili muscles. The eccrine coil is always embedded in adipocytes (dermal fat tissue). The middle drawing shows how the 1 mm punch is introduced into the scalp skin with the angle parallel to the hair shaft direction, in order to avoid FU transection. The incision made with the circular punch releases the FU from the surrounding dermis, allowing it to be removed with fine tip forceps (drawing on the right). (This figure was published in Exp Dermatol 2018; 27: 679 and is reproduced with permission of John Wiley & Sons. Inc. [9])

One of the difficulties encountered with EG isolation is that EGs, as opposed to sebaceous glands or hair follicles, are not visible under the stereomicroscope, which makes dissection impossible. To circumvent this problem, the gland can be stained prior to dissection with the supravital dye methylene blue (MB) [16, 17] or with neutral red (NR) [15]. The MB and neutral red staining method have been shown to be nontoxic at low concentrations, and the stained glands grow in culture just as successfully as controls [9, 15].

Another way to isolate glands is by enzymatic digestion of skin samples with collagenase. This method yields a larger number of glands [18–20], but the exposure of cells to collagenase is potentially damaging for physiological functional studies.

2 Materials

2.1 Local Anesthetic Infiltration

1 vial of 10 ml of lidocaine 1%.

1 vial of 1 ml of adrenaline 1/1000.

Luer lock syringes of 1 ml.

30-G needles.

2.2 Surgical Equipment for Follicular Unit Harvesting (See Note 1)

Punches: Sharp titanium tipped round punches (Ellis Instruments, Robbins Instruments, Cole Instruments, A to Z Instruments). Diameter of 1.00 mm. Beginners can use bigger punches up to 1.25 mm in diameter in order to diminish the risk of FU transection (*see* **Note 2**).

Punch handle for manual punch: Versi-Handle (Robbins Instruments) or CIT Manual Hand Punch (Cole Instruments).

Forceps for removal of the FUs previously isolated with the punch: FUE extracting forceps (Ertip Medical, Robbins, Ellis, A to Z).

Ancillary equipment: Surgical loupes (Zeiss; Akura) in the range of 3.5× to 5× are recommended for FU harvesting.

Supplementary material: Sterile gauze pads, physiological saline 0.9%, antibiotic ointment (Fucidin or Bactroban type).

2.3 Equipment for Eccrine Gland Microdissection

Stereomicroscope for microscopic tissue dissection.

0.02% of methylene blue (MB) vital dye in physiologic saline or PBS.

0.2% of neutral red (NR) vital dye in physiologic saline or PBS.

Material for microdissection: Scalpel handle, #15 blades, fine-tip forceps, 30-G needles, physiological saline.

3 Methods

3.1 Anesthetic Infiltration

1. Add 0.1 cc of adrenaline 1/1000 into a 10 cc vial of lidocaine 1%. This will produce a solution containing 10.1 cc of lidocaine 1% with adrenaline 1/100,000, which is the anesthetic used during the whole procedure.
2. Inject the anesthetic very superficially and slowly using a 1 cc Luer lock syringe and a 30-G needle in the area of the scalp where the FUs are going to be harvested (*see* **Note 3**).

3.2 Follicular Unit Harvesting

1. Trim with scissors the hairs of the scalp that will be harvested to 1 mm of length in order to facilitate the incisions with the punch.
2. Punch incision around the FU: This technique should be performed with appropriate loupe magnification. Select the FU to harvest (*see* **Note 4**). The 1 mm punch must be perfectly centered on the exit path of the hairs. Perform the incision by rotating the punch back and forth between the thumb and index finger as it enters the skin (*see* **Note 5**). The punch is advanced parallel to the direction of the hair shaft in order to avoid transection (*see* **Note 6**) (Fig. 1 and Electronic Supplementary Video 1).
3. Extraction of the FU: Release the FU from the subcutaneous fat at the bottom of the graft using forceps (*see* **Note** 7).
4. Immediately immerse the extracted FU grafts in physiological saline which is used as the holding solution until the microscopic dissection procedure of the eccrine coils begins (*see* **Note 8**).
5. Place an antibiotic ointment on top of the wound, which should be reapplied by the patient 2–3 times daily for 3–4 days (*see* **Note 9**).

3.3 Microscopic Dissection of the Eccrine Coil

1. *Visualization of the eccrine glands with MB or with NR*: Immerse the FU completely in a solution *of MB 0.02% or NR 0.2%* for 10–30 min. Both MB and NR give a very clear delineation of the EG, providing a distinct blue and red coloring of the EG, respectively, easily visible against the background of the dermal collagen (*see* **Note 10**) (Figs. 2 and 3).
2. *Isolation of the eccrine coil (see* ***Note 11****)*: Microdissect the stained eccrine coil from the FU under the stereomicroscope using fine tip forceps, #15 scalpel blades, or 30-G needles (*see* **Note 12**) (Fig. 2c).

The eccrine coils isolated by this methodology can be cultured ex vivo in a mixture of supplemented Williams E and F12 media [9] or can be used for functional studies [4].

4 Notes

1. All equipment and punches must be fully cleaned and sterilized.
2. Follicular units from the scalp can be harvested in two ways: by strip excision or by FUE (follicular unit excision). For the purpose of harvesting a limited number of FUs, the FUE technique, which is the technique described in this paper, is

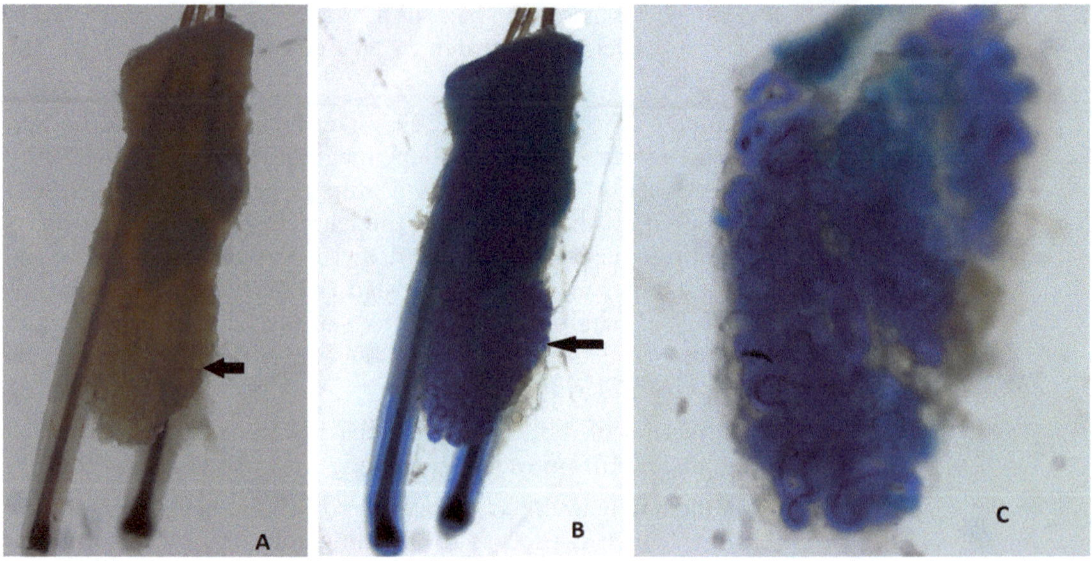

Fig. 2 (**a**) Follicular unit immediately after harvesting from the scalp with 1 mm punch containing two terminal anagen follicles. The eccrine coil cannot be identified but its presence is suspected when a mass of adipocytes is seen in this location (arrow). (**b**) Methylene Blue (MB) staining of the same FU allows for the visualization of the eccrine coil (arrow). Note this typical anatomic location of the eccrine coil, adjacent to the inferior portion of anagen hair follicles. (**c**) The eccrine coil after its isolation by manual dissection under the stereomicroscope

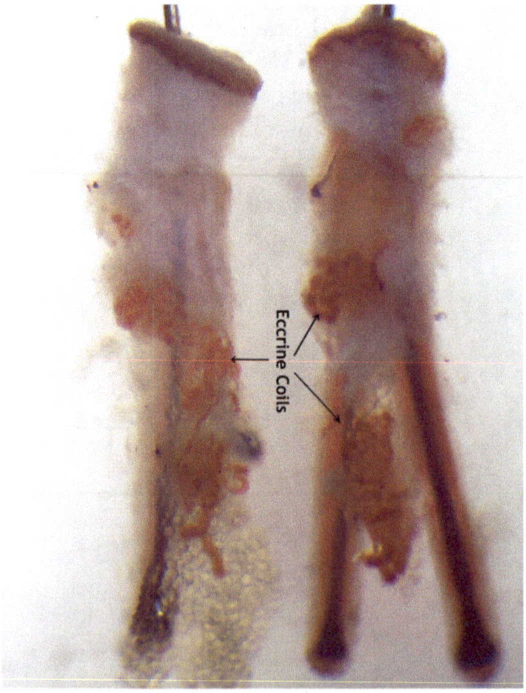

Fig. 3 Neutral red staining of two follicular units harvested from the scalp showing the presence of eccrine coils

preferable. FUE can be performed with manual or motorized punches. Manual punches are cheaper and less complicated for the beginner surgeon. However, when harvesting hundreds or even thousands of FU grafts in hair transplantation, most surgeons prefer to use the motorized punch technique because it is faster and less labor intensive (Electronic Supplementary Video 1).

3. After anesthetic infiltration, wait at least 15 min before harvesting so that the adrenaline achieves its maximal hemostatic effect.
4. The FUs can be harvested from any area of the scalp, but in hair transplant surgery they are normally harvested from the occipital or lateral sides of the scalp. Choose FUs with thick terminal hair groupings since they can be removed more easily than miniaturized hairs. In the author's experience around 70–80% of the FUs harvested from the scalp contain eccrine coils.
5. The punch incision around the FU is the most critical step. Here, the experience and feeling of the surgeon is very important to avoid significant damage through follicular transection. Use the minimum depth required to release the graft. Set Initial depth at 2–3 mm and adjust according to graft quality and hair transection. Stopping the advance of the punch at 3–4 mm below the skin is critical because follicles below this level start to splay apart, increasing the chance of hair follicle transection. Precise depth control is required. Ideally the punch should not penetrate deeper than 3–4 mm in order to minimize follicular unit transection. The end point of this punch incision maneuver is to core out or isolate an intact FU, releasing it from the surrounding dermal tissue. If done at the proper depth, the FU can be easily extracted with the forceps.
6. The success of this isolation method is very technique dependent. Obtaining an intact FU with a tiny punch only 1 mm in diameter is very challenging for the beginner surgeon. In non-experienced hands, the possibility of FU transection and damage is so high that this EG isolation method will not be efficient. For this reason, the author recommends that researchers interested in obtaining fresh human EGs from the scalp establish a collaboration with an experienced hair transplant surgeon who does this type of harvesting procedure on an almost daily basis, and who can provide scalp FUs to the researcher to perform the dyeing and EG microdissection.
7. The limited bleeding that can occur during FU harvesting can be easily managed with gauze compression for a few minutes.
8. Ideally, the eccrine coils should be dissected as soon as possible after FU harvesting.

9. The circular wounds left on the scalp after the FU punch harvesting will close naturally in 2–4 days (by second intention). Patient postoperative care is very simple. The patient can shampoo their hair the day after the FU harvesting procedure.
10. During the process of dyeing the FU with MB or NR, check under the stereomicroscope regularly to see if the eccrine gland is visible and present within the FU. Try to avoid using too high a dye concentration, and do not leave the FU too long in the dye as the goal is to make the EG visible with minimal background staining of the FU.
11. Note that the ductal excretory portion cannot be isolated with this method because at some point the duct is transected by the 1 mm punch.
12. Take care to keep the FU moist during microscopic dissection by adding a few drops of physiological saline as needed.

References

1. Lu C, Fuchs E (2014) Sweat gland progenitors in development, homeostasis, and wound repair. Cold Spring Harb Perspect Med 4: a0152223
2. Sato K, Kand WH, Saga K, Sato KI (1989) Biology of sweat glands and their disorders. I. Normal sweat gland function. J Am Acad Dermatol 20:537–563
3. Kurata R, Futaki S, Nakano I et al (2017) Three-dimensional cell shapes and arrangements in human sweat glands as revealed by whole-mount immunostaining. PLoS One 12: e0178709
4. Bovell D (2015) The human eccrine sweat gland: structure, function and disorders. J Local Glob Heath Sci 2015. https://doi.org/10.5339/jlghs.2015.5
5. Rittié L, Sachs DL, Orringer JS, Voorhees JJ, Fisher GJ (2013) Eccrine sweat glands are major contributors to reepithelialization of human wounds. Am J Pathol 182:163–171
6. Schittek B, Hipfel R, Sauer B et al (2001) Dermcidin: a novel human antibiotic peptide secreted by sweat glands. Nat Immunol 2:1133–1137
7. Poblet E, Jiménez-Acosta F, Hardman JA, Escario E, Paus R (2016) Is the eccrine gland an integral, functionally important component of the human scalp pilosebaceous unit? Exp Dermatol 25:149–150
8. Poblet E, Jimenez F, Escario-Travesedo E et al (2018) Eccrine sweat glands associate with the human hair follicle within a defined compartment of dermal white adipose tissue. Br J Dermatol 178:1163–1171
9. Jimenez F, Alam M, Hernandez I, Poblet E, Hardman JA, Paus R (2018) An efficient method for eccrine gland isolation from human scalp. Exp Dermatol 27:678–681
10. Vogel JE, Jimenez F, Cole J et al (2013) Hair restoration surgery: the state of the art. Aesthet Surg J 33:128–151
11. Jiménez-Acosta F, Ponce-Rodríguez I (2017) Follicular unit extraction for hair transplantation: an update. Actas Dermosifiliogr 108:532–537
12. Headington JT (1984) Transverse microscopy anatomy of the human scalp. Arch Dermatol 120:449–456
13. Sato K, Sato F (1983) Individual variations in structure and function of human eccrine sweat gland. Am J Phys 245:R203–R208
14. Lee CM, Jones CJ, Kealey T (1984) Biochemical and ultrastructural studies of human eccrine sweat glands isolated by shearing and maintained for seven days. J Cell Sci 72:259–274
15. Brayden DJ, Fitzpatrick J (1995) Cultured human sweat gland epithelia. Isolation of glands using Neutral Red. Pharm Res 12:171–175
16. Wolfe S, Cage G, Epstein M et al (1970) Metabolic studies of isolated human eccrine sweat glands. J Clin Invest 49:1880–1884
17. Mangos J (1973) Transductal fluxes of Na, K, and water in the human eccrine sweat gland. Am J Phys 224:1235–1240

18. Okada N, Kitano Y, Morimoto T (1983) Isolation of a viable eccrine sweat gland by dispase. Arch Dermatol Res 275:130–133
19. Kealy T (1983) The metabolism and hormonal responses of human eccrine sweat glands isolated by collagenase digestion. Biochem J 212:143–148
20. Hubka KM, Wu D, Harrington DA et al (2015) Dissociative and nondissociative models for culture of human eccrine glands for toxicology testing and tissue engineering applications. Appl In Vitro Toxicol 1:187–197

Chapter 15

Identification of Long Noncoding RNA by In Situ Hybridization Approaches

Mara Mancini, Anna Maria Lena, and Eleonora Candi

Abstract

In situ hybridization (ISH) and fluorescence in situ hybridization (FISH) techniques enable us to detect the expression of a specific RNA in fixed cells or tissue sections. Here, we describe in detail two procedures adjusted to reveal specifically lncRNAs in normal human keratinocytes and in skin tissue samples. Examples of the results obtained by the two different approaches are also shown.

Key words Noncoding RNA, Long non coding RNA, In situ hybridization, Fluorescence in situ hybridization, Skin, Keratinocytes

1 Introduction

About 80% of the human genome is transcribed, but only 2% of the transcripts codifies for proteins [1–3]. The majority of the transcripts is represented by noncoding RNAs (ncRNAs) [4, 5]. Among them, there are some "housekeeping ncRNAs" such as small nuclear RNAs (snRNAs), players in RNA splicing [6, 7]; transfer RNAs (tRNAs) and ribosomal RNAs (rRNAs), essential for mRNAs' translation into proteins [8]; and small nucleolar RNAs (snoRNAs) modifiers of rRNAs and tRNAs [7], also important for translation (Fig. 1a). Most of the other ncRNAs (miRNAs and long noncoding RNAs) were extensively studied during the last 10 years as key regulators of cellular physiology and pathology [4, 9], as diagnostic and prognostic markers in different cancers [10–15] and also as new putative therapeutic targets [16, 17].

Long ncRNAs are classified as RNA molecules of more than 200 nucleotides in length. Over 10,000 lncRNAs have been annotated in human transcriptome and their genes have been positioned inter- or intragenes in the genome [18, 19], but only few of them have been fully characterized. In general, they are gene expression regulators acting in different ways: as antisense to coding mRNAs

Natalia V. Botchkareva and Gillian E. Westgate (eds.), *Molecular Dermatology: Methods and Protocols*, Methods in Molecular Biology, vol. 2154, https://doi.org/10.1007/978-1-0716-0648-3_15,

© Springer Science+Business Media, LLC, part of Springer Nature 2020

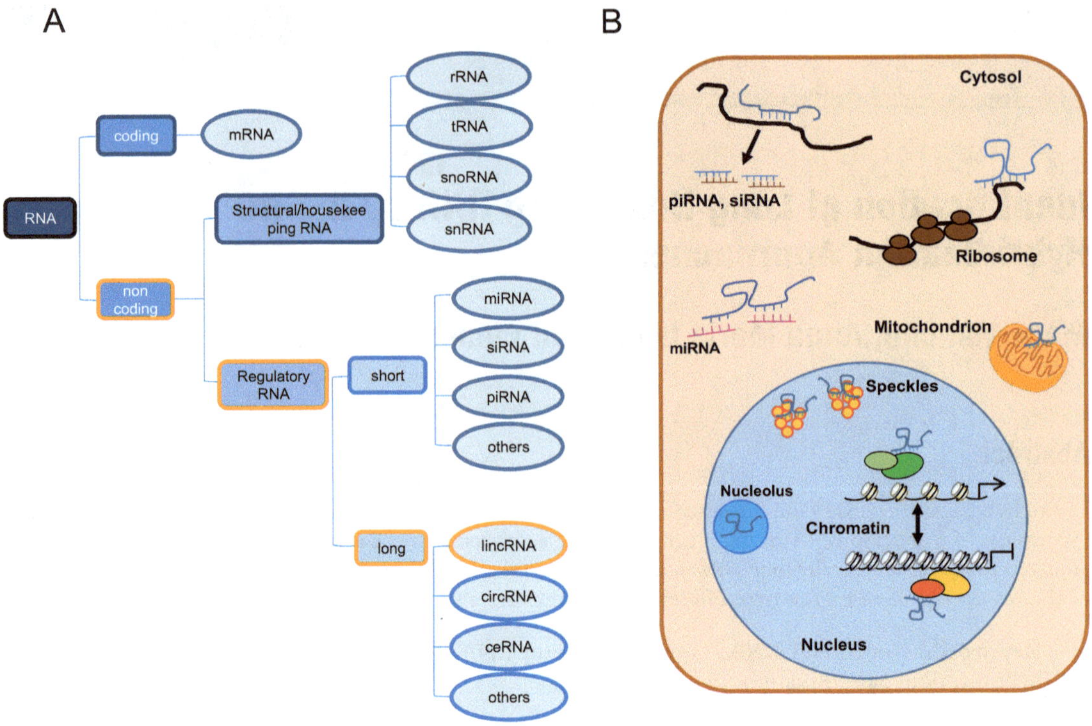

Fig. 1 Coding and noncoding RNAs. (**a**) Noncoding RNA classification. (**b**) Noncoding RNA functions and subcellular localization

[20], as precursors or sponges for smaller RNAs (circRNAs, ceRNA) [21], directly interacting with transcription regulating and chromatin remodeling complexes to address their action to specific loci [22] or affecting RNA/proteins localization and stability [20] (Fig. 1a, b). LncRNAs' roles in epidermal development, keratinocyte differentiation, and skin homeostasis as well as in skin malignancies and pathologies were also recently described [23, 24]. Two skin-specific lncRNAs are the prodifferentiation TINCR and the antidifferentiation ANCR lncRNAs [25, 26]. TINCR is localized in the cytoplasm of differentiated keratinocytes in association with differentiation specific mRNAs to increase their stability [27, 28]. On the other hand, ANCR is downregulated during keratinocyte differentiation and it is able to maintain their proliferative potential repressing epidermal differentiation program [25, 26]. In squamous cell carcinoma and basal cell carcinoma the uncontrolled keratinocyte proliferation and the repression of differentiation pathways are the main players of the neoplastic transformation. Very few studies investigated how lncRNAs are involved in these malignancies. Interestingly, the lncRNA TINCR is downregulated in squamous cell carcinoma supporting the finding of its role in keratinocyte differentiation [29]. Over 70% of melanomas present mutations in the BRAF

oncogene. Different studies describe the lncRNA expression profiles in melanoma cell lines compared to normal melanocytes or in melanocytes overexpressing the mutant BRAF oncogene or in BRAF mutant melanoma specimens. Of particular interest is a novel lncRNA transcript, BANCR (BRAF activated noncoding RNA) because its depletion affects motility of melanoma cells [30]. SPRY4-IT1 lncRNA was highly upregulated in melanoma cell lines and clinical samples and its knockdown led to cell death, decreased migration and invasion ability [31–33]. Another melanoma long intergenic noncoding RNAs (lincRNA) called SAMMSON, interacts with p32, a pro-oncogenic mitochondrial metabolism regulator. SAMMSON silencing abrogates crucial mitochondrial functions in a cancer-cell-specific manner [34]. An upregulated psoriasis susceptibility-related RNA gene induced by stress (PRINS) is involved in the hyperproliferative skin disorder psoriasis [35]. This lncRNA and some others were identified comparing gene expression profiles of epidermis from healthy subjects and psoriasis patients [36].

Using deep sequencing a dynamic modulation of lncRNA expression was also recently shown during skin UV-induced damage, wound healing, and physiological aging [37–41], indicating that lncRNAs play an important role also in skin pathologies.

All the described studies are based mainly on new technical approaches like microarrays, DNA and RNA deep sequencing, and chromatin immunoprecipitation (ChIP) sequencing [42]. Although, these wide screening data generated a huge amount of information about lncRNA expression profiles, different approaches are needed to elucidate their functions. The first step to try to clarify their physiological role is acquire knowledge of their cellular or tissue spatial localization in cells and in epidermis, a complex stratified organ that includes different cell types (Fig. 1b).

In situ hybridization (ISH) and Stellaris® fluorescence in situ hybridization (FISH) techniques adjusted specifically to lncRNA detection are described here in detailed step-by-step protocols (Fig. 2), and examples of ISH and FISH results experimentally obtained in human skin (Fig. 4a, b) and normal human keratinocytes are also shown (Fig. 4c).

To study the expression of lncRNA by ISH we have successfully used Exiqon in situ probes that include a novel class of modified nucleotide termed LNA. Structurally, in the LNA probes, the ribose ring is connected by a methylene bridge between the 2′-O and 4′-C atoms, thus "locking" the ribose ring in the ideal conformation for Watson–Crick binding. When incorporated into a DNA or RNA oligonucleotide LNA makes the pairing with a complementary nucleotide strand more rapid and increases the stability of the resulting duplex.

The protocol below is designed for human skin cryosections and for digoxigenin (DIG)-conjugated LNA probes. The final

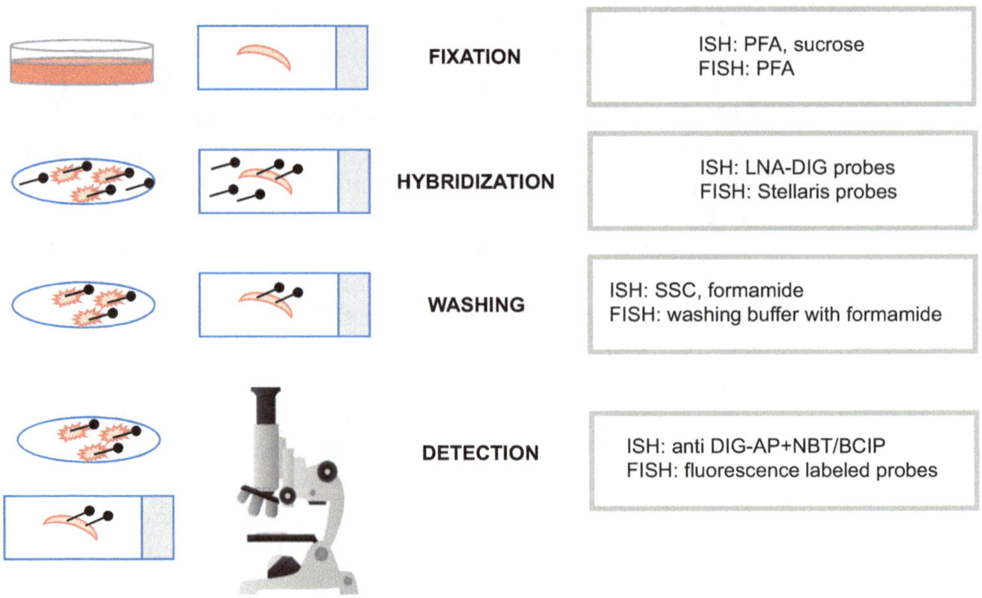

Fig. 2 Schematic representation of ISH and FISH experiment flowchart. The flowchart illustrates the main phases of the ISH and FISH procedures (*see* details in the text)

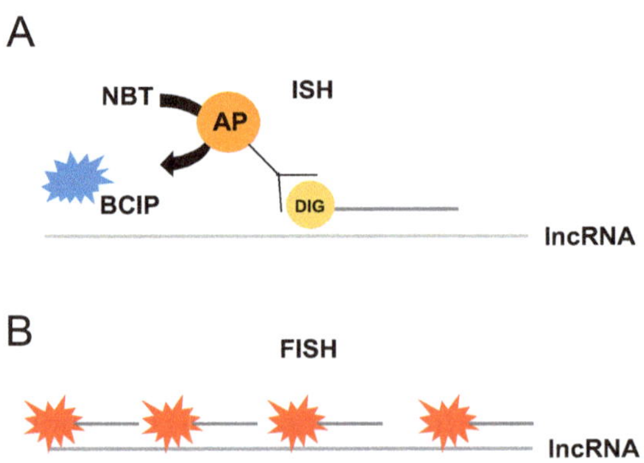

Fig. 3 ISH and FISH probes. (**a**) ISH probe is an oligonucleotide DIG labeled at 5′ or 3′ end or both. DIG tag is then detected by an AP conjugated antibody and revealed by NBT/BCIP staining; (**b**) FISH probes are a mix of short oligonucleotides complementary to ncRNA sequence and directly conjugated with a fluorophore

section staining is performed with an alkaline phosphatase color reaction where nitro-blue-tetrazolium and 5-bromo-4-chloro-3-indolyl phosphate (NBT/BCIP) are used to enhance the signal strength (Fig. 3a) [43].

Stellaris® RNA FISH is an RNA visualization method that allows simultaneous detection and localization of individual RNA molecules at the cellular and tissue level in fixed samples using

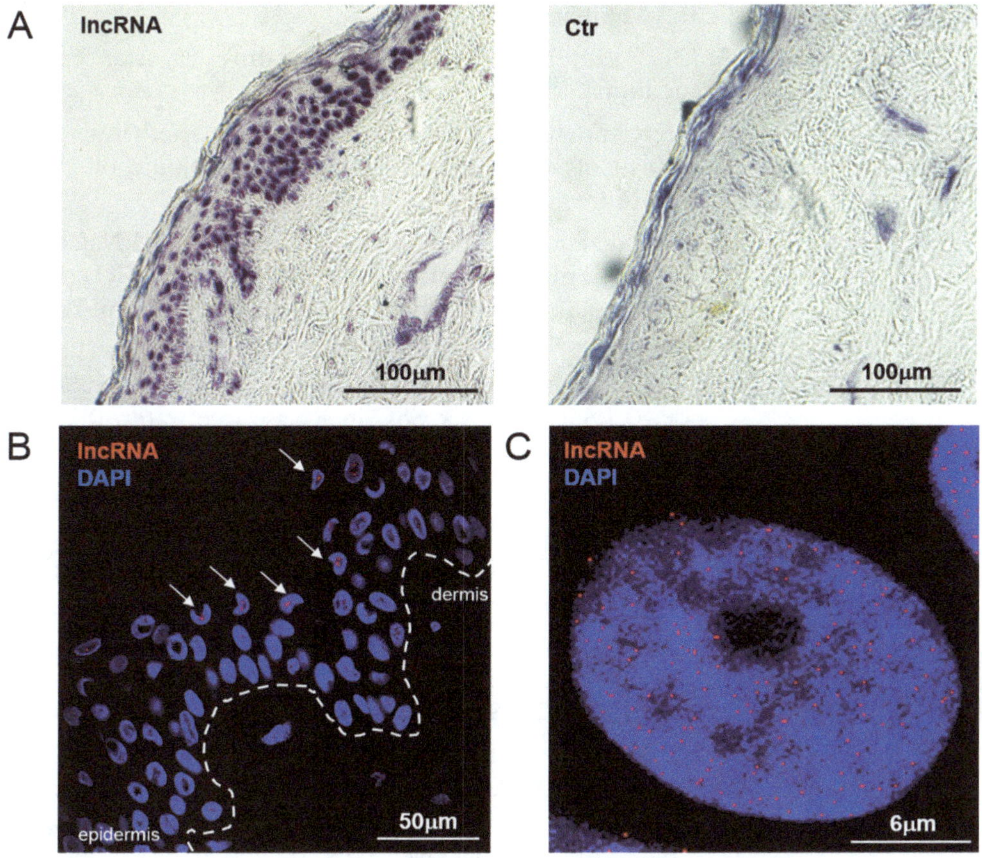

Fig. 4 ISH and FISH examples. (**a**) Nuclear lncRNA detection in human skin section by ISH. Left panel, lncRNA-specific probe; right panel, scramble control probe. (**b**) Nuclear lncRNA detection in human skin section by FISH. (**c**) Nuclear lncRNA detection in normal human keratinocytes by FISH

fluorescence microscopy. A set of Stellaris® RNA FISH Probes is composed of a pool of up to 48 unique probes, each labeled with a fluorophore, that collectively bind along an RNA target transcript to produce a diffraction limited spot (Fig. 3b). It is possible to use Stellaris® custom fish probes that are designed against lncRNA of interest by utilizing the Stellaris® RNA FISH Probe Designer (Biosearch Technologies, Inc., Petaluma, CA) available online at www.biosearchtech.com/stellarisdesigner. The lncRNA were hybridized with the Stellaris® RNA FISH Probe set labeled with Qasar 570 (Biosearch Technologies, Inc.).

2 Materials

2.1 *In Situ Hybridization (ISH)*

Prepare all the solutions and buffers in DNase/RNase-free water.

1. 4% Paraformaldehyde (PFA)/PBS (Phosphate Buffered Saline).

2. 0.5 M Sucrose/PBS.
3. OCT (Optimal Cutting Temperature cryostat embedding medium).
4. Acetylation solution: add 2.33 ml of triethanolamine and 500 μl of acetic anhydride to a final volume of 200 ml of RNase-free water.
5. Hybridization solution: 50% formamide, 0.3 M NaCl, 20 mM, Tris–HCl pH 8.0, 5 mM EDTA, 10 mM NaH_2PO_4 pH 8, 10% dextran sulfate, 1× Denhart's solution, 0.5 mg/ml Yeast RNA, and RNase-free water up to 20 ml.
6. Washing solutions: 5× SSC in RNase-free water, 50% formamide, 0.1% Tween, 1× SSC in RNase-free water, 0.2× SSC in RNase-free water, 1× PBS in RNase-free water.
7. Blocking solution: 1× PBS/10% goat serum and 0.1% Tween 20.
8. Antibody: Alkaline phosphatase (AP)-conjugated anti-digoxigenin (DIG) Fab fragment antibody.
9. NBT/BCIP developer solution: 3.4 μl 100 mg/ml NBT, 3.5 μl 50 mg/ml BCIP, 2.4 μl 24 mg/ml Levamisole, 5 μl of 10% Tween 20, 986 μl of B3 buffer.
10. B3 buffer: 3.35 ml 1.5 M Tris–HCl pH 9.5, 1 ml 5 M NaCl, 2.4 ml 1 M $MgCl_2$, and RNase-free water up to 50 ml.
11. LNA probe: 19–23 nucleotide long LNA probes are designed with approximately 30% of LNA content and the position of LNA incorporations is determined by the supplier (Exiqon, miRCURY probes). The epitope used in our protocol is DIG and the probes were ordered from Exiqon directly DIG labeled (Fig. 3a). We use to include a scramble probe to discriminate the specificity of the signal.

2.2 Fluorescent In Situ Hybridization (FISH)

1. Fixation buffer: 1 ml 37% formaldehyde, 1 ml 10× PBS, and 8 ml of RNase-free water.
2. Hybridization buffer: mix 900 μl Stellaris RNA FISH Hybridization Buffer (Biosearch Technologies) and 100 μl Formamide.
3. Stellaris FISH probes: Stellaris® custom fish probes are designed against lncRNA of interest by utilizing the Stellaris® RNA FISH Probe Designer at www.biosearchtech.com/stellarisdesigner (Biosearch Technologies, Inc., Petaluma, CA). The RNA is hybridized with the Stellaris RNA FISH Probe set labeled with Qasar 570 (Biosearch Technologies, Inc.) (Fig. 3b).

4. Wash buffer A: 2 ml Stellaris RNA FISH Wash Buffer A (Biosearch Technologies), 1 ml formamide, and 7 ml RNase-free water.
5. Wash buffer B (Biosearch Technologies).
6. 4′,6-Diamidino-2-phenylindole (DAPI).

3 Methods

The main points of in situ hybridization and fluorescence in situ hybridization methods are summarized in Fig. 2.

3.1 In Situ Hybridization (ISH)

3.1.1 Tissue Preparation and Embedding

1. Prefix dissected tissue in 4% PFA at 4 °C for an overnight (*see* **Note 1**).
2. After fixation, to minimize freeze fracturing of the tissue, place the specimen directly into 0.5 M sucrose/1× PBS and incubate overnight at 4 °C *see* **Note 2**).
3. Remove the specimen from the sucrose.
4. Freeze the tissue in OCT embedding medium and store blocks at −80 °C until ready to use.

3.1.2 Sectioning and Tissue Fixation

1. Cut sections of 14 μm on cryostat and collect them on Superfrost PLUS slides.
2. Air-dry sections at least 20 min but not longer than 3 h.
3. Fix the dried slides in 4% PFA in PBS for 10 min at 4 °C.

3.1.3 Acetylation of the Tissue

1. Wash the slides three times for 5 min with 1× PBS at room temperature.
2. Acetylate the slides for 10 min in acetylation solution. Acetylation positively charges amino groups of proteins, thus reducing background binding of the negatively charged probe to the tissue sections.
3. Wash the slides three times for 5 min with 1× PBS.

3.1.4 Hybridization and Slides Washing

1. Prepare 150 μl of denaturizing hybridization buffer by adding the LNA DIG-labeled probe with a final concentration of 20–40 nM.
2. Denature probes by heating them at 65 °C for 5 min and chill on ice.
3. Add the hybridization solution to the slides and cover the sections with RNase-free plastic coverslips.
4. Incubate the slides overnight at a temperature that is 20–22 °C below melting temperature (Tm) of the miRCURY LNA probe in a humidified chamber.

5. Wash the slides once with 5× SSC at room temperature for 20 min.
6. Wash the slides two times for 30 min at the same temperature of the hybridization step in 50% formamide, 0.1% Tween 20, and 1× SSC.
7. Wash the slides for 15 min in 0.2× SSC at room temperature.
8. Wash the slides for 15 min in 1× PBS at room temperature.

3.1.5 Antibody Incubation and Detection

1. Incubate the slides for 1 h at room temperature with blocking solution composed of 10% heat-inactivated goat serum, 0.1% Tween 20, and 1× PBS.
2. Incubate the slides for 2–3 h at room temperature with blocking solution preincubated for 1 h with AP-conjugated anti-DIG Fab fragment diluted 1:200 (*see* **Note 3**). During incubation place the slides in a humidified chamber.
3. Wash the slides 2 times for 30 min with 1× PBS–0.1% Tween 20.
4. Prepare 150 μl of developer solution for each slide and develop at room temperature in the dark ranging from 10 min up to 4 days, depending on the Noncoding RNA expression levels. The reaction can be monitored using a light microscope, terminate when a strong blue staining is observed.
5. Stop the color reaction by washing the slides three times for 10 min in 1× PBS.
6. Mount the slides in glycerol or any water-soluble mounting media.
7. Visualize the staining by using standard light microscope (*see* **Note 4**).

3.2 Fluorescent In Situ Hybridization (FISH)

3.2.1 Fixation of Frozen Tissue

1. Slice frozen tissue at a thickness of 4–10 μm using a cryostat and mount onto a microscope slide.
2. Thaw the slide-mounted tissue section to room temperature.
3. Immerse the slide in fixation buffer for 10 min at room temperature (*see* **Note 1**).
4. Wash the slides twice with 1× PBS for 5 min.
5. Immerse the slide in 70% (vol/vol) ethanol for at least 1 h at room temperature. The slides can be stored at +2 to +8 °C in 70% ethanol up to a week before hybridization.

3.2.2 Fixation of Adherent Cells

1. Grow cells on round cover glass in a 12-well cell culture plate.
2. Aspirate growth medium and wash with 1 ml of 1× PBS.
3. Add 1 ml of fixation buffer.
4. Incubate at room temperature for 10 min.

5. Wash twice with 1 ml of 1× PBS.
6. Immerse cells in 1 ml of 70% ethanol for at least 1 h at +2 to +8 °C. Cells can be stored at +2 to +8 °C in 70% ethanol up to a week before hybridization.

3.2.3 Hybridization in Adherent Cells and/or Frozen Tissue

1. Prepare the hybridization buffer by adding 1–2 μl of probe stock solution to a final volume of 100 μl of hybridization buffer. This creates a working probe solution of 125 nM.
2. Wash the cover glass containing adherent cells or slide mounted tissue with wash buffer A for 5 min.
3. Assemble the humidified chamber and dispense 100 μl of the hybridization buffer containing probe on the cover glass containing cells or slide mounted tissue.
4. Incubate in the dark at 37 °C for at least 4 h. Incubation can be continued up to 16 h.
5. Wash cover glass or slides with wash buffer A for 30 min at 37 °C.
6. Incubate cover glass or slides for 30 min at 37 °C in the dark with wash buffer A containing 5 ng/ml of DAPI to counterstain the nuclei.
7. Wash cover glass or slides with wash buffer B for 5 min at room temperature.
8. Add a small drop of mounting medium onto the tissue slides and cover with a clean cover glass. For cover glass containing cells add a small drop of mounting medium onto a microscope slide, and mount cover glass onto the slide.
9. Proceed to imaging using fluorescence or confocal microscope.

4 Notes

1. For optimal fixation it may be critical to use fresh formaldehyde solutions.
2. Place tissue directly into sucrose without washing. Residual 4% paraformaldehyde does not affect the detection.
3. It is possible to use a PAP Pen (liquid blocker super PAP Pen) to draw a liquid-repellent barrier around section on slide to decrease amount of antibody solution used per slide.
4. Adapted from Heller et al. [44].

References

1. Alexander RP, Fang G, Rozowsky J, Snyder M, Gerstein MB (2010) Annotating non-coding regions of the genome. Nat Rev Genet 11:559–571
2. The ENCODE Project Consortium, Dunham I, Kundaje A, Aldred SF, Collins PJ, Davis CA, Doyle F, Epstein CB, Frietze S, Harrow J et al (2012) An integrated encyclopedia of DNA elements in the human genome. Nature 489:57–74
3. Derrien T, Johnson R, Bussotti G, Tanzer A, Djebali S, Tilgner H, Guernec G, Martin D, Merkel A, Knowles DG et al (2012) The GENCODE v7 catalog of human long noncoding RNAs: analysis of their gene structure, evolution, and expression. Genome Res 22:1775–1789
4. Mercer TR, Dinger ME, Mattick JS (2009) Long non-coding RNAs: insights into functions. Nat Rev Genet 10:155–159
5. Fatica A, Bozzoni I (2014) Long non-coding RNAs: new players in cell differentiation and development. Nat Rev Genet 15:7–21
6. Wassarman D, Steitz J (1992) Interactions of small nuclear RNA's with precursor messenger RNA during in vitro splicing. Science 257:1918–1925
7. Tollervey D, Kiss T (1997) Function and synthesis of small nucleolar RNAs. Curr Opin Cell Biol 9:337–342
8. Mattick JS, Makunin IV (2006) Non-coding RNA. Hum Mol Genet 15:R17–R29
9. Anastasiadou E, Jacob LS, Slack FJ (2018) Non-coding RNA networks in cancer. Nat Rev Cancer 18:5–18
10. Chandra Gupta S, Nandan Tripathi Y (2017) Potential of long non-coding RNAs in cancer patients: from biomarkers to therapeutic targets. Int J Cancer 140:1955–1967
11. Formosa A, Lena AM, Markert EK, Cortelli S, Miano R, Mauriello A, Croce N, Vandesompele J, Mestdagh P, Finazzi-Agrò E et al (2013) DNA methylation silences miR-132 in prostate cancer. Oncogene 32:127–134
12. Formosa A, Markert EK, Lena AM, Italiano D, Finazzi-Agro' E, Levine AJ, Bernardini S, Garabadgiu AV, Melino G, Candi E (2013) MicroRNAs, miR-154, miR-299-5p, miR-376a, miR-376c, miR-377, miR-381, miR-487b, miR-485-3p, miR-495 and miR-654-3p, mapped to the 14q32.31 locus, regulate proliferation, apoptosis, migration and invasion in metastatic prostate cancer cells. Oncogene 33:5173–5182
13. Amelio I, Lena AM, Viticchiè G, Shalom-Feuerstein R, Terrinoni A, Dinsdale D, Russo G, Fortunato C, Bonanno E, Spagnoli LG et al (2012) Mir-24 triggers epidermal differentiation by controlling actin adhesion and cell migration. J Cell Biol 199:347–363
14. Rivetti di Val Cervo P, Lena AM, Nicoloso M, Rossi S, Mancini M, Zhou H, Saintigny G, Dellambra E, Odorisio T, Mahe C et al (2012) p63-microRNA feedback in keratinocyte senescence. Proc Natl Acad Sci U S A 109:1133–1138
15. Viticchiè G, Lena AM, Cianfarani F, Odorisio T, Annicchiarico-Petruzzelli M, Melino G, Candi E (2012) MicroRNA-203 contributes to skin re-epithelialization. Cell Death Dis 3:e435
16. Matsui M, Corey DR (2017) Non-coding RNAs as drug targets. Nat Rev Drug Discov 16:167–179
17. Chen L, Dzakah EE, Shan G (2018) Targetable long non-coding RNAs in cancer treatments. Cancer Lett 418:119–124
18. Wilusz JE, Sunwoo H, Spector DL (2009) Long noncoding RNAs: functional surprises from the RNA world. Genes Dev 23:1494–1504
19. Iyer MK, Niknafs YS, Malik R, Singhal U, Sahu A, Hosono Y, Barrette TR, Prensner JR, Evans JR, Zhao S et al (2015) The landscape of long noncoding RNAs in the human transcriptome. Nat Genet 47:199–208
20. Rashid F, Shah A, Shan G (2016) Long non-coding RNAs in the cytoplasm. Genomics Proteomics Bioinformatics 14:73–80
21. Thomson DW, Dinger ME (2016) Endogenous microRNA sponges: evidence and controversy. Nat Rev Genet 17:272–283
22. Peschansky VJ, Wahlestedt C (2014) Non-coding RNAs as direct and indirect modulators of epigenetic regulation. Epigenetics 9:3–12
23. Hombach S, Kretz M (2013) The non-coding skin: exploring the roles of long non-coding RNAs in epidermal homeostasis and disease. Bioessays 35:1093–1100
24. Botchkareva NV (2017) The molecular revolution in cutaneous biology: noncoding RNAs: new molecular players in dermatology and cutaneous biology. J Invest Dermatol 137: e105–e111
25. Kretz M, Webster DE, Flockhart RJ, Lee CS, Zehnder A, Lopez-Pajares V, Qu K, Zheng GXY, Chow J, Kim GE et al (2012) Suppression of progenitor differentiation requires the

long noncoding RNA ANCR. Genes Dev 26:338–343

26. Lopez-Pajares V, Qu K, Zhang J, Webster DE, Barajas BC, Siprashvili Z, Zarnegar BJ, Boxer LD, Rios EJ, Tao S et al (2015) A LncRNA-MAF:MAFB transcription factor network regulates epidermal differentiation. Dev Cell 32:693–706
27. Kretz M, Siprashvili Z, Chu C, Webster DE, Zehnder A, Qu K, Lee CS, Flockhart RJ, Groff AF, Chow J et al (2013) Control of somatic tissue differentiation by the long non-coding RNA TINCR. Nature 493:231–235
28. Xu T, Liu X, Xia R, Yin L, Kong R, Chen W, Huang M, Shu Y (2015) SP1-induced upregulation of the long noncoding RNA TINCR regulates cell proliferation and apoptosis by affecting KLF2 mRNA stability in gastric cancer. Oncogene 34:5648–5661
29. Zhou W, Zhang S, Li J, Li Z, Wang Y, Li X (2019) lncRNA TINCR participates in ALA-PDT-induced apoptosis and autophagy in cutaneous squamous cell carcinoma. J Cell Biochem 120:13893–13902
30. Li R, Zhang L, Jia L, Duan Y, Li Y, Bao L, Sha N (2014) Long non-coding RNA BANCR promotes proliferation in malignant melanoma by regulating MAPK pathway activation. PLoS One 9:e100893
31. Khaitan D, Dinger ME, Mazar J, Crawford J, Smith MA, Mattick JS, Perera RJ (2011) The melanoma-upregulated long noncoding RNA SPRY4-IT1 modulates apoptosis and invasion. Cancer Res 71:3852–3862
32. Zhao W, Mazar J, Lee B, Sawada J, Li J-L, Shelley J, Govindarajan S, Towler D, Mattick JS, Komatsu M et al (2016) The long noncoding RNA SPRIGHTLY regulates cell proliferation in primary human melanocytes. J Invest Dermatol 136:819–828
33. Zhao J, Yin M, Deng H, Jin FQ, Xu S, Lu Y, Mastrangelo MA, Luo H, Jin ZG (2016) Cardiac Gab1 deletion leads to dilated cardiomyopathy associated with mitochondrial damage and cardiomyocyte apoptosis. Cell Death Differ 23:695–706
34. Leucci E, Vendramin R, Spinazzi M, Laurette P, Fiers M, Wouters J, Radaelli E, Eyckerman S, Leonelli C, Vanderheyden K et al (2016) Melanoma addiction to the long non-coding RNA SAMMSON. Nature 531:518–522
35. Sonkoly E, Bata-Csorgo Z, Pivarcsi A, Polyanka H, Kenderessy-Szabo A, Molnar G, Szentpali K, Bari L, Megyeri K, Mandi Y et al (2005) Identification and characterization of a novel, psoriasis susceptibility-related noncoding RNA gene, PRINS. J Biol Chem 280:24159–24167
36. Gupta R, Ahn R, Lai K, Mullins E, Debbaneh M, Dimon M, Arron S, Liao W (2016) Landscape of long noncoding RNAs in psoriatic and healthy skin. J Invest Dermatol 136:603–609
37. Zheng L-L, Li J-H, Wu J, Sun W-J, Liu S, Wang Z-L, Zhou H, Yang J-H, Qu L-H (2016) deepBase v2.0: identification, expression, evolution and function of small RNAs, LncRNAs and circular RNAs from deep-sequencing data. Nucleic Acids Res 44: D196–D202
38. Williamson L, Saponaro M, Boeing S, East P, Mitter R, Kantidakis T, Kelly GP, Lobley A, Walker J, Spencer-Dene B et al (2017) UV irradiation induces a non-coding RNA that functionally opposes the protein encoded by the same gene. Cell 168:843–855.e13
39. Herter EK, Xu Landén N (2017) Non-coding RNAs: new players in skin wound healing. Adv Wound Care 6:93–107
40. Kour S, Rath PC (2016) Long noncoding RNAs in aging and age-related diseases. Ageing Res Rev 26:1–21
41. Kim C, Kang D, Lee EK, Lee J-S (2017) Long noncoding RNAs and RNA-binding proteins in oxidative stress, cellular senescence, and age-related diseases. Oxidative Med Cell Longev 2017:1–21
42. Kashi K, Henderson L, Bonetti A, Carninci P (2016) Discovery and functional analysis of lncRNAs: methodologies to investigate an uncharacterized transcriptome. Biochim Biophys Acta 1859:3–15
43. Obernosterer G, Martinez J, Alenius M (2007) Locked nucleic acid-based in situ detection of microRNAs in mouse tissue sections. Nat Protoc 2:1508–1514
44. Heller S, Sheane CA, Javed Z, Hudspeth AJ (1998) Molecular markers for cell types of the inner ear and candidate genes for hearing disorders. Proc Natl Acad Sci U S A 95:11400–11405

Chapter 16

Detection of MicroRNAs by In Situ Hybridization in Skin

Maximilian E. Pickup and Mohammed I. Ahmed

Abstract

MicroRNAs (miRNAs) are a family of small noncoding RNAs (~19–24 nt) playing a key role in the execution of gene expression programs in various cells and tissues. Many technical challenges have been encountered when investigating miRNAs, in particular, determining the spatiotemporal expression pattern of miRNAs in cells and tissues. We describe here a well-established in situ hybridization protocol for the detection and analysis of spatiotemporal expression patterns of miRNAs in skin and its appendages such as the hair follicle in both frozen and paraffin-embedded tissue sections. We describe in detail the different steps that are associated with utilizing in situ hybridization procedure on either frozen or paraffin-embedded tissues for miRNAs localization. Postfixation, tissues are hybridized with LNA double labeled probes with digoxygenin. Detection of hybridized probes is performed by using an alkaline phosphatase coupled antibody against digoxygenin. The final step involves the use of substrates to develop the color of alkaline phosphatase-LNA-probe structure leading to identification of the spatiotemporal location of target miRNAs in target tissue and cells. We also discuss two options for substrate color development in these procedures: (1) NBT/BCIP and (2) BM Purple. This method is a simple and convenient way of determining the spatiotemporal expression pattern of miRNAs, which has been a challenge since their discovery, due to their relatively small size. Knowledge gained from in situ hybridization is crucial for better understanding of the roles of individual miRNA(s) during distinct stages of development in various cells and tissues. These protocols will be beneficial to the wider scientific community.

Key words In situ hybridization, MicroRNAs, Skin and hair follicles

1 Introduction

In situ hybridization (ISH) is a technique used to identify the spatiotemporal expression of transcripts of coding and more recently noncoding RNA, including microRNAs, in either fixed tissue samples or in cells, using labeled probes complimentary to target RNA. This technique is widely used in research labs and provides a visual map of genetic expression in aged, mutated or diseased samples in various animal cells and tissues. This expression map coupled with visualizing cellular structures (by counterstaining or immunofluorescence staining) can give possible explanations as to the phenotype seen in different samples affected by aging,

Natalia V. Botchkareva and Gillian E. Westgate (eds.), *Molecular Dermatology: Methods and Protocols*,
Methods in Molecular Biology, vol. 2154, https://doi.org/10.1007/978-1-0716-0648-3_16,
© Springer Science+Business Media, LLC, part of Springer Nature 2020

diseases, or genetic abnormalities [1]. Here we aim to demonstrate a clear methodical approach for detection of specifically small, noncoding RNA known as microRNAs (miRNAs, miRs). MiRNAs are approximately ~19–24 nt nucleotides in length, which makes them more challenging to detect than long gene transcripts. MiRNAs function as gene regulators at posttranslational level, mainly in gene silencing [2–4]. MiRNAs' function as a regulatory level of genetic expression at distinct time points of tissue development and/or tissue regeneration.

ISH can be utilized on tissues and cells for better understanding of specific roles of individual miRNAs at distinct developmental time points. Here we demonstrate an optimized protocol for the detection of miRNAs in frozen and paraffin-embedded tissue samples, in particular, in the skin (Fig. 1; [2, 3, 5, 6]. We will describe separately a step-by-step procedure for performing in situ hybridization in frozen and paraffin-embedded tissue sections.

2 Materials

Prepare all the solutions and buffers in DNase/RNase-free water.

1. UltraPure water or DEPC-treated water (*see* **Note 1**).
2. UltraPure PBS.
3. 4% paraformaldehyde (PFA) in PBS.
4. Acetic anhydride–TEA solution: 300 mM triethanolamine, 0.25% acetic anhydride.
5. Permeabilization buffer: 1% Triton X-100.
6. Hybridization buffer: formamide DI (50%), 20× SSC pH 4.5 (5× SSC), yeast RNA (50 μg/ml), 20% SDS solution (1% SDS), heparin (50 μg/ml).
7. Wash buffer 1: formamide (50%), 20× SSC pH 4.5 (5× SSC); 20% SDS solution (1% SDS).
8. Wash buffer 2: formamide (50%), 20× SSC pH 4.5 (5× SSC).
9. 1× TBST buffer solution: TBST (1% Tween 20 in TBS), levamisole (1 mg/ml).
10. Blocking solution: blocking reagent (2%), heat-inactivated sheep serum (5%), Tween 20 (1%), and levamisole (1 mg/ml) in 1× TBST.
11. Modified blocking solution: heat-inactivated sheep serum (5%), Tween 20 (1%), and levamisole (1 mg/ml) in 1× TBST.
12. NTMT buffer: 5 M NaCl (100 mM), 1 M Tris–HCl pH 9.5 (100 mM), 2 M $MgCl_2$ (50 mM), Tween 20 (1%).
13. Proteinase K (1 μg/ml in PBS).
14. Glycine (2 mg/ml in PBS).

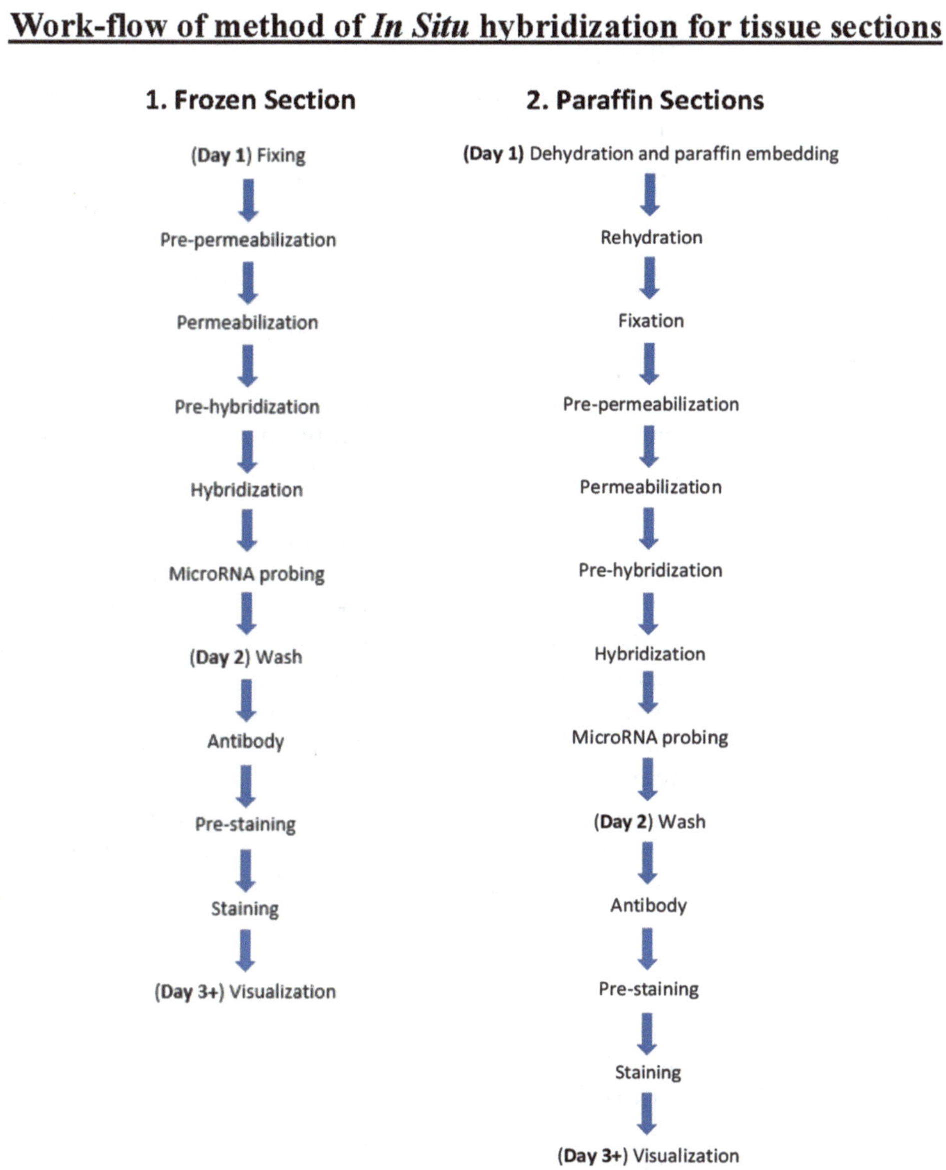

Fig. 1 In situ hybridization workflow on frozen and paraffin-embedded tissue sections. Each work-flow illustrates the main steps of conducting in situ hybridization (ISH) for both frozen and paraffin embedded tissue sections. ISH on frozen sections consists of the following key steps: tissue permeabilization, hybridization, probing, and visualization. ISH on paraffin-embedded sections follows similar key steps with additional steps of dehydration and hydration of tissue and prepermeabilization of tissue sections

3 Methods

3.1 In Situ Hybridization on Frozen Tissue Sections

1. Prepare 5–10 μm tissue sections either on coated RNase-free or Superfrost/plus slides. Freeze the slides on dry ice immediately after cutting sections and store the slides at −80 °C until ready to use.

2. Air-dry the slides at room temperature for approximately 10 min.
3. Place the slides in a slide mailer and incubate in 4% PFA for 10 min at room temperature.
4. Wash the slides in PBS for 5 min.
   (a) Incubate the tissue samples in acetic anhydride–TEA solution for 10 min at room temperature.
5. Wash the tissue samples in PBS for 5 min.
6. Incubate the tissue samples in permeabilization buffer for 10 min at room temperature.
7. Incubate the tissue samples in hybridization buffer for 1 h at predetermined hybridization temperature (*see* **Note 2**). Complete the following steps at predetermined hybridization temperature.
8. *Probe preparation*: Prepare the miRCURY double DIG-labeled LNA™ miRNA probe (Exiqon, Germany) (*see* **Notes 3** and **16**) solution at 2 pmol. We recommend miRNA concentration optimization prior to this stage (*see* **Note 4**). Denature the probe in hybridization buffer at 80 °C for 5 min after which, immediately chill it on ice for 5 min. Vortex (30 s) and centrifuge probe solution (full speed for 30 s), and keep at room temperature.
9. Place the slides horizontally in a humidified chamber. Flood a humidified chamber with ddH_2O prior to adding slides to prevent the slides drying out during subsequent overnight incubation.
10. Draw a hydrophobic circle around the tissue samples with a PAP pen. Pipette the hybridization buffer containing the miRNA probe (100 μl) directly on the tissue section and carefully cover it with a clean coverslip to evenly distribute the microRNA probe and hybridization solution (*see* **Note 5**).
11. Incubate the slides at predetermined hybridization temperature overnight or for a minimum of 16 h.

Complete the following washing steps at hybridization temperature (from **step 12**; *see* **Note 6**).

12. Pour off hybridization buffer containing the miRNA probe from the slides and place the slides back into a slide mailer (*see* **Note 7**).
13. Wash the slides three times with Wash Buffer 1 for 15 min per wash at hybridization temperature.
14. Wash the slides three times with Wash Buffer 2 for 15 min per wash at hybridization temperature.

The following steps should be carried out at room temperature:

15. Wash three times with TBST buffer solution for 10 min per wash.
16. Incubate the samples in Blocking Solution (*see* **Note 8**) for 60 min.
17. *Antibody incubation*: add anti–digoxygenin (DIG)-alkaline phosphatase (AP) antibody into Modified Blocking Solution in dilution range from 1:1000 to 1:2500 (*see* **Note 9**) and incubate at room temperature for 2 h.
18. *Substrate color development*: The following steps are carried out at room temperature in the slide mailer. Wash the slides four times with TBST buffer solution for 15 min per wash.
19. Wash three times with freshly made NTMT buffer for 15 min per wash.
20. Incubate the slides in BM purple or NBT/BCIP substrate solution (*see* **Note 10**) in the dark for 12–96 h at room temperature or until a color reaction has occurred. We recommend to initially observe the development of signal for 4–6 h to prevent over development of signal taking place (*see* **Note 11**).
21. Wash the slides three times with PBS for 10 min per wash.
22. Counterstain with nuclear fast red solution for 30 s to 2 min (*see* **Note 12**).
23. Wash three times with water for 1 min per wash.
24. Air-dry the slides and then dip once in xylene to prepare tissue for coverslip.
25. Pipet a few drops of DPX mounting medium onto the slides and coverslip (*see* **Note 13**) and let dry overnight.

3.2 In Situ Hybridization of Paraffin-Embedded Tissue Sections

1. *Tissue processing*: Harvested tissue must be processed for tissue sectioning. This involves dehydrated of tissue by immersing it into a series of aqueous alcohol solutions gradually moving to pure alcohol (ranging from 50% to 100% absolute alcohol in ddH_2O). The tissue can then be embedded in paraffin wax, which enables the cutting of sections of between 8 and 10 μm thickness of skin tissue (*see* **Note 14**).
2. *Rehydration of tissue sections*: Place slides of tissue sections into a slide mailer and clear tissue using Histo-Clear II solution for 10 min, after which, a further two times with Histo-Clear II solution for 2 min per wash. Remove the Histo-Clear II solution and incubate the slides in aqueous alcohol solutions gradually decreasing in concentration moving to PBS (ranging from 100%, 90%, 80%, 70%, and 50% alcohol, then PBS) each step for 5 min.

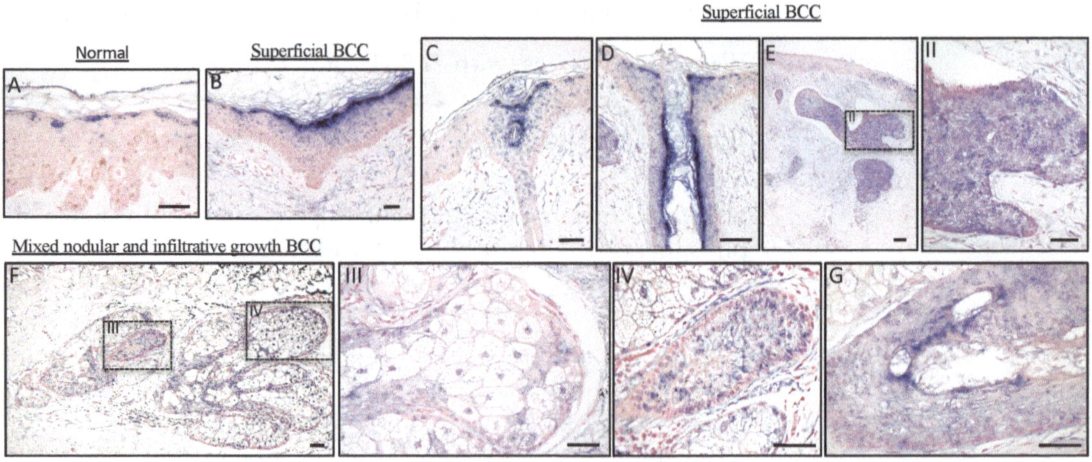

Fig. 2 MicroRNA-21 spatiotemporal expression pattern in human basal cell carcinoma (BCC). (**a**) Healthy human skin: miR-21 is expressed at relative low levels with expression restricted to the superficial/differentiated layers of the skin. (**b**, **c**) Superficial BCC: band-like lymphoid infiltrate in opposition to the basaloid nests (arrow). (**e–g**) Mixed nodular and infiltrative growth BCC/Nodular BCC manifests multiple variably sized and rounded nodules: miR-21 expression is elevated with expression restricted to the growth of BCC/Nodular regions of carcinoma (arrow). Scale bars, 50 μm

3. Incubate the tissue slides with 4% PFA for 15 min at room temperature.
4. Wash slides with PBS for 5 min.
5. Incubate the slides in the proteinase K (*see* **Note 15**) for 5 min.
6. Incubate the samples in glycine for 5 min.
7. Wash the slides in PBS for 5 min.
8. Incubate the slides in previously used 4% PFA (from **step 3**) for 15 min.
9. Continue to follow **step 5** of Subheading 3.1 beginning with '*Acetylation of tissue*' to complete in situ hybridization for paraffin sections.
10. Image capture using bright field microscope (Fig. 2).

4 Notes

1. We recommend using UltraPure water or PBS commercially bought to prevent any degradation of RNA targets during the in situ hybridization process. We also suggest using DEPC-treated water as an alternative to commercially purchasing UltraPure water/PBS.
2. We use miRCURY double DIG-labeled LNA™ probes from Qiagen/Exiqon. The use of LNA™ technology and the double DIG-labeled probes on either side of the sequence

improves the specificity, sensitivity, and detection of miRNA. Melting temperature of each probe can be found on the Exiqon/Qiagen datasheet provided with probe purchased or online. From our experience a reduction of 18–20 °C from the predicted RNA melting temperature is a good starting point. We also advise to optimize the hybridization temperature on tissue sections with high level of expression of miRNA of interest (as a positive control) vs. tissue samples with low or absence of the expression of targeted miR to confirm specificity (negative control; comparison of background vs. signal staining).

3. Nonradioisotope labeling compounds include DIG, fluorescein, biotin, and bromodeoxyuridine (BrdU; Jensen, 2014). Biotin and digoxigenin are the most commonly used methods and are visualized by indirect detection from a fluorescently labeled or enzyme linked protein or antibody. Incorporation of these compounds is commonly performed by nick translation, PCR, or oligonucleotide 3′/5′ labeling. These labeling methods provide a higher resolution compared to the isotope equivalent and can generally be conducted in a much shorter space of time compared to radiolabeled probes [7].

   We recommend using miRCURY double DIG-labeled LNA™ probes for detection of miRNAs as they provide improved specificity, stability of probe, and low background to signal staining. These labeled probes can be visualized by alkaline phosphatase (AP)-linked antibody. Visualization of target probe is achieved by the addition of BM purple or NBT/BCIP substrates (*see* **Note 10**) that leads to a precipitation of color development and identification of the spatiotemporal expression and localization of miRNA(s) on cells or tissues.

4. As with predetermining hybridization temperature for each probe, it is important to also determine the amount of probe used per tissue. As mentioned in **Note 2**, we use miRCURY double DIG-labeled LNA™ probes. The use of LNA™ technology and the double DIG-labeled probes on either side of the target miRNA sequence provides opportunity to apply reduced concentration of miRNA probe per tissue section/slide.

5. We recommend using a new clean coverslip which will cover the entire area of interest (where tissue section(s) are located). This will require steady hands and also great care not to damage the tissue. We have, however, performed the above experiment with and without coverslips. If carrying out experiments without coverslips, the humidifying chamber requires sufficient amount of buffer/water to prevent slides drying out overnight.

6. Due to the use of miRCURY double DIG-labeled LNA™ probes in our protocols, it allows to wash tissue sections at higher temperatures, usually around the hybridization step. Washing tissue sections at the hybridization temperature will lead to the efficient removal of nonattached probe, which can result in nonspecific signal. Washing at hybridization temperature also does not reduce specificity or lead to loss of signal detection of miRNA(s) in our tissue samples.
7. You can reuse the slide mailer used on day one of ISH, on the second day without the concerns of RNA degradation.
8. We have noticed that the Blocking Solution will take some time to dissolve completely at room temperature. We suggest heating the Blocking Solution at the washing temperature for approximately 1 h, giving a rigorous shake every 10–15 min until completely dissolved. You can only use it once it is completely dissolved.
9. Anti-digoxygenin-alkaline phosphatase (anti-DIG-AP) requires dilution optimization. We recommend using range from 1:1000 (for low expressing miRNAs) to 1:2500 for highly expressed miRNAs. Dilution factor will also impact; (1) rate of substrate development, and (2) rate of development of specific signal to background staining on your tissue section.
10. For miRNA ISH in skin, we have utilized two different substrates for color development of miRNA signal; these are BM purple and/or NBT/BCIP solution. BM purple is a substrate for alkaline phosphatase (AP) and will produce a deep purple color precipitate upon reaction with anti-DIG-AP antibody bound to a specific target. NBT/BCIP also uses AP for visualization. BCIP reaction with AP causes a production of insoluble NBT dimer product ranging from reddish to purple in color and, like BM purple, can be observed visually under a light microscope. Tissue section can be counterstained with nuclear fast red using either substrate [8] and is highly recommended.

    In our application of ISH in skin, BM Purple is chromogenic substrate specifically for AP-linked antibodies resulting in low background to signal staining (Fig. 2). Over 12–96 h, BM purple will change color from light-yellow (initial color) to a deep purple precipitate (complete development of signal), allowing for visualization of the specific target miRNA sequence, which can be seen under a light microscope. BM purple detection option is recommended for targets that are expressed at low to medium levels. An advantageous characteristic of developing your miRNA signal with BM purple at room temperature, is that it allows the user to incubate the tissue sections for longer periods of time without the risk of over-development or background staining (between 12 and 96 h), resulting in a distinct and sharp staining of target miRNA (s) with counterstain of tissue.

The alternative to BM purple is NBT/BCIP substrate solution. In our experience, NBT/BCIP detection option is suitable for miRNAs that are abundantly expressed in the target tissues. NBT/BCIP develops considerably faster at room temperature than BM purple, usually around 6–12 h. We suggest frequent checking for signal development during incubation to prevent overstaining leading to increased background to signal staining. Therefore, for visualizing of low to mid-level expressing miRNAs in skin we recommend BM purple, while abundant miRNAs can be detected using either BM purple or NBT/BCIP.

11. From our experience we have noticed that shorter incubation time is required for more abundant miRNAs compared to miRNAs expressed at lower level. BM Purple substrate development can be accelerated when incubated at 30 °C but care must be taken as to not to over develop the reaction as it could lead to increased background to signal staining.
12. We recommend optimizing nuclear fast red staining as this can vary depending on tissue type being stained.
13. Remove any bubbles by gently pressing the coverslip with forceps; it should be done carefully as to not damage the tissue sections during this process.
14. Tissue thickness should range between 8 and 10 μm for skin. If thicker tissue sections are used, we recommend increasing incubation times of each step during the procedure.
15. Treatment of paraffin-embedded tissue sections with Proteinase K will break the protein–protein links. This will lead to the unmasking of target antigens, epitopes or in our case miRNAs resulting in enhanced detection.
16. We highly recommend using controls (either tissue and/or probe) to determine the specificity of a miRNA-target probe signal being generated. *Negative controls*: these help to identify any issues with nonspecific signal development, which may invalidate the experimental data. A 'scrambled' probe sequence can be used as the negative control. This probe will not bind to any specific miRNA [3, 9].

    Alternatively, positive probe control can be used as an indicator of successful completion of ISH. We suggest using the probes for detection of the noncoding small nuclear RNA either U6 or SnoRNA that are abundantly expressed in all tissues we have investigated. This will validate the success of the hybridization protocol. Conducting positive and negative controls alongside target miRNAs on the same tissues samples is essential to ensure that the signal observed is specific to your miRNA of interest.

Acknowledgments

I would like to thank my mentors Dr. Natalia Botchkareva and Dr. Andrei Mardaryev for their help and support when first introducing in situ hybridization to me as young Ph.D. student. This technique has been crucial for us to understand and delineate the role of microRNAs in skin.

References

1. Jensen E (2014) Technical review: in situ hybridization. Anat Rec (Hoboken) 297 (8):1349–1353. https://doi.org/10.1002/ar.22944
2. Ahmed MI, Alam M, Emelianov VU, Poterlowicz K, Patel A, Sharov AA, Mardaryev AN, Botchkareva NV (2014) MicroRNA-214 controls skin and hair follicle development by modulating the activity of the Wnt pathway. J Cell Biol 207(4):549. https://doi.org/10.1083/jcb.201404001
3. Ahmed MI, Pickup ME, Rimmer AG, Alam M, Mardaryev AN, Poterlowicz K, Botchkareva NV, Botchkarev VA (2019) Interplay of microRNA-21 and Satb1 in epidermal keratinocytes during skin aging. J Investig Dermatol 139:2538. https://doi.org/10.1016/j.jid.2019.04.022
4. Botchkareva NV (2012) MicroRNA/mRNA regulatory networks in the control of skin development and regeneration. Cell Cycle 11 (3):468–474. https://doi.org/10.4161/cc.11.3.19058
5. Ahmed MI, Mardaryev AN, Lewis CJ, Sharov AA, Botchkareva NV (2011) MicroRNA-21 is an important downstream component of BMP signalling in epidermal keratinocytes. J Cell Sci 124(Pt 20):3399–3404. https://doi.org/10.1242/jcs.086710
6. Mardaryev AN, Ahmed MI, Vlahov NV, Fessing MY, Gill JH, Sharov AA, Botchkareva NV (2010) Micro-RNA-31 controls hair cycle-associated changes in gene expression programs of the skin and hair follicle. FASEB J 24 (10):3869–3881
7. Komminoth P, Merk FB, Leav I, Wolfe HJ, Roth J (1992) Comparison of 35S- and digoxigenin-labeled RNA and oligonucleotide probes for in situ hybridization. Expression of mRNA of the seminal vesicle secretion protein II and androgen receptor genes in the rat prostate. Histochemistry 98(4):217–228
8. Horwitz JP, Chua J, Noel M, Donatti JT, Freisler J (1966) Substrates for cytochemical demonstration of enzyme activity. II. Some dihalo-3-indolyl phosphates and sulfates. J Med Chem 9 (3):447. https://doi.org/10.1021/jm00321a059
9. Hanna JA, Wimberly H, Kumar S, Slack F, Agarwal S, Rimm DL (2012) Quantitative analysis of microRNAs in tissue microarrays by in situ hybridization. Biotechniques 52 (4):235–245. https://doi.org/10.2144/000113837

Chapter 17

Chromatin Immunoprecipitation of Low Number of FACS-Purified Epidermal Cells

Carmit Bar, V. Julian Valdes, and Elena Ezhkova

Abstract

Chromatin immunoprecipitation followed by sequencing (ChIP-seq) is a method designed to detect interactions between chromatin and the proteins bound to it. This method has been widely used for characterizing epigenetic landscapes in many cell types; however, a limiting factor has been the requirement of a high number of cells. Here, we describe a protocol for ChIP in epidermal cells from a newborn mouse, purified by fluorescence-activated cell sorting (FACS). This protocol has been optimized specifically for prefixed, low cell numbers, resulting in enough immunoprecipitated DNA suitable for genome-wide analysis.

Key words ChIP-seq, Epidermal cells, Low cell number ChIP, FACS purified cells

1 Introduction

Epigenetic modifications include a plethora of chemical modifications in DNA bases such as methylation and hydroxylation of cytosine or modifications of histone proteins, including acetylation, methylation, and phosphorylation [1–4]. These modifications affect the chromatin architecture and regulate access of the transcription machinery to the DNA. These epigenetic marks are established, maintained, and interpreted by proteins able to write, erase or read DNA and histone modifications [5]. These modifications are highly dynamic and may vary according to the cellular context. Thus, they have a direct influence in gene expression and are critical in formation of cellular diversity [6]. Detailed characterization of diverse epigenetic modifications at a genome-wide level has been a hallmark to study complex biological processes such as cell differentiation and tissue development [6, 7].

Differentiation of epidermal progenitor cells during stratification of epithelial tissues, such as the skin and the oral mucosa, is a highly regulated process where epigenetic regulation plays

Natalia V. Botchkareva and Gillian E. Westgate (eds.), *Molecular Dermatology: Methods and Protocols*,
Methods in Molecular Biology, vol. 2154, https://doi.org/10.1007/978-1-0716-0648-3_17,
© Springer Science+Business Media, LLC, part of Springer Nature 2020

important roles both in maintenance of the basal progenitor population and proper acquisition of epidermal cell fate. Our group's previous studies have shown that Polycomb repressive complexes (PRC) 1 and 2, catalyzing the H2AK119ub and H3K27me3 histone modifications, respectively [8–12] are critical regulators of hair follicle and epidermal development, and Merkel cell differentiation in the skin epidermis [13–15]. In the tongue epithelium, PRC1 is critical for the maintenance of fungiform papillae, housing taste buds, and filiform papillae formation [16]. Other epigenetic complexes such as histone acetylases, deacetylases, DNA methyltransferases, and chromatin remodelers have also been linked to biological processes in the epidermis [17–20].

One of the most powerful tools to assess the epigenome of a cell is chromatin immunoprecipitation followed by sequencing (ChIP-seq), designed to study specific interactions between DNA and the protein complexes bound to it [21]. During ChIP, cells are first cross-linked to preserve the interactions of proteins with the DNA, nuclei are extracted, and genome is fragmented by sonication. Genome segments bound to a protein are then isolated by specific antibody binding and recovered by precipitation. Next, the cross-linking is reversed, and the DNA is purified. To map these interactions at a genome-wide level, ChIP-purified DNA can be sequenced by next-generation sequencing methods. To achieve high-resolution ChIP sequencing, a high number of cells is required, often presenting a limitation for ChIP experiments [21]. Here, we present a detailed and optimized protocol to successfully perform chromatin immunoprecipitation in low-number samples of FACS-purified epidermal cells, resulting in a sufficient quantity of immunoprecipitated DNA from a few thousand cells. This protocol can be adapted to other cell populations or epigenetic regulators to characterize the epigenetic landscape of a developing tissue at a genome-wide scale.

2 Materials

2.1 Skin Tissue Processing

1. Sterile 1× Dulbecco's phosphate-buffered saline (DPBS) without calcium and magnesium.
2. Dispase, diluted to a final concentration of 1.26 U/mL.
3. Trypsin (0.25%) supplemented with 2.21 mM EDTA.
4. E-media for keratinocyte culture [22] supplemented with 0.05 mM $CaCl_2$.
5. 6-Well plates or 6 mm tissue culture dishes.
6. 15 mL conical tubes.

2.2 Viability Staining and Cell Fixation

1. Zombie violet viability detection kit (Biolegend).
2. Formaldehyde 16% (w/v), methanol-free, store at room temperature (RT).
3. Glycine, prepare 2.5 M stock solution. Final working concentration, 125 mM. Stock solution is stable indefinitely at RT.

2.3 Purifying Epidermal Cells by FACS

1. Sterile 1× Hank's Balanced Salt Solution (HBSS), without calcium and magnesium, supplemented with 2% fetal bovine serum (FBS).
2. Cell strainers: FACS tubes with strainer caps, and 0.45 μm cell strainers fitting conical 50 mL tubes.
3. Antibodies:
   - EpCAM-APC (BioLegend, RRID: AB_1134102), working dilution 1:200.
   - Sca-1-PerCP-Cy5.5 (Biolegend, RRID: AB_893619), working dilution 1:200.
   - Integrinα6-FITC (eBiosciences, RRID: AB_11150059), working dilution 1:100.

2.4 ChIP

1. cOmplete™ Mini Protease Inhibitor Tablets (Roche). Dissolve one tablet in 1 mL of water for a 50× stock. Store aliquots at −20 °C.
2. ChIP DNA Clean & Concentrator Kit™ (Zymo Research).
3. Bovine serum albumin (Sigma-Aldrich).
4. Proteinase K, recombinant, PCR Grade (Roche). Prepare 20 mg/mL stock solution. Dissolve powder in 10 mM Tris–HCl pH 7.5, 20 mM $CaCl_2$, 50% glycerol. Store 500 μL aliquots at −20 °C.
5. RNase A, from bovine pancreas (Roche). Prepare at 10 mg/mL dissolved in 10 mM Tris–HCl pH 7.5, 15 mM NaCl. Heat at 100 °C for 20 min to inactivate DNases. Store 1 mL aliquots at −20 °C.
6. Nonidet P 40 Substitute (NP-40).
7. Na-deoxycholate.
8. *N*-Lauroylsarcosine.
9. Triton X-100.
10. Protein G Dynabeads (Life Technologies).
11. Magnetic rack, Dynamag 2.
12. LoBind tubes (Eppendorf).
13. Qubit dsDNA HS Assay kit and Qubit tubes (Thermo).

2.5 Stock Solutions

Prepare the following stock solutions prior to buffer preparation. Use ultrapure water (resistivity >18.18 MΩ cm at 25 °C). Stock solutions are stable at least 1 year at room temperature. Also *see* **Note 1**.

1. 1 M HEPES–KOH, pH 7.5.
2. Tris–HCl pH 8.0.
3. 5 M NaCl.
4. 0.5 M EDTA.
5. 1 M EGTA.
6. 50% Glycerol.
7. 10% NP40.
8. 10% Triton X-100.
9. 5% Na-Deoxycholate.
10. 10% *N*-Lauroylsarcosine.
11. 10% SDS.
12. 2.5 M LiCl (MW 42.4 g/mol).

2.6 Cell Lysis Buffers

Prepare 50–500 mL of each lysis buffer. Use ultrapure water (resistivity >18.18 MΩ cm at 25 °C) to dilute to final volume and filter (0.44 μm). Store at 4 °C. Buffers are stable up to 1 year.

1. *Lysis Buffer I.* 50 mM HEPES–KOH pH 7.5, 140 mM NaCl, 1 mM EDTA, 10% glycerol, 0.5% NP-40, 0.25% Triton X-100.
2. *Lysis Buffer II.* 10 mM Tris–HCl pH 8.0, 200 mM NaCl, 1 mM EDTA, 0.5 mM EGTA.
3. *Lysis Buffer III.* 10 mM Tris–HCl pH 8.0, 100 mM NaCl, 1 mM EDTA, 0.5 mM EGTA, 0.1% Na-deoxycholate, 0.5% *N*-lauroylsarcosine. Keep this stock without protease inhibitors. At the day of experiment, add one tablet of cOmplete™ protease inhibitors into each 50 mL of Lysis Buffer III.

2.7 Wash and Elution Buffers

Prepare 50–500 mL of each lysis buffer. Use ultrapure water (resistivity >18.18 MΩ cm at 25 °C) to dilute to final volume and filter (0.44 μm). Store Wash and TE buffers at 4 °C. Store elution buffer at room temperature. Buffers are stable up to 1 year.

1. *Low Salt Wash Buffer.* 20 mM Tris–HCl pH 8.0, 150 mM NaCl, 2 mM EDTA, 0.1% SDS, 1% Triton X-100.
2. *High Salt Wash Buffer.* 20 mM Tris–HCl pH 8.0, 500 mM NaCl, 2 mM EDTA, 0.1% SDS, 1% Triton X-100.
3. *LiCl Wash Buffer.* 20 mM Tris–HCl pH 8.0, 250 mM LiCl, 1 mM EDTA, 1% Na-deoxycholate, 1% NP-40.

4. *TE Buffer*. 10 mM Tris–HCl pH 8.0, 1 mM EDTA.
5. *Elution Buffer*. 50 mM Tris–HCl pH 8.0, 10 mM EDTA, 1% SDS.

3 Methods

3.1 Dissection of Mouse Newborn (P0) Back Skin Samples, Preparation of Cell Suspension, and Prefixing

The described procedure is for one skin sample. For multiple samples, prepare an appropriate master mix. Centrifuge should be cooled to 4 °C.

1. Prepare a 6-well plate, or a 6 mm cell culture dish with 2 mL of dispase solution, diluted in sterile DPBS, to a final concentration of 1.26 U/mL.
2. Gently remove the skin from the back of P0 mouse. Peeled skin contains dermis and epidermis (Fig. 1a). If necessary, remove excess adipose tissue underneath the dermis with blunt-edged forceps.
3. Place the skin in the dispase solution, with the dermis facing down.
4. Incubate overnight at 4 °C (for alternatives, *see* **Note 2**).
5. Warm 0.25% trypsin–EDTA solution at 37 °C bath for 10 min.
6. Gently peel the dermis from the epidermis using sharp forceps. Discard the dermis (Fig. 1a).
7. Place the peeled epidermis in a new plate/well containing 2 mL of warm 0.25% trypsin–EDTA. Incubate for 10 min at 37 °C with gentle shaking.
8. Using a p1000 micropipette, gently pipet under the stratum corneum to release the epidermal cells from the tissue. Transfer the cell suspension to a new 15 mL conical tube.
9. Add 2 mL of ice-cold E-media supplemented with calcium (*see* Subheading 2) to the dish containing stratum corneum to release as many cells as possible. Transfer the media to the conical tube containing the trypsinized cell suspension and mix well by pipetting. Do not vortex.
10. Spin the cell suspension for 10 min, 300 rcf.
11. Decant the supernatant carefully and resuspend the pellet in 10 mL 1×DPBS.
12. Spin the cell suspension for 10 min, 300 rcf.
13. Decant the supernatant carefully and resuspend the pellet in 10 mL 1×DPBS.
14. Spin the cell suspension for 10 min, 300 rcf.
15. While cell suspension is in the centrifuge, dilute Zombie dye 1:100 in 1×DPBS, final volume of 300 μL per sample.

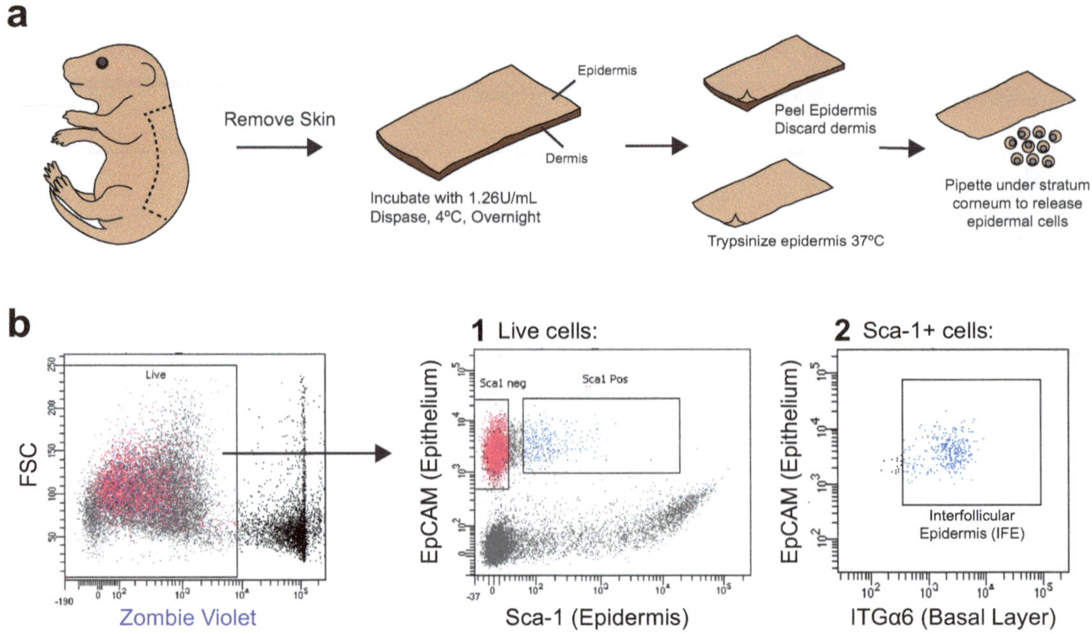

Fig. 1 Isolation of epidermal cells from a newborn mouse. (**a**) Skin dissection and isolation of the epidermis by enzymatic digestion. Dispase breaks hemidesmosomes, allowing separation of the dermis and epidermis. Trypsin separates the epidermal cells. (**b**) An example of a FACS plot to isolate basal interfollicular epidermis (IFE). Zombie violet negative cells are alive. From living cells, EpCAM is a general marker of epithelial cells. Sca-1 marks epidermal cells, excluding hair follicles. Integrinα6 (ITGα6) is a marker of basal cells only

16. Resuspend cells in 300 μL of Zombie dye solution. Incubate on ice for 10 min. Pipet to mix after 5 min of incubation.
17. Add 10 mL of 1×DPBS to the cells and spin 10 min, 300 rcf.
18. Decant supernatant carefully and add 10 mL 1×DPBS.
19. Spin the cell suspension for 10 min, 300 rcf.
20. Decant supernatant carefully and resuspend the pellet well in 1 mL 1×DPBS.
21. Fill the volume to final 4.5 mL with 1×DPBS.
22. Add 300 μL of 16% formaldehyde, methanol-free, directly from the ampule, to the diluted cell suspension. Final dilution is 1% (*see* **Notes 3–5**).
23. Incubate cell suspension at room temperature for 10 min, with rocking.
24. Add 250 μL of 2.5 M glycine stock solution directly to the cell suspension to stop the fixation. Final glycine concentration is 125 mM. Incubate at room temperature for 5 min, with rocking.
25. Add 5 mL of 1×DPBS to wash cells and spin for 5 min, 930 rcf (*see* **Notes 6** and 7).

26. Carefully decant the supernatant and add 10 mL of 1×DPBS.
27. Spin the cell suspension for 5 min, 930 rcf.
28. Decant supernatant carefully and continue directly to antibody staining and FACS-sorting.

    Optional STOP: After the second wash, flash-freeze cells in liquid nitrogen and store cells at −80 °C for future use. Cells can be stored at −80 °C for several months.

3.2 Antibody Staining and FACS Sorting

If using a combined sample of frozen cell suspensions, *see* **Note 8**.

1. Resuspend cells in 1 mL 1×DPBS and pass through BD FACS strainer.
2. Spin 5 min, 930 rcf.
3. Decant supernatant.
4. Stain cells with antibodies diluted with 1×HBSS, supplemented with 2% FBS (*see* antibody dilutions in Subheading 2) in a final volume of 200 μL for 30 min on ice. For larger samples, *see* **Note 9**.
5. Add 1 mL 1×HBSS, supplemented with 2% FBS to the stained cells.
6. Spin 5 min, 930 rcf.
7. Decant supernatant and wash again with 1 mL 1×HBSS supplemented with 2% FBS.
8. Decant supernatant and resuspend in 300 μL 1×HBSS supplemented with 2% FBS.
9. Sort cells by FACS (Fig. 1b) and collect into 15 mL conical tubes containing 3 mL 1×DPBS.
10. For best results, begin the ChIP protocol immediately after sorting.

    Optional STOP: Freeze sorted cells in liquid nitrogen and store at −80 °C for future use (*see* **Note 10**).

3.3 Nuclei Isolation

Fixed cells should be permeabilized to eliminate cytoplasmic proteins and improve immunoprecipitation of chromatin. All steps should be carried out at 4 °C. *See* **Notes 11–14** before starting the experiment.

1. If the samples are frozen, thaw them on ice (10–15 min) and gently tap the tubes to disaggregate the cell pellet.
2. For each 100,000–500,000 fixed cells, add 1 mL of Lysis Buffer I and disaggregate the pellet using a micropipette. To the resuspended cells, add 4 more mL of Lysis Buffer I and vortex 3 s. Rotate the tube for 10 min at 4 °C.
3. Spin for 7 min, 930 rcf at 4 °C and decant the supernatant with one gentle movement.

4. Tap the tube to disaggregate the cell pellet and resuspend it in 5 mL of Lysis Buffer II. Gently vortex the tube and rotate the sample for 10 min at 4 °C.
5. Spin the tube for 7 min, 930 rcf at 4 °C and decant the supernatant with one gentle movement.
6. Repeat the spin at 930 rcf for 2 min. Carefully remove the supernatant with a micropipette.
7. Resuspend the permeabilized cell nuclei in 3 mL of Lysis Buffer III supplemented with 1× Protease Inhibitors. Use a micropipette to gently mix the sample.

3.4 Optimization of Chromatin Sonication

For ChIP-seq, fragmented chromatin should range from 150 to 400 bp with a peak in 300 bp. Bigger fragments, up to 600 bp, can be used for ChIP-qPCR. Samples should be kept at 4 °C throughout during sonication to prevent protein degradation. *See* **Notes 15–21** for more details and possible troubleshooting suggestions.

1. For an ultrasonicator with a probe, place the 15 mL conical tube with the permeabilized nuclei (in 3 mL of Lysis Buffer III with protease inhibitors) in a 500 mL beaker with ice water. Use a 1/8″ microtip for 15 mL conical tubes adjusting the height, such that the probe is fully submerged in the sample but 5–7 mm above the bottom of the tube, to avoid contact between the sonicator tip and the tube walls. Adjust the voltage of the ultrasonicator to 20–25% power and program cycles of 30 s (1 s ON/1.5 s OFF). Let the sample to cool down for at least 90 s and repeat the sonication cycle 5–20 times. After sonication, place the sample on ice.

   ***Steps 2–6** refer to optimization of sonication process. Once the sonication process has been optimized, immediately proceed to immunoprecipitation (Subheading 3.5).
2. To assess the shearing of the chromatin, collect 5% of the volume every 5 cycles and substitute the same volume with Lysis Buffer III. Save 100 μL as a presonication sample.
3. Leave the samples overnight at 65 °C to reverse the cross-linking.
4. Add 2 μL of RNase A (10 mg/mL) and incubate for 1 h at 37 °C.
5. Add 2 μL of Proteinase K (20 mg/mL) and incubate 1 h at 55 °C.
6. Purify 100 μL of the sonicated samples by phenol–chloroform extraction or by affinity columns. Elute the sonicated DNA in 15 μL and assess the shearing of the chromatin on a 2% agarose gel with a 100 bp DNA marker to determine the fragment size.

7. Evaluate the shearing of the chromatin to select sonication settings and cycles that results in reproducible shearing between 150 and 400 bp with a peak at 300 bp.

3.5 Immunoprecipitation

Once sonication of the fixed chromatin has been optimized, immunoprecipitation can be performed. Different epigenetic marks may require different amounts of sheared chromatin. *See* **Notes 22–27** for more details and possible troubleshooting suggestions.

1. Transfer the 3 mL of sonicated chromatin to three DNA LoBind tubes coated with 0.05% BSA in 1×DPBS (*see* **Notes 11–13**). Spin for 10 min at maximal speed (~5000 rcf) at 4 °C.
2. Transfer the supernatant to a new 15 mL conical tube coated with 0.05% BSA in 1×DPBS and add 300 μL of 10% Triton X-100. Gently invert the tube several times to mix the solution. This sheared chromatin can be kept at 4 °C for up to 24 h. Alternatively, flash-freeze the sheared chromatin in liquid nitrogen and store at −80 °C for later use. Samples can be stored for several months.
3. Antibody-tube preparation: Set up one LoBind DNA tube for each sample to be immunoprecipitated. *See* **Note 12** for BSA coating instructions. Pipet 1–3 μL of each antibody directly into the remaining 0.05% BSA in DPBS and leave tubes on ice until use.
4. Calculate the number of cells per mL of sonicated chromatin from **step 2** and add the volume equivalent to 5000–10,000 cells to each tube with diluted antibodies for ChIP-qPCR and at least 50,000 cells to each tube with diluted antibodies for ChIP-seq. If necessary, adjust the volume to have at least 300 μL/IP tube. Keep the volume constant between samples for reproducible results. *See* also **Note 25**.
5. Save 1–5% of the volume of one IP sample in a new LoBind tube to use as the input. Store input at 4 °C for future use.
6. Rotate the IP tubes (chromatin + antibody) overnight at 4 °C. (*See* **Note 25** for alternatives.)
7. After antibody incubation, equilibrate the Protein G magnetic Dynabeads: for each μL of antibody, use 5 μL of Dynabeads by pipetting the *total volume necessary for all IP tubes.* Put the tube on a magnet and wait 30–60 s until all the beads adhere to the magnet. Do not vortex the magnetic beads.
8. Wash the beads: Aspirate the supernatant and add 1 mL of 0.05% BSA in 1×DPBS. Remove the tube from the magnet and invert it gently several times until all the magnetic beads are resuspended. Avoid vortexing. Place the tube again on the magnet and wait for the bead to adhere to the magnet. Aspirate the supernatant and repeat the wash with 1 mL of 0.05% BSA in

DBPS for a total of three washes. Leave the beads in 1 mL 0.05% BSA in DBPS and leave it rotating at 4 °C until use (*see* **Note 26**).

3.6 Washes and Elution

Immunoprecipitated chromatin bound to magnetic beads should be astringently washed to increase specificity and reduce background. *See* **Note 27**.

1. After the overnight incubation of the IP samples, place the tubes on the magnet until all the beads adhere to the magnet side. Discard supernatant or save it for Western blot analysis or other IP experiments.
2. While the tubes are still on the magnet, add 1 mL of low salt buffer to the beads. Remove the sample from the magnet and invert several times to solubilize the beads. Rotate the tubes for 10 min at 4 °C. Place the tube again on the magnet and wait 30–60 s until all magnetic beads adhere to the magnet. Then, aspirate the supernatant.
3. Wash once with 1 mL of high salt buffer as in **step 2**.
4. Wash twice with 1 mL of LiCl wash buffer as in **step 2**.
5. Wash once with TE buffer at room temperature as in **step 2**.
6. Aspirate the TE buffer after the last wash while the tubes are still on the magnet and add 100 μL of elution buffer. Remove the tubes from the magnet and vortex for 10 s to disaggregate the beads.
7. Add elution buffer to the input sample from **step 5** of Subheading 3.5 to a total volume of 100–160 μL.
8. Leave the washed beads and the input sample at 65 °C overnight to reverse the cross-linking. Do not incubate longer than 16 h.

3.7 DNA Purification and Quantification

1. After reverse cross-linking, vortex the IP tubes, and quickly spin them on a bench centrifuge. Place the tubes on a magnet and let them stand for 60 s or until all the beads adhere to the magnet.
2. Transfer the supernatant to a new DNA LoBind tube.
3. Wash the remaining beads with 100 μL of TE with 2 μL of RNase A (10 mg/mL). Vortex the tubes and place them on the magnet.
4. Collect the *supernatant* into the same tube from **step 2**.
5. Add 40 μL of TE with 2 μL of RNase A (10 mg/mL) to the input tube.
6. Incubate the bead-purified solution and the input tubes for 2 h at 37 °C.

7. Add 3 μL of Proteinase K (10 mg/mL) to each IP tube and input sample, vortex briefly and incubate for 2 h at 55 °C. Then, proceed immediately to DNA purification or store the samples at −20 °C for later DNA purification.
8. Proceed with DNA purification using ChIP DNA Clean & Concentrator kit (Zymo) or a similar affinity column following manufacturer's instructions. Equilibrate all buffers to room temperature prior to purification.
9. Before elution, centrifuge the column 1 min at maximal speed on a dry collection tube.
10. Elute the DNA in 7–14 μL of 10 mM Tris pH 8 prewarmed to 37 °C.
11. Repeat the elution step with 7–14 μL of 10 mM Tris pH 8 warmed at 37 °C for a total volume of 14–28 μL.
12. Quantify the IP-DNA using the HS dsDNA Qubit kit.

For ChIP-library construction, several commercial options are available. Most kits can construct libraries for next-generation sequencing from 1–3 ng of DNA from ChIP samples. Some kits require DNA samples in a maximal volume of 5 μL to initiate the protocol. In this case, elute the ChIP samples in a reduced volume on **step 10**. Table 1 presents an example of the total amount of purified ChIP DNA material obtained from different number of FACS-purified epidermal basal cells.

3.8 ChIP-qPCR and Assessment of Input Recovery Percentage

Before ChIP-seq NGS library construction, it is essential to corroborate that the ChIP protocol worked successfully and assess the background level in the sample. ChIP for nonspecific IgG can be used as a negative control to initially optimize the duration and astringency of the washes, and more important, amplification of negative regions to test by qPCR that will give a direct measurement of the background signal on each ChIP (*see* **Notes 28–32**).

Table 1
Total ng of DNA recovered after ChIP

FACS-purified cells	10% input (ng)	ChIP H3K27me3 (ng)	ChIP H2AK199Ub (ng)
1000	13.44	0.83	1.15
5000	10.27	1.23	1.17
15,000	47.04	1.63	2.24
50,000	153.60	2.80	7.34
500,000	616.80	5.42	70.32

The indicated number of cells were collected in individual tubes and the full ChIP protocol was performed until purification. Concentration was quantified by Qubit HS dsDNA kit. ChIP antibodies used: H3K27me3: Millipore 07-449, RRID:AB_310624; H2A119Ub: Cell Signaling 8240S, RRID:AB_10891618

Table 2
A list of control primers that can be used for mouse genome

	Loci	Forward	Reverse
Negative regions	Intergenic Chr3	TGTCTGGAATGTGGTGG TTTGA	GCCCACTGCTATAA TTAGGAAGGA
	Intergenic Chr5	CCCTCATCACAGACCCAC TTCT	GTGGGAGTGGATGTATCTC TGACTT
Active	ACTB promoter	TCCAACCAGAGTTCTGC TTCCAAGT	TGCTAGCAA TAGCCGGAAAGCCA
Polycomb targeted	*HoxB13*	TGTGTGTGCATGGTGTTTGT	TTCAGCCACTCCTCCAAACT
	HoxD13	TCGGCAACGGTTACTACAGC	TGTACTTCTCCACCGGGAAA

1. Dilute the ChIP-purified DNA and input sample 1:100–1:500 in 10 mM Tris pH 8.0 with 0.05% Tween 20. Use the same dilution for all samples independently of the concentration (*see* **Note 28**).
2. Assemble qPCR reactions in a duplicate or triplicate with 4–6 μL of diluted ChIP-purified DNA and perform 45 cycles of amplification.
3. Primers: We have found that intergenic regions are usually devoid of transcription factors, transcriptionally active marks, and Polycomb group proteins. Therefore, they can be used as negative controls for each ChIP sample. In the mouse genome, we consistently found the intergenic regions in chromosome 3 and 5 as solid negative controls (*see* primer list on Table 2). For positive controls for ChIP of transcription factors or transcriptionally active associated histone marks, promoters of housekeeping genes, such as Actin B can be used. Primers for positive controls should be located approximately 100 bp upstream of the TSS or centered to reported NGS peaks if available. All primers, experimental or positive and negative control, should have a Tm of 65 °C and amplify products of 100 ± 10 bp.
4. After the qPCR, calculate input recovery percentage for each primer/IP. In this method, the signal obtained for each ChIP is divided by the theoretical signal of the undiluted input to calculate the IP efficiency. To adjust the dilution of the input, first calculate the dilution factor. For example, if 1% of the material were saved as input, the dilution factor (DF) is 100. Divide the signal of the input by the dilution factor using the following formula:

$$\text{Adjusted Input} = \text{Ct of Input} - (\text{Log2 of DF})$$

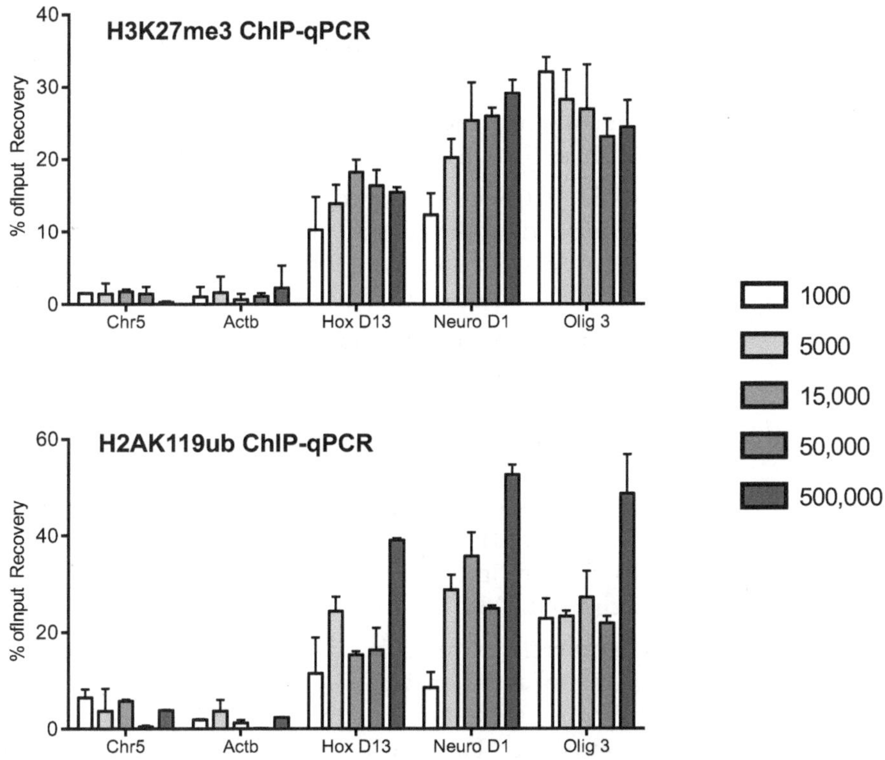

Fig. 2 ChIP-qPCR of H3K27me3 and H2AK119Ub from different numbers of FACS-purified epidermal cells. The percentage of input recovery was calculated for the negative regions (chromosome 5 and Actin B promoter) as well as Polycomb targeted genes (*HoxD13*, *NeuroD1*, and *Olig3*)

To calculate the input recovery, divide the Adjusted Input value by the ChIP signal and express it as percentage; transform the values from Log2 to a linear scale using the following formula:

% of input recovery = 100 × 2^[(Adjusted input) – (Ct of ChIP)]

5. Alternatively, once the Percentage of Input Recovery has been calculated for each primer, the signal can be normalized to the signal of the negative control to calculate the fold enrichment.

An example calculating the percentage of input recovery using different number of FACS purified cells is presented in Fig. 2.

4 Notes

1. Preparation of ChIP wash buffers can be time-consuming. Therefore, we recommend preparing all stock solutions in advance and preparing 50–500 mL of each buffer prior to the ChIP protocol. Use ultrapure water (Resistivity >18.18 MΩ cm

at 25 °C) for all stock solutions and buffers. Once prepared, all buffers should be filtered (0.44 μm) and stored at 4 °C up to 1 year.

2. The minimal incubation time at 4 °C is 4 h. Alternatively, incubate for 1 h at 37 °C with gentle shaking.
3. Adding 16% formaldehyde directly to the sample is critical. Resuspending the cells in a prediluted master mix of 1% formaldehyde, or in prechilled formaldehyde, will result in improper fixation and poor ChIP quality.
4. The fixative agent and fixation time may be optimized depending on the targeted immunoprecipitated protein. A 10-min fixation with formaldehyde at room temperature is a good starting point for histones and histone modifications. For non-histone proteins, especially transcription factors, longer fixation times and secondary fixative agents such as disuccinimidyl glutarate (DSG) or ethylene glycol bis-succinimidylsuccinate (EGS) can improve ChIP results but should be carefully optimized [23].
5. Commercial formaldehyde solutions may include up to 15% of methanol as a stabilizer. The presence of methanol will result in methanol-fixated chromatin products that cannot be reverted or shredded. This effect is particularly evident when fixed cells are stored, even at −80 °C. Using recently opened methanol-free formaldehyde is recommended to avoid this problem.
6. Increasing centrifugation speed to *930 rcf* is critical to precipitate the cells after fixation.
7. Cell pellets become transparent after fixation.
8. In the case of combining several samples of frozen and fixed cell suspensions, pass the cells through 0.45 μm cell strainers to remove large cell clumps often generated by fixation. Spin at 930 rcf for 10 min.
9. A larger sample (e.g., a combined suspension from many animals) may be stained in a larger volume.
10. Freezing does not affect histone ChIP but is not recommended for ChIP of proteins that are rarer than histones, such as transcription factors. In this case, freezing may result in a low ChIP yield.
11. For low cell-number samples, we recommend minimizing tube transfers and using the same 15 mL conical tube for collection of FACS-purified cells up to the sonication step. If changing the tube is necessary, coating the new tube with 0.05% BSA in 1×DPBS helps to reduce cell loss. Micropipette tips can also be coated with the 0.05% BSA solution by pipetting the maximum volume for a few seconds before pipetting the cell suspension. When working with a very low number of cells

(1000–50,000 cells), the micropipette tip has to be coated with 0.05% BSA before pipetting the solution with cells but using the same volume of the Lysis buffers used for 100,000 cells.

12. Coat each tube by pipetting 1 mL of 0.05% BSA in DPBS and vortex for 3 s; incubate the tubes 3–5 min to coat. Decant or recover the 0.05% BSA, but leave 30–50 μL of solution in each tube.
13. The solution of 0.05% BSA in 1×DBPS can be easily contaminated with bacteria, so it is essential to prepare this solution fresh and not store it for more than 48 h at 4 °C. To avoid waste, frozen aliquots of 0.05% BSA can be stored indefinitely. Quality and testing of BSA are important to avoid the presence of proteinase or nuclease activity.
14. Some protocols include SDS in the sonication buffer to improve disruption of the cells, but presence of SDS is incompatible with antibodies and therefore must be diluted and neutralized before immunoprecipitation. However, when working with a low number of cells (less than one million), this will result in overdilution of the chromatin and failure of the ChIP. Here we have optimized lysis and sonication buffers to avoid the use of SDS to keep chromatin concentration at maximum to improve immunoprecipitation.
15. Different commercial equipment can be used to successfully shear chromatin. To optimize sonication, we recommend only varying the cycle number/time of sonication and analyzing chromatin size on an agarose gel.
16. Some cells are particularly difficult to sonicate. Detergents as SDS can be added to the sonication buffer from 0.1% to 1% to improve cell disruption. Sonication power has to be adjusted to prevent foam formation, and the sample has to be diluted up to ten times before immunoprecipitation to prevent antibody denaturalization by the detergent. When working with a low number of cells, dilution of the sample will negatively affect immunoprecipitation efficiency, so it is better to optimize sonication conditions to avoid the use of SDS.
17. To proper evaluate shearing of the chromatin on an agarose gel, it is necessary to perform the cross-linking reversal followed by Proteinase K and RNase A treatments. The presence of RNA can be easily confused for or hide the shredded chromatin size, whereas improper reversal of the cross-linking will prevent migration on the agarose gel.
18. When using a probe sonicator, we noticed that polypropylene tubes tend to stay in place more easily than polystyrene tubes. To help keep tube in place, set a tape ring around the sonicator tip exactly at the height of the tube. This will not interfere with chromatin shearing and will prevent the tip from touching the tube walls. It is also convenient to use a clear glass beaker as a

sonication ice water bath: place the tube close to the wall of the beaker to have a good view of the sonicator tip inside the tube.

19. Avoid formation of foam during sonication because this will prevent breaking of the cells and chromatin shearing. The appearance of white spots during sonication is normal and expected. It is convenient to set a trial sonication test only with a sonication buffer. Assay the power of the equipment and set the power output that does not generate any foam. Also test for overheating of the sample; some warming of the tube is normal, but if overheating is observed, longer incubation times between sonication samples is recommended.

20. Always choose the shortest number of sonication cycles/time. Oversonication will result in degradation of protein complexes and failure of the experiment. Analysis of the sample proteins by Western blot after sonication is a good way to evaluate the integrity of the protein to be immunoprecipitated. Depending on the equipment and voltage, oversonication may result in chromatin fragments of about 100 bp. Therefore, when optimizing sonication, a titration time curve should be performed to observe a ladder size of the shredded chromatin.

21. In general, immunoprecipitation of histone is more efficient and will require lower amounts of chromatin, whereas non-abundant proteins or transcription factors may require more starting material.

22. Proteins G and A have different affinities toward different isotypes of antibodies. Use magnetic beads with protein G or A accordingly. Sepharose G/A beads can be used for ChIP but require blocking before antibody conjugation and centrifugation steps between washes.

23. Some protocols recommend using an IgG control to evaluate background signal, but only the same isotype of antibody and preimmune serum from the same animal would be indicative of a background signal. Using a "beads only" control without any antibody would be more informative in evaluating noise and optimizing washes and incubation times. However, an anti-H3 antibody will give an extremely robust signal to any target and can be used as a control when optimizing the ChIP protocol for the first time.

24. We have found that using 1–3 μL of antibody per ChIP sample gives good results. Using higher amounts of antibody may result in increased unspecific signal unless the number of cells is accordingly increased. However, we have observed that substantially increasing the number of cells (over five million) may increase the background signal using this protocol. As currently ChIP-seq libraries can be constructed with low amounts of DNA (0.5–1 ng), we favor the use of fewer cells per antibody to obtain a specific signal and lower background in the ChIP-seq.

25. For immunoprecipitation, incubate the antibody and chromatin for at least 4 but no more than 16 h. To promote efficient immunoprecipitation and not overdilute antibodies, maintain a volume of at least 200 μL but no more than 600 μL per IP tube.
26. Washed beads can be kept rotating in 0.05% BSA in 1XDPBS up for 16 h before use.
27. The duration of each wash can be increased up to 40 min if unspecific signal is observed. All solutions are kept at 4 °C, except for the TE and elution buffer.
28. For qPCR, keep the dilution factor of input and ChIP samples constant. Do not store diluted samples; they are not stable even at −20 °C. If new primers are to be tested, prepare fresh dilutions for consistent and comparable qPCR results.
29. In the same sense, the amount of recovered ChIP-ed material is not indicative of a successful ChIP or of a higher percentage of input recovery. These only reflect the amount of starting material and antibody concentration. Using a lower number of cells tends to result in higher input recovery, but low recovery is not indicative of failure.
30. The only way to assess the quality of a ChIP and decision to construct ChIP-seq libraries is evaluating negative and positive controls and the ratio between them by qPCR.
31. Before proceeding to sequencing, the ratio of negative/positive controls can be evaluated in the ChIP-seq libraries that should be similar to the original ChIP samples.
32. Although evaluating the ChIP-qPCR by calculating the "fold enrichment" or bulk-H3 normalization may help to normalize the background signal, this may result in overnormalization and masking biological results. We recommend presenting the input recovery percentage for each primer/IP without normalization to assess the biological differences and successes of the ChIP protocol.

Acknowledgments

We thank members of the Ezhkova and Valdes labs for discussion and suggestions. E.E. is supported by the National Institute of Arthritis and Musculoskeletal and Skin Diseases of the National Institutes of Health under award number R01AR069078 and the Tisch Cancer Institute P30 Cancer Support Grant. C.B. is a Merksamer Fund scholar. J.V. is supported by the PAPIIT-UNAM IA202118, CONACYT 284867 and SECTEI/277/2019 grants.

References

1. Chen Z, Li S, Subramaniam S, Shyy JY, Chien S (2017) Epigenetic regulation: a new frontier for biomedical engineers. Annu Rev Biomed Eng 19:195–219. https://doi.org/10.1146/annurev-bioeng-071516-044720
2. Cubenas-Potts C, Matunis MJ (2013) SUMO: a multifaceted modifier of chromatin structure and function. Dev Cell 24(1):1–12. https://doi.org/10.1016/j.devcel.2012.11.020
3. Farrelly LA, Thompson RE, Zhao S, Lepack AE, Lyu Y, Bhanu NV, Zhang B, Loh YE, Ramakrishnan A, Vadodaria KC, Heard KJ, Erikson G, Nakadai T, Bastle RM, Lukasak BJ, Zebroski H 3rd, Alenina N, Bader M, Berton O, Roeder RG, Molina H, Gage FH, Shen L, Garcia BA, Li H, Muir TW, Maze I (2019) Histone serotonylation is a permissive modification that enhances TFIID binding to H3K4me3. Nature 567(7749):535–539. https://doi.org/10.1038/s41586-019-1024-7
4. Strahl BD, Allis CD (2000) The language of covalent histone modifications. Nature 403 (6765):41–45. https://doi.org/10.1038/47412
5. Kouzarides T (2007) Chromatin modifications and their function. Cell 128(4):693–705. https://doi.org/10.1016/j.cell.2007.02.005
6. Hemberger M, Dean W, Reik W (2009) Epigenetic dynamics of stem cells and cell lineage commitment: digging Waddington's canal. Nat Rev Mol Cell Biol 10(8):526–537. https://doi.org/10.1038/nrm2727
7. Chen T, Dent SY (2014) Chromatin modifiers and remodellers: regulators of cellular differentiation. Nat Rev Genet 15(2):93–106. https://doi.org/10.1038/nrg3607
8. Secundino NF, Araujo MS, Oliveira GH, Massara CL, Carvalho OS, Lanfredi RM, Pimenta PF (2002) Preliminary description of a new entomoparasitic nematode infecting Lutzomyia longipalpis sand fly, the vector of visceral leishmaniasis in the New World. J Invertebr Pathol 80(1):35–40
9. Margueron R, Reinberg D (2011) The Polycomb complex PRC2 and its mark in life. Nature 469(7330):343–349. https://doi.org/10.1038/nature09784
10. de Napoles M, Mermoud JE, Wakao R, Tang YA, Endoh M, Appanah R, Nesterova TB, Silva J, Otte AP, Vidal M, Koseki H, Brockdorff N (2004) Polycomb group proteins Ring1A/B link ubiquitylation of histone H2A to heritable gene silencing and X inactivation. Dev Cell 7(5):663–676. https://doi.org/10.1016/j.devcel.2004.10.005
11. Wang H, Wang L, Erdjument-Bromage H, Vidal M, Tempst P, Jones RS, Zhang Y (2004) Role of histone H2A ubiquitination in Polycomb silencing. Nature 431 (7010):873–878. https://doi.org/10.1038/nature02985
12. Zhang LD, Yu D, Qu Y, Wang Z, Gao Y, Pan Q, Wang J, Guan HQ (2011) Effect of eye-acupuncture on cerebral intracellular adhesion molecule-1 expression in rats with acute cerebral ischemia-reperfusion injury. Zhen Ci Yan Jiu 36(6):409–413
13. Bardot ES, Valdes VJ, Zhang J, Perdigoto CN, Nicolis S, Hearn SA, Silva JM, Ezhkova E (2013) Polycomb subunits Ezh1 and Ezh2 regulate the Merkel cell differentiation program in skin stem cells. EMBO J 32 (14):1990–2000. https://doi.org/10.1038/emboj.2013.110
14. Dauber KL, Perdigoto CN, Valdes VJ, Santoriello FJ, Cohen I, Ezhkova E (2016) Dissecting the roles of Polycomb repressive complex 2 subunits in the control of skin development. J Invest Dermatol 136:1647. https://doi.org/10.1016/j.jid.2016.02.809
15. Cohen I, Zhao D, Menon G, Nakayama M, Koseki H, Zheng D, Ezhkova E (2019) PRC1 preserves epidermal tissue integrity independently of PRC2. Genes Dev 33(1–2):55–60. https://doi.org/10.1101/gad.319939.118
16. Bar C, Cohen I, Zhao D, Pothula V, Litskevitch A, Koseki H, Zheng D, Ezhkova E (2019) Polycomb repressive complex 1 controls maintenance of fungiform papillae by repressing sonic hedgehog expression. Cell Rep 28 (1):257–266.e255. https://doi.org/10.1016/j.celrep.2019.06.011
17. Boxer LD, Barajas B, Tao S, Zhang J, Khavari PA (2014) ZNF750 interacts with KLF4 and RCOR1, KDM1A, and CTBP1/2 chromatin regulators to repress epidermal progenitor genes and induce differentiation genes. Genes Dev 28(18):2013–2026. https://doi.org/10.1101/gad.246579.114
18. LeBoeuf M, Terrell A, Trivedi S, Sinha S, Epstein JA, Olson EN, Morrisey EE, Millar SE (2010) Hdac1 and Hdac2 act redundantly to control p63 and p53 functions in epidermal progenitor cells. Dev Cell 19(6):807–818. https://doi.org/10.1016/j.devcel.2010.10.015

19. Rinaldi L, Datta D, Serrat J, Morey L, Solanas G, Avgustinova A, Blanco E, Pons JI, Matallanas D, Von Kriegsheim A, Di Croce L, Benitah SA (2016) Dnmt3a and Dnmt3b associate with enhancers to regulate human epidermal stem cell homeostasis. Cell Stem Cell 19 (4):491–501. https://doi.org/10.1016/j.stem.2016.06.020

20. Bao X, Tang J, Lopez-Pajares V, Tao S, Qu K, Crabtree GR, Khavari PA (2013) ACTL6a enforces the epidermal progenitor state by suppressing SWI/SNF-dependent induction of KLF4. Cell Stem Cell 12(2):193–203. https://doi.org/10.1016/j.stem.2012.12.014

21. Park PJ (2009) ChIP-seq: advantages and challenges of a maturing technology. Nat Rev Genet 10(10):669–680. https://doi.org/10.1038/nrg2641

22. Rheinwald JG, Green H (1977) Epidermal growth factor and the multiplication of cultured human epidermal keratinocytes. Nature 265(5593):421–424. https://doi.org/10.1038/265421a0

23. Tian B, Yang J, Brasier AR (2012) Two-step cross-linking for analysis of protein-chromatin interactions. Methods Mol Biol 809:105–120. https://doi.org/10.1007/978-1-61779-376-9_7

Chapter 18

3D-FISH Analysis of the Spatial Genome Organization in Skin Cells in Situ

Andrei N. Mardaryev and Michael Y. Fessing

Abstract

Spatial genome organization in the cell nucleus plays a crucial role in the control of genome functions. Our knowledge about spatial genome organization is relying on the advances in gene imaging technologies and the biochemical approaches based on the spatial dependent ligation of the genomic regions. Fluorescent in situ hybridization using specific fluorescent DNA and RNA probes in cells and tissues with the spatially preserved nuclear and genome architecture (3D-FISH) provides a powerful tool for the further advancement of our knowledge about genome structure and functions. Here we describe the 3D-FISH protocols allowing for such an analysis in mammalian tissue in situ including in the skin. These protocols include DNA probe amplification and labeling; tissue fixation; preservation and preparation for hybridization; hybridization of the DNA probes with genomic DNA in the tissue; and post-hybridization tissue sample processing.

Key words Epigenetics, Spatial genome organization, Fluorescent in situ hybridization, 3-D FISH analysis

1 Introduction

Spatial genome organization in the cell nucleus plays a crucial role in the control of genome functions, including DNA replication, repair and transcription. For example, spatial contact between gene promoters and distal gene regulatory elements, such as gene enhancers and silences, is important to assure proper gene transcription levels [1, 2]. Changes in the cellular gene transcription programs drive stem and progenitor cell activity and differentiation underlying tissue and organ development, homeostasis, and regeneration. Advances in our understanding of the spatial genome organization and its role in the developmental control of gene expression are based on the studies using imaging of the labeled genome regions [2, 3] and biochemical techniques based on the spatial proximity dependent ligation of the genomic DNA fragments (chromosome conformation capture and its variations) [2, 4].

Natalia V. Botchkareva and Gillian E. Westgate (eds.), *Molecular Dermatology: Methods and Protocols*, Methods in Molecular Biology, vol. 2154, https://doi.org/10.1007/978-1-0716-0648-3_18,
© Springer Science+Business Media, LLC, part of Springer Nature 2020

The visualization of specific genomic regions (genes, gene loci, chromosome subregions, and whole chromosomes) has been usually performed using fluorescent in situ hybridization (FISH) analysis with fluorescently labeled DNA or RNA probes [3, 5, 6]. More recently, the alternative techniques have been developed employing specific DNA binding fluorescent proteins or enzymatically inactive fluorophore conjugated CAS9 (dCAS9) protein with specific guide RNAs to the genomic regions of interest [7].

The FISH analysis involves harsh tissue treatment that can lead to changes of the nuclear architecture and spatial genome organization. To overcome this problem, FISH techniques using cells and tissues with spatially preserved nuclei were developed, and is called 3D-FISH [5]. These techniques could also be combined with immune-fluorescent analysis, providing the tools to study spatial genome organization in relation to the different nuclear bodies [8, 9]. We were the first group who employed 3D-FISH to demonstrate the role of spatial genome organization in the control of developmentally regulated gene expression in skin tissue in situ by using fluorophore and hapten-labeled complex DNA probes with confocal laser scanning microscopy [8, 10, 11]. These studies led us to uncover the developmentally associated changes in the genome locus condensation [8, 10], gene positioning relative to the nuclear center [8, 10], chromosomes [8], nuclear bodies [8], and other genes [8, 10, 12]. Examples of some images representing single optical tissue sections after 3D-FISH are presented in Fig. 1.

Here we describe the 3D-FISH technology using frozen tissue sections with the 3D-tissue architecture preserved and include

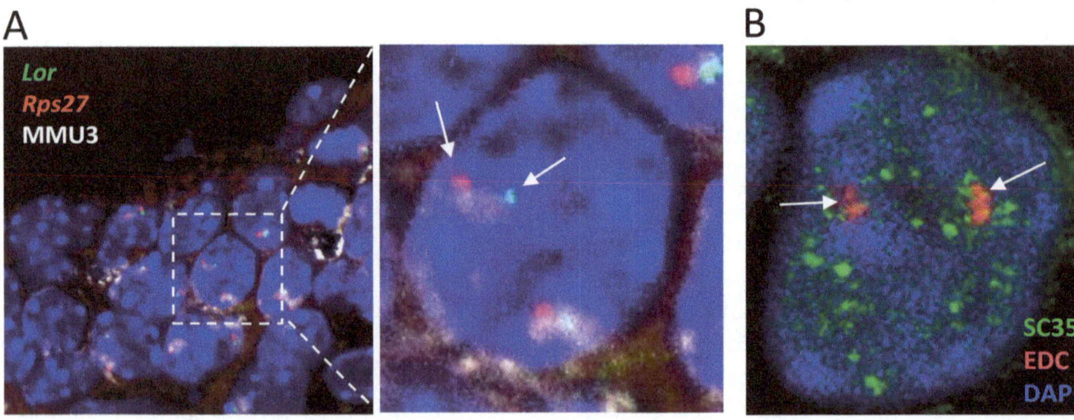

Fig. 1 Detection of genes, gene loci, and chromosome territories in the skin tissue using in situ 3D-FISH. (**a**) A single optical section of a confocal microscopy image using fluorescent dye-labeled DNA probes and showing localization of *Lor* (green) and *Rps27* (red) genes in relation to the corresponding chromosome territory 3 (MMU3) in epidermal keratinocytes of mouse embryonic skin (E14). (**b**) Coapplication of 3D-FISH and immunofluorescent detection of the EDC gene locus (red) and SC35-enriched nuclear speckles (green), respectively

gene/gene locus specific and whole chromosome probe preparation; tissue preservation and embedding; tissue section processing and preparation for hybridization; probe hybridization with genomic DNA in the tissue, and posthybridization tissue processing. The important image acquisition and image analysis techniques depend on the available microscopy equipment and the specific research question and these are widely covered elsewhere [8–10, 13] and are not described here.

2 Materials

2.1 Amplification and Labeling of the DNA Probes by DOP-PCR

1. Whole chromosome paint after the first or second amplification of the flow-sorted chromosome (*see* **Note 1**).
2. 10×PCR buffer: 200 mM Tris–HCl (pH 8.4), 500 mM KCl.
3. 25 mM magnesium chloride.
4. 100 μM 6 MW primer: 5′-CCGACTCGAGNNNNNNATG TGG-3′, where N is an equal mix of A, T, C, and G in the oligonucleotide preparation.
5. dNTP mix: 2.5 mM dATP, 2.5 mM dTTP, 2.5 mM dCTP, 2.5 mM dGTP.
6. DNA free Taq polymerase: 5 U/μl DNA free Taq polymerase.
7. dACG mix: 2 mM dATP, 2 mM dCTP, 2 mM dGTP.
8. dTTP: 2 mM dTTP.
9. 1 mM hapten or fluorophore-conjugated dUTP (*see* **Note 2**).
10. Agarose gel: 1% agarose gel in TAE buffer (40 mM Tris–acetate, 1 mM EDTA, pH 8.3) containing 0.5 mg/ml ethidium bromide.
11. TAE buffer: 40 mM Tris–acetate, 1 mM EDTA, pH 8.3.
12. DNA loading buffer.
13. DNA molecular weight marker in 100 bp–5 kb range.
14. Thermocycler.
15. Bucket with ice.
16. Microcentrifuge.
17. Tank and power supply for the agarose gel electrophoresis.
18. Gel documentation system.

2.2 DNA Amplification Utilizing Bacteriophage Phi29 DNA Polymerase

1. GE Healthcare Life Sciences Illustra GenomiPhi V2 DNA Amplification Kit or other Phi29 DNA polymerase-based DNA amplification system.
2. Agarose gel: 0.8% agarose gel in TAE buffer (40 mM Tris–acetate, 1 mM EDTA, pH 8.3) containing 0.5 mg/ml ethidium bromide.

3. TAE buffer: 40 mM Tris–acetate, 1 mM EDTA, pH 8.3.
4. DNA loading buffer.
5. DNA molecular weight marker in 1–10 kb range.
6. TAE buffer: 40 mM Tris–acetate, 1 mM EDTA, pH 8.3.
7. Thermocycler or water bath at 30 °C.
8. Bucket with ice.
9. Microcentrifuge.
10. Tank and power supply for the agarose gel electrophoresis.
11. Gel documentation system.

2.3 Labeling the BAC or Phagemid DNA Probes by Nick Translation

1. 10×NT buffer: 0.5 M Tris–HCl pH 7.5, 50 mM $MgCl_2$, 0.05% BSA.
2. NT dNTP mix: 0.5 mM dATP, 0.5 mM dGTP, 0.5 mM dCTP, 0.1 mM dTTP.
3. 0.1 M mercaptoethanol.
4. DNase I stock solution: 1 mg/ml DNase I (grade II, from bovine pancreas) in 150 mM NaCl, 10 mM $CaCl_2$, 50% glycerol (*see* **Note 3**).
5. E. *coli* DNA Polymerase I for NT: 5 U/μl DNA polymerase I (*see* **Note 3**).
6. Hapten or fluorophore conjugated dUTP: 1 mM hapten or fluorophore-conjugated dUTP (*see* **Note 2**).
7. NT stop mix: 100 mM NaCl, 20 mM Tris–HCl pH 7.5, 20 mM EDTA, 0.1% bromophenol blue, 0.5% dextran blue (*see* **Note 4**).
8. Agarose gel: 1% agarose gels in TAE buffer (40 mM Tris–acetate, 1 mM EDTA, pH 8.3) containing 0.5 mg/ml ethidium bromide.
9. TAE buffer: 40 mM Tris–acetate, 1 mM EDTA, pH 8.3.
10. DNA loading buffer.
11. DNA molecular weight makers in 100 bp–1 kb range.
12. Thermocycler or water baths.
13. Bucket with ice.
14. Microcentrifuge.
15. Tank and power supply for the agarose gel electrophoresis.
16. Gel documentation system.

2.4 Tissue Fixation and Cryopreservation

1. 10×PBS: 1.5 M NaCl, 30 mM Na_2HPO_4, 10.5 mM KH_2PO_4, pH 7.4.
2. 4% formaldehyde in 1×PBS, pH 7.4 (*see* **Note 5**).

3. 0.1 M phosphate buffer: 81 mM Na_2HPO_4, 19 mM NaH_2PO_4, pH 7.4.
4. 50 mM NH_4Cl, 1×PBS, pH 7.4.
5. 5% sucrose, 0.1 M phosphate buffer.
6. 12.5% sucrose, 0.1 M phosphate buffer.
7. 20% sucrose, 0.1 M phosphate buffer.
8. Embedding medium (*see* **Note 6**).
9. Sodium citrate buffer: 10 mM sodium citrate pH 6.0.
10. Molds for tissue embedding (*see* **Note 7**).
11. Rotator for tubes.
12. Bucket with dry ice.

2.5 Preparation of the Tissue Sections for Hybridization

1. Sodium citrate buffer: 10 mM sodium citrate, pH 6.0.
2. 2×SSC buffer prepared by diluting 20×SSC buffer: 3 M NaCl, 0.3 M sodium citrate, pH 7.0.
3. 50% formamide in 2×SSC (*see* **Note 8**).
4. Positively charged glass slides.
5. Heat-resistant Coplin jars.
6. Cryostat blades.
7. Microwave oven (*see* **Note 9**).
8. Water bath.
9. Cryostat.

2.6 Preparation of the DNA Probe Solution for Hybridization

1. Cot-1 DNA.
2. 2 M sodium chloride.
3. 95% ethanol.
4. Formamide (*see* **Note 8**).
5. 2 × hybridization mix (20% dextran sulfate, 4×SSC).
6. Microcentrifuge with cooling set at 4 °C.
7. Vortex mixer.
8. Water bath.
9. −80 °C freezer.

2.7 Hybridization Setup

1. 0.17 mm thick glass coverslips.
2. Diamond cutter.
3. Transparent nail polish.
4. Rubber cement.
5. Fine forceps.
6. Floating metal box with a lid.

7. Water bath at +37 °C.
8. Hot plate at +85 °C (*see* **Note 10**).

2.8 Posthybridization Tissue Processing

1. 4×SSC, 2×SSC and 0.1×SSC prepared from 20×SSC buffer: 3 M NaCl, 0.3 M sodium citrate, pH 7.0.
2. Blocking buffer: 4% BSA, 4×SSC, 0.1% Triton X-100, 0.1% Saponin.
3. Fluorophore-conjugated avidin, streptavidin, and/or anti-hapten antibodies (*see* **Note 11**).
4. DNA counterstaining solution: 25 ng/μl DAPI or 1 μM TO-PRO-3, 2×SSC.
5. Embedding medium (*see* **Note 12**).
6. 22 × 22 mm 0.17 mm thick glass coverslips.
7. Fine forceps.
8. Transparent nail polish.
9. Moisturizing chamber for the immunostaining.

3 Methods

3.1 Labeling the DNA Probes

For the 3D-FISH spatial genome organization analysis, labeled DNA probes could be generated using whole chromosomal DNA or DNA for specific subchromosomal (pericentromeric, telomeric, or gene/gene locus) regions. We describe here the amplification and labeling of the Whole Chromosome Paints (WCPs) using specific whole chromosome DNA by degenerate oligonucleotide-primed PCR (DOP-PCR) and the gene/gene locus specific probes using bacterial artificial chromosomes (BACs) or phasmids by amplification utilizing bacteriophage Phi29 DNA polymerase, followed by nick-translation labeling. Set up all the enzymatic reactions on ice. We have routinely used these methods for probe generation [8, 10, 12]. The probe quality is crucial for the successful 3D-FISH experiment. The quality of the probes and the detection strategy should be checked in 2D-FISH analysis with the metaphase chromosome spreads [14] before attempting the 3D-FISH experiment.

3.1.1 Whole Chromosome Paint Amplification by DOP-PCR

1. In the PCR tube combine 10 μl 10×PCR buffer, 8 μl 25 mM magnesium chloride, 2 μl 100uM 6 MW primer (Telenius), 8 μl 2.5 mM each dNTP mix, 70 μl water, 1 μl (5 U/μl) DNA-free Tag polymerase, 1 μl (30–200 ng/μl) specific whole chromosome DNA (after the first or the second amplification, *see* **Note 1**).
2. Gently mix the contents of the tube and collect onto the bottom of the tube using a microcentrifuge.

3. Place the tube in the PCR machine and run the amplification program: +94 °C for 3 min, followed by 36 cycles of +94 °C for 1 min, +56 °C for 1 min and +72 °C for 2 min, final extension +72 °C for 5 min, held at +4 °C.
4. Assess the size distribution and yield of the amplified DNA by analyzing 5 μl of the resulting reaction mixture in 1% agarose–TAE gel. The amplified products should be visible as smear ranging from approximately 200 bp to 1.5 kb. The concentration of the amplified products usually ranges from 30 to 200 ng/μl.
5. Store the remaining amplified products at −20 °C where it can be stored for several years.

3.1.2 Whole Chromosome Paint Labeling by DOP-PCR

1. In the PCR tube, combine 10 μl 10×PCR buffer, 8 μl 25 mM magnesium chloride, 2 μl 100 μM 6 MW primer (Telenius), 5 μl dACG mix (2 mM each), 1 μl 2 mM dTTP, 58 μl water, 6 μl 1 mM hapten or fluorophore-conjugated dUTP (*see* **Note 2**), 1 μl specific whole chromosome DNA after the second or third amplification (30–200 ng/μl), and 0.5 μl DNA-free Taq polymerase (5 U/μl).
2. Gently mix the contents of the tube and collect onto the bottom of the tube using a microcentrifuge.
3. Place the tube in the PCR machine and run the amplification program: +94 °C for 3 min, followed by 25 cycles of +94 °C for 1 min, +56 °C for 1 min and +72 °C for 30 s, final extension +72 °C for 5 min, hold at +4 °C.
4. Assess the size distribution and yield of the labelled DNA by analyzing 5 μl of the resulting reaction mixture in 1% agarose/TAE gel. The amplified products should be visible as smear ranging from approximately 200 bp to 1.5 kb. The concentration of the amplified products usually ranges from 30 to 200 ng/μl.
5. Store the remaining amplified products at −20 °C where it can be stored for several years.

3.1.3 BAC or Phasmid DNA Amplification Utilizing Bacteriophage Phi29 DNA Polymerase

We routinely amplify BAC or phasmid DNA by isothermal strand displacement synthesis catalyzed by the bacteriophage Phi29 DNA polymerase using GE Healthcare illustra™ GenomiPhi V2 DNA Amplification Kit. For a bacteriophage Phi29 DNA polymerase–based amplification system from other suppliers, follow the corresponding supplier's protocol. This method does not require highly purified DNA (crude purification from the E. *c*oli using routine alkaline lysis followed by precipitation with isopropanol without phenol–chloroform extraction provides DNA of sufficient purity and concentration).

1. Combine 9 μl of sample buffer and 1 μl (10–30 ng) DNA.
2. Heat at +95 °C for 3 min and chill at +4 °C (*see* **Note 13**).
3. Prepare master mix by combining 9 μl reaction buffer and 1 μl enzyme mix. Mix gently.
4. Combine the sample DNA after the **step 2** and 10 μl of the master mix prepared according to the **step 3**. Mix gently.
5. Incubate at +30 °C for 1.5 h followed by incubation at +65 °C for 10 min and cooling at +4 °C (*see* **Note13**).
6. Assess the size distribution and yield of the amplified DNA by analyzing 2 μl of the amplification products in 0.8% agarose/TAE gel with the appropriate DNA molecular weight ladder. The amplified DNA will appear as a smear around 10 kb. The expected yield is 4–7 μg (200–350 ng/μl).

3.1.4 Labeling the BAC or Phagemid DNA Probes by Nick Translation

1. Dilute 1 μl of DNase I stock solution in 250 μl of water on ice. Mix gently (*see* **Note 14**).
2. Combine 1 μg DNA and water to the final volume of 31 μl. Add 5 μl NT buffer, 5 μl 0.1 M mercaptoethanol, 5 μl NT dNTP mix, 2 μl labeled dUTP (*see* **Note 2**), 1 μl diluted DNase I (from the **step 1**), and 1 μl E. *c*oli DNA polymerase I. Mix gently.
3. Incubate at +15 °C for 1.5 h. Place on ice.
4. Analyze 5 μl of the nick translation products in 1% agarose/TAE gel with the appropriate DNA molecular weight ladder. The probe DNA should appear as a smear between 200 and 600 bp. If the most of the probe DNA fragments are larger than 600 bp, add 1 μl of the freshly diluted DNase I (*see* **step 1**) and incubate at room temperature for 5 min.
5. Add equal volume of the NT stop solution to the NT reaction mixture. Mix well. The concentration of the DNA probe would be about 10 ng/μl. Store the labeled probe for up to a year at −20 °C.

3.2 Preparation of the Frozen Tissue Sections

3.2.1 Tissue Fixation and Cryopreservation

1. Rinse freshly isolated tissue (or animal embryos) with 1×PBS (pH 7.4) at room temperature for 5 min two times (*see* **Note 15**).
2. Fix the tissue in 4% formaldehyde in 1×PBS (pH 7.4) at +4 °C overnight on the rotator.
3. Carefully remove the fixative solution avoiding touching the tissue. Add 50 mM ammonium chloride solution in 1×PBS, carefully mix the solution by inverting the tube several times and incubate the tissue at room temperature for 5 min.
4. Wash the tissue twice in 0.1 M phosphate buffer (pH 7.4) at room temperature for 10 min each.

5. Incubate the tissue in 5% sucrose/0.1 M phosphate buffer with gentle rotation at room temperature for 1 h.
6. Incubate the tissue in 12.5% sucrose–0.1 M phosphate buffer with gentle rotation at room temperature for 1 h.
7. Incubate the tissue in 20% sucrose–0.1 M phosphate buffer with gentle rotation at 4 °C overnight.
8. Fill the embedding dish with the embedding medium and carefully place the tissue in it. Make sure that the tissue is completely covered and avoid bubbles.
9. Put the embedding dish on the dry ice and let the medium completely freeze. It will take several minutes.
10. Keep the fixed frozen tissue samples at −80 °C for up to 5 years.

3.2.2 Preparation of the Tissue Sections for Hybridization

1. Cut the tissue sections at 20 μM thickness using a cryostat at −30 °C and mount them on the positively charged glass slides. The slides could be stored at −80 °C for up to a year.
2. Remove the slide(s) with the frozen sections from the storage and dry them at room temperature for 1 h.
3. Place the slide(s) in a Coplin jar filled with 10 mM sodium citrate buffer (pH 6.0) and incubate them at room temperature for 5 min.
4. Heat the slides in the microwave oven until the first sign of boiling. Do not allow the buffer to boil, as it will destroy the nuclear architecture and tissue morphological appearance.
5. Place the Coplin jar with the slide(s) in the water bath at +90 °C and incubate for 10 min (*see* **Note 16**).
6. Cool the slide(s) in the buffer at room temperature for 5 min.
7. Transfer the slide(s) into 2×SSC buffer and incubate at room temperature for 10 min.
8. Transfer the slide(s) into 50% formamide in 2×SSC and incubate at room temperature for at least 2 h (*see* **Note 17**).

3.3 DNA Hybridization

3.3.1 Preparation of the DNA Probe Solution for Hybridization

1. Combine the labeled BAC, phagemid, and/or whole chromosome probes to achieve final concentration of 30–50 ng/μl for each probe in the hybridization mixture (*see* **Note 18**). Add 10 μg Cotl DNA for the first 250 ng of the total probe DNA and 5μg Cotl DNA for each additional 250 ng of the total probe DNA (*see* **Note 19**). Add 2 M sodium chloride solution to the final sodium chloride concentration of 50 mM (*see* **Note 20**).
2. Add 2.5 volumes of the ice cold 95% ethanol to the probe mixture. Mix well and incubate at −80 °C for 30 min.

3. Collect the DNA pallet by centrifugation at 14,000 × *g* for 15 min at +4 °C.
4. Remove the supernatant and air dry the DNA pallet for 10–15 min.
5. Dissolve the DNA pallet in the half of the final hybridization mixture volume (*see* **Note 21**) of the formamide by incubating the probe at +55 °C and vortexing every 2 min. It will take about 10–15 min.
6. Add equal volume of 2× hybridization mix and mix well.

3.3.2 Hybridization Set Up

1. Take the slide from 50% formamide/2×SSC and remove the excess buffer. Cover the tissue with the hybridization glass chamber. Add the hybridization mixture with the dissolved probes under the glass chamber avoiding bubbles (Fig. 2) (*see* **Note 22**).
2. Seal the chamber on the slide with the rubber cement and let the cement completely dry.
3. Incubate the sealed slide at +37 °C for at least 2 h (*see* **Note 23**).
4. Denature the tissue and probe DNA by incubating the slide on the hot plate at +85 °C for 5 min (*see* **Note 24**).
5. Incubate the slide at 37 °C for 2–3 days in a metal box floating in the water bath.

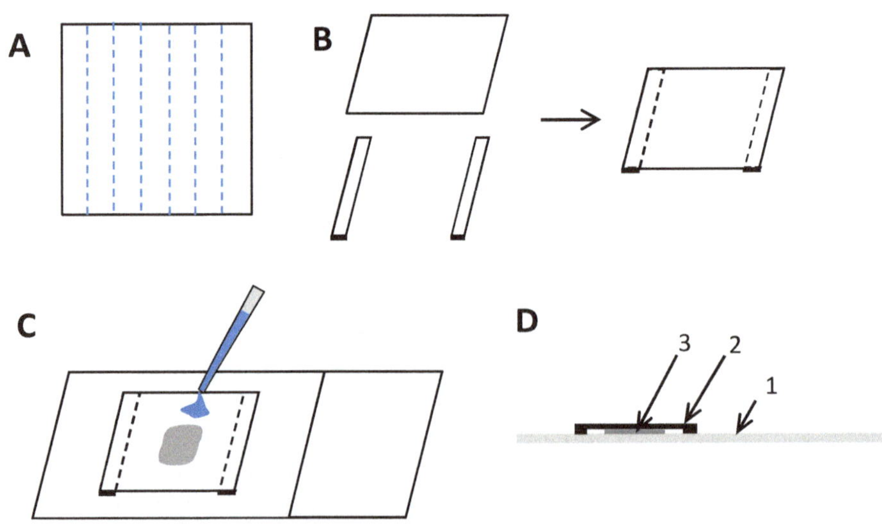

Fig. 2 Glass chamber preparation for hybridization of probes on skin tissue sections. (**a**) Using a diamond cutter, cut coverslips into several strips (2–3 mm wide); (**b**) Assemble a glass chamber by attaching two strips onto a coverslip using nail polish; (**c**) Apply the glass chamber over a tissue section on a slide and load the hybridization mixture over the tissue under the glass chamber using a pipette tip; (**d**) A side view of the assembled slide (1) with the glass chamber (2) and tissue section (3)

3.4 Post-Hybridization Tissue Processing

1. Carefully peel off the rubber cement and remove the glass chamber from the slide using the forceps. Put the slide in the Coplin jar with 2×SSC at +37 °C and incubate for 10 min with gentle agitation on the rocking plate (*see* **Note 25**).
2. Repeat washes with 2×SSC at +37 °C for 10 min twice more.
3. Wash the slides with 0.1×SSC at +60 °C for 10 min.
4. If only the fluorophore-labeled probes were used, proceed to the DNA counterstaining and slide embedding (**step 11**). Otherwise transfer the slide into 4×SSC at room temperature.
5. Apply the blocking buffer (4% BSA, 4×SSC, 0.1% Triton X-100, 0.1% Saponin) to the tissue sections and incubate at room temperature for 15 min.
6. Put the 22 × 22 mm glass chamber, similar to the hybridization chamber used for the hybridization, around the tissue section on the slide. Apply the required anti-hapten antibodies and/or fluorophore-conjugated avidin (or streptavidin) in the blocking buffer to the section under the chamber (*see* **Note 26**).
7. Incubate the slide at +4 °C overnight in the moisturized chamber.
8. Carefully remove the glass chamber and wash the slide three times in 4×SSC for 10 min with gentle agitation.
9. If only the fluorophore-labeled avidin or streptavidin was used in **step 6**, proceed to the DNA counterstaining and slide embedding. Otherwise stain the tissue sections with the appropriate secondary fluorophore-conjugated antibodies diluted in the blocking buffer at +37 °C for 4 h or at +4 °C overnight (*see* **Note 11**).
10. Wash the slides three times in 4×SSC at +37 °C for 10 min with gentle agitation.
11. Incubate the tissue with 25 ng/μl DAPI or 1 μM TO-PRO-3 in 2×SSC at room temperature for 30 min.
12. Embed using the counter-stain free embedding medium and seal the coverslip on the slide with the nail polish.
13. The tissue is ready for microscopy (*see* **Note 27**). The slides can be stored for up to a year if DAPI counter-stain is used or for up to 4 weeks if TO-PRO-3 counterstain is used.

4 Notes

1. Whole Chromosome Paints (flow-sorted and DOP-PCR amplified chromosomes) could be obtained from flow-sorting facilities or other commercial sources. The quality of the paints drops after forth DOP-PCR rounds of amplification.

2. Hapten and fluorophore-conjugated dUTP could be obtained from different suppliers or prepared as described elsewhere [15]. The preparation of the conjugated dUTP in-house provides very significant saving (more than tenfold). We have routinely prepared and used biotin-dUTP (Bio-dUTP), digoxigenin-dUTP (Dig-dUTP), fluorescein isothiocyanate-dUTP (FITC-dUTP), and cyanine 3-dUTP (Cy3-dUTP). The choice of fluorophores depends on the confocal microscope that is used for imaging.
3. DNase I and DNA polymerase I are very temperature sensitive. Store and transport them in the benchtop cooler at −20 to −30 °C and do not allow warming above −20 °C.
4. Bromophenol blue and dextran blue dye inclusion helps to better monitor the DNA probe pellet, the probe dissolving in the hybridization mixture and the hybridization mixture during the hybridization set up.
5. Methanol-free nonoxidized formaldehyde should be used. We prepared 4% formaldehyde solution in 1×PBS by dissolving paraformaldehyde. The solution could be aliquoted and kept at −20 °C for up to half a year before use. Alternatively, we use 32% formaldehyde solution packed in the inert atmosphere in sealed glass ampules, available from several suppliers, to prepare the working fixation solution.
6. We routinely use Tissue-Tek™ O.C.T. medium. Similar embedding medium from other suppliers could be used.
7. We routinely use the peel away disposable plastic embedding molds.
8. Use the molecular biology grade or deionized formamide kept at +4 °C for up to a year or at −20 °C for several years.
9. Use microwave oven at 700–800 W with the transparent door.
10. Check the plate temperature using a thermometer. The proper temperature is crucial.
11. The choice of the fluorophores depends on the confocal microscope that will be used for the imaging. We routinely use Alexa 488, FITC, Cy3, and Cy5-conjugated antibodies and DAPI for the DNA counterstaining. Use the antibodies with the lower molecular weight to improve the antibody penetration inside the tissue.
12. We routinely use Vectashield™ embedding medium without DAPI from Vector Laboratories. A similar medium is available from other suppliers.
13. We use a thermocycler for the incubation.
14. Prepare the DNase I diluted solution from the stock solution just before use. Discard the unused diluted solution.

15. Volumes of all the solutions should be in vast excess (at least tenfold) of the tissue volume.
16. This time of incubation works well for the whole thickness mouse and human skin or mouse embryos. The time should be optimized for the specific tissue sections if necessary.
17. Incubation for at least 12 h at +4 °C improves the probe penetration into the tissue. The tissue could be stored at +4 °C for up to a week.
18. This probe concentrations work well in our hands in most cases. They could be increased up to 100 ng/μl if the signals after the hybridization are weak.
19. Cot-1 DNA suppresses hybridization of the repetitive sequences in the probe to the repetitive genomic DNA sequences in the tissue.
20. Nick-translation stop solution in the BAC and/or phasmid probes usually contains enough sodium chloride for the probe DNA precipitation.
21. The final volume of the hybridization mixture depends on the hybridization chamber size required to cover the tissue. Fifteen microliters is required for the 12 × 12 mm chamber and 25 μl for the 18 × 18 mm chamber.
22. We load the probe using small pipette tips.
23. The incubation for up to 16 h will improve the probe penetration into the tissue.
24. Temperature and timing are crucial.
25. Protect from light if the fluorophore-labeled probes are used.
26. At this step the immune-fluorescent staining with the antibodies specific for the different nuclear proteins can be combined with the tissue processing after the 3D-FISH hybridization. In this case use the primary antibodies in the **step 6**, relevant fluorophore conjugated secondary antibodies in the **step 9** and substitute 4×SCC with 1×PBS in all blocking and washing buffers.
27. We use confocal microscope and scan the samples at the voxel size of 100 × 100 × 200 nm. The imaging and image analysis depends on the specific scientific question and is widely describe elsewhere [8–10, 13].

Acknowledgments

We greatly appreciate the invaluable help and support of Thomas Cremer, Irina Solovei, and Boris Joffe (Ludwig-Maximilians University Munich, Germany) in establishing 3D-FISH technology and analysis in our laboratories. We extend special thanks to Prof.

Vladimir Botchkarev (Boston University School of Medicine, Boston, MA, USA) and all our colleagues and collaborators with whom we have had the pleasure to study the role of nuclear architecture and spatial genome organization in skin biology.

References

1. Benabdallah NS, Bickmore WA (2015) Regulatory domains and their mechanisms. Cold Spring Harb Symp Quant Biol 80:45. https://doi.org/10.1101/sqb.2015.80.027268
2. Dekker J, Misteli T (2015) Long-range chromatin interactions. Cold Spring Harb Perspect Biol 7(10):a019356. https://doi.org/10.1101/cshperspect.a019356
3. Cremer T, Cremer M (2010) Chromosome territories. Cold Spring Harb Perspect Biol 2 (3):a003889. https://doi.org/10.1101/cshperspect.a003889
4. Kim TH, Dekker J (2018) 3C-based chromatin interaction analyses. Cold Spring Harb Protoc 2018(9). https://doi.org/10.1101/pdb.top097832
5. Cremer M, Grasser F, Lanctot C, Muller S, Neusser M, Zinner R, Solovei I, Cremer T (2008) Multicolor 3D fluorescence in situ hybridization for imaging interphase chromosomes. Methods Mol Biol 463:205–239. https://doi.org/10.1007/978-1-59745-406-3_15
6. Fauth C, Speicher MR (2001) Classifying by colors: FISH-based genome analysis. Cytogenet Cell Genet 93(1–2):1–10. https://doi.org/10.1159/000056937
7. Knight SC, Tjian R, Doudna JA (2018) Genomes in focus: development and applications of CRISPR-Cas9 imaging technologies. Angew Chem Int Ed Engl 57(16):4329–4337. https://doi.org/10.1002/anie.201709201
8. Mardaryev AN, Gdula MR, Yarker JL, Emelianov VU, Poterlowicz K, Sharov AA, Sharova TY, Scarpa JA, Joffe B, Solovei I, Chambon P, Botchkarev VA, Fessing MY (2014) p63 and Brg1 control developmentally regulated higher-order chromatin remodelling at the epidermal differentiation complex locus in epidermal progenitor cells. Development 141 (1):101–111. https://doi.org/10.1242/dev.103200. 141/1/101 [pii]
9. Solovei I, Kreysing M, Lanctot C, Kosem S, Peichl L, Cremer T, Guck J, Joffe B (2009) Nuclear architecture of rod photoreceptor cells adapts to vision in mammalian evolution. Cell 137(2):356–368. https://doi.org/10.1016/j.cell.2009.01.052
10. Fessing MY, Mardaryev AN, Gdula MR, Sharov AA, Sharova TY, Rapisarda V, Gordon KB, Smorodchenko AD, Poterlowicz K, Ferone G, Kohwi Y, Missero C, Kohwi-Shigematsu T, Botchkarev VA (2011) p63 regulates Satb1 to control tissue-specific chromatin remodeling during development of the epidermis. J Cell Biol 194(6):825–839. https://doi.org/10.1083/jcb.201101148
11. Gdula MR, Poterlowicz K, Mardaryev AN, Sharov AA, Peng Y, Fessing MY, Botchkarev VA (2013) Remodeling of three-dimensional organization of the nucleus during terminal keratinocyte differentiation in the epidermis. J Invest Dermatol 133(9):2191–2201. https://doi.org/10.1038/jid.2013.66. jid201366 [pii]
12. Poterlowicz K, Yarker JL, Malashchuk I, Lajoie BR, Mardaryev AN, Gdula MR, Sharov AA, Kohwi-Shigematsu T, Botchkarev VA, Fessing MY (2017) 5C analysis of the Epidermal Differentiation Complex locus reveals distinct chromatin interaction networks between gene-rich and gene-poor TADs in skin epithelial cells. PLoS Genet 13(9):e1006966. https://doi.org/10.1371/journal.pgen.1006966
13. Ronneberger O, Baddeley D, Scheipl F, Verveer PJ, Burkhardt H, Cremer C, Fahrmeir L, Cremer T, Joffe B (2008) Spatial quantitative analysis of fluorescently labeled nuclear structures: problems, methods, pitfalls. Chromosome Res 16(3):523–562. https://doi.org/10.1007/s10577-008-1236-4
14. Macoska JA (2001) Fluorescence in situ hybridization (FISH) to metaphase and interphase chromosomes. Methods Mol Med 53:101–123. https://doi.org/10.1385/1-59259-144-2:101
15. Muller S, Neusser M, Kohler D, Cremer M (2007) Preparation of complex DNA probe sets for 3D- FISH with up to six different fluorochromes. CSH Protoc 2007:pdb.prot4730

Chapter 19

Method to Study Skin Cancer: Two-Stage Chemically Induced Carcinogenesis in Mouse Skin

Andrei N. Mardaryev

Abstract

Two-stage chemical carcinogenesis method is widely used to elucidate genetic and molecular changes that lead to skin cancer development, as well as to test chemotherapeutic properties of novel drugs. This protocol allows researchers to reliably induce benign papilloma development and their conversion to squamous cell carcinoma in the skin of susceptible mouse strains in response to a single dose of carcinogen 2,4-dimethoxybenzaldehyde (DMBA) and repetitive applications of tumor promoter 12-O-tetradecanoyl-phorbol 13-acetate (TPA).

Key words Skin, Chemical carcinogenesis, DMBA, TPA, Papilloma, Squamous cell carcinoma

1 Introduction

Skin cancer is the most common form of human malignancy with a rapidly increasing incidence worldwide. In vivo animal models provide invaluable tools for dissecting molecular changes leading to skin cancer initiation and progression, and may also lead to development of new strategies for skin cancer treatment and prevention.

The mouse skin model of two-stage chemical carcinogenesis is widely accepted as the best-characterized and versatile system for the study of sequential molecular events that lead to formation of epithelial tumors [1, 2]. In this model, skin tumors initiated using a single topical subcarcinogenic dose of a genotoxic agent such as 7,12-dimethylbenz[a]anthracene (DMBA), followed by repeated application of a tumor-promoting (nongenotoxic) irritant such as the phorbol ester, 12-O-tetradecanoylphorbol-13-acetate (TPA) (Fig. 1) [3]. By doing so, the two-stage chemical carcinogenesis protocol provides several advantages: (1) it is possible to distinguish molecular events or effects related to the tumor initiation versus tumor promotion stages; (2) the tumor development can be

Natalia V. Botchkareva and Gillian E. Westgate (eds.), *Molecular Dermatology: Methods and Protocols*,
Methods in Molecular Biology, vol. 2154, https://doi.org/10.1007/978-1-0716-0648-3_19,
© Springer Science+Business Media, LLC, part of Springer Nature 2020

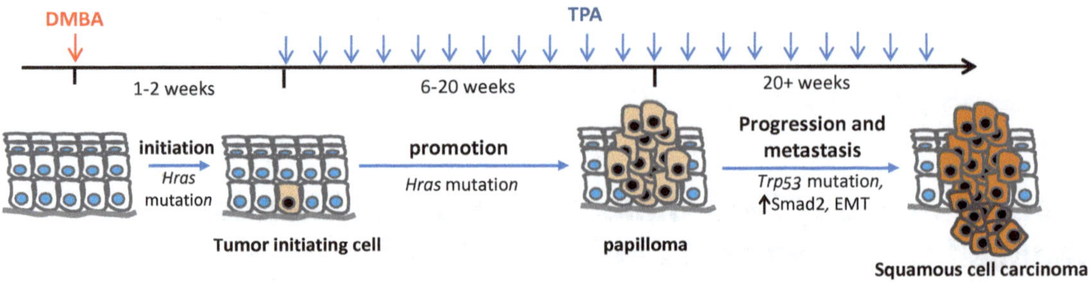

Fig. 1 Molecular changes occur in response to a single subcarcinogenic dose of DMBA and repetitive application of TPA leading to papilloma development and their progression to SCC

monitored visually and using noninvasive evaluation methods throughout the life span of the animal, while tissue collection and pathological assessment are performed at the end of the experiment; (3) it allows for assessment of the effect of genetic manipulations, therapeutic and chemopreventive agents, and dietary and environmental factors in the tumor initiation and promotion [1, 3].

At the first stage (initiation) of the chemically induced skin carcinogenesis model, a chemical mutagen causes mutations of key genes in skin epithelial cells. Activating mutations in *Hras1* and *Kras* genes are frequently associated with the tumor initiation upon exposure to chemical carcinogens, including the widely used tumor initiator DMBA [2, 4, 5]. *Hras1* mutations can be detected in epidermal keratinocytes as early as 1 day following the initial single dose of DMBA and present in the majority of epithelial tumors developed after tumor promoter treatment [6]. Recent studies suggest that quiescent epithelial stem and progenitor cells located in the basal layer of the epidermis and hair follicle bulge (K15+/CD34+ cells) are primarily targeted to transform into tumor-initiating cells during the initiation stage [7–9].

Repeated topical application of chemical irritants leads to expansion of mutated cells (tumor promotion stage) and increase of epidermal thickness (hyperplasia). Due to acquired mutations, initiated cells have a growth advantage allowing for their selective expansion. The widely used tumor promoter TPA was shown as a potent activator of protein kinase C (PKC), proinflammatory cytokines, and Wnt/β-catenin signaling pathways, which are involved in cell proliferation and survival [10, 11].

The tumors that develop as a result of exposure to DMBA and TPA are mainly papillomas and to a much less extent squamous cell carcinomas (SCC). Papillomas may progress to invasive SCC as early as 20 weeks after application of tumor promoter TPA. The frequency of malignant conversion depends on genetic background and the dose of DMBA and TPA used [3, 12], and is mechanistically linked to chromosomal abnormalities [13, 14], Trp53 mutations [13, 15], activation of TGFβ/Smad2 signaling, and enhanced epithelial–mesenchymal transition (EMT) [16–20].

Despite the technical simplicity of the DMBA/TPA protocol, careful design and attention to details are required for the proper execution of a two-stage skin carcinogenesis study. In this section, the DMBA/TPA skin carcinogenesis experimental design and execution considerations will be discussed.

2 Materials

1. Animals: Female FVB/n mice, 7–9 weeks of age, with dorsal skin in the resting phase of hair cycle (*see* **Note 1**).
2. 2,4-Dimethoxybenzaldehyde (DMBA) dissolved in acetone at 100 mg per 100 ml (0.1%, w/v or 390 nmol per 100-μl application) (*see* **Note 2**).

   The 0.1% DMBA stock solution should be made fresh prior to each experiment, or stored at −20 °C. Protect all DMBA solutions from light by wrapping vials in foil or using brown glass.
3. 12-O-tetradecanoylphorbol 13-acetate (TPA) dissolved in acetone at 5 mg per 50 ml (for a 0.01%, w/v solution or 16.2 nmol per 100-μl application) (*see* **Note 2**).

   Protect the 0.025% TPA stock solution from light, and store it at −20 °C for up to 6 months.
4. Surgical clippers and electric shaver.
5. Disposable 15 and 50 ml tubes.
6. Pipettes and 100 μl tips.
7. Calipers or other measuring devices (e.g., Vernier calipers).

3 Methods

1. Perform sample size calculation and randomize the mice into treatment and control groups (*see* **Note 3**).
2. Remove the dorsal hair from each mouse using a surgical clipper and shaver, while gently restraining mice by the tail (*see* **Note 4**).
3. Two days after hair removal, apply 0.1 ml of DMBA to the shaved area. Treat control mice with 0.1 ml of acetone.
4. After 1 week, apply 0.1 ml of TPA to the DMBA-treated area. Perform two applications of TPA per week, maintaining 3–4 day intervals between treatments (*see* **Note 5**). Treat control mice with 0.1 ml of acetone.
5. Continue applying TPA for up to 52 weeks (*see* **Note 6**).

6. Measure palpable tumors using caliper and count papillomas of 1 mm in diameter and greater on a weekly basis. Keep record of the number of tumors per mouse and the number of mice remaining in the experiment (*see* **Note 7**).
7. Note visual changes in papillomas and their conversion to SCCs (*see* **Note 8**).

Keep record of mice that die unexpectedly during the experiment and use approved methods of killing when reaching humane endpoints or mice becoming critically unwell.

Harvest skin and tumors for histological analysis and further studies (*see* **Note 9**). Tumors and adjacent skin may be snap-frozen, cryopreserved in optimum cutting temperature (OCT) compound, or fixed in formalin and embedded in paraffin.

4 Notes

1. The choice of genetic background has a significant impact on the frequency of tumor development and malignant conversion. For instance, a low-dose regimen typically results in the papilloma incidence of ~100% in FVB mice with up to 50% of papillomas converting to SCCs, while C57Bl/6 strain is more resistant to tumor promotion and SCC conversion and require higher doses [21]. Age-matched females are preferred for the experiments as male mice may fight and cannot be grouped together. Number of animals in cages should be determined according to approved institutional animal care guidelines.
2. For the lower-dose regimens listed in Table 1, dilute the 0.1% DMBA and 0.01% (w/v) TPA solutions in acetone to yield the appropriate concentrations. The high doses of DMBA and TPA may lead to increased tumor burden and may affect the health and well-being of mice, which in turn may affect the results and conclusions drawn from the experiment. Furthermore, the higher doses my mask a potential chemopreventing effect of drugs and genetic manipulations. Depending on the anticipated effect, a pilot experiment is recommended to perform dose-response analyses with a range of DMBA or TPA doses.

Table 1
Low- and high-dose regimens for DMBA/TPA two-step carcinogenesis

Low-dose	6.4 μg (25 nmol) DMBA; 1 μg (1.6 nmol) TPA twice per week 25 μg (97.5 nmol) DMBA; 2 μg (3.2 nmol) TPA twice per week 50 μg (195 nmol) DMBA; 4 μg (6.4 nmol) TPA twice per week
High-dose	50 μg (195 nmol) DMBA; 10 μg (16.2 nmol) TPA twice per week 100 μg (390 nmol) DMBA; 10 μg (16.2 nmol) TPA twice per week

For an even milder promotion regimen, TPA may be applied once a week or once every 2 weeks

3. Sample size calculation predicts the number of mice needed per group to allow sufficient statistical power to compare the response of control and test groups [22]. Sample size can be calculated using statistical software packages (e.g., SPSS Sample Power, G*Power or Quanto) or online resources (e.g., http://powerandsamplesize.com/Calculators/). For randomization process, a random number generator is recommended to randomly assign age- and gender-matched animals to treatment groups.
4. It is important to remove back hair to ensure uniform application of DMBA. If necessary, provide short anesthesia with inhaled anesthetic substances such as isoflurane to reduce the risk of damaging the skin by clipper. Ensure proper training and obtaining gas anesthesia machine (calibrated within the last 12 months) with adequate gas scavenging system or filter, and an induction chamber constructed of a see-through material (glass, polycarbonate, etc.). Once shaved, the skin of the mice should be examined to determine the approximate phase of the hair cycle. Mice in telogen (resting) stage of hair cycle should be selected for the experiment. The skin of mice in telogen appears pink and thin, while becoming thickened and white (in FVB) or dark grey (in pigmented mice) during the hair growth phase (anagen).
5. Time of the first TPA application following DMPA treatment can be extended for several weeks without a negative impact on tumor yield [3].
6. TPA is applied to the skin until the tumor response reaches a plateau. For highly and moderately susceptible strains, such as FVB, TPA application is typically performed for 20 weeks.
7. To demonstrate tumor response to DMAB/TPA treatment, a tumor incidence and overall tumor burden can be calculated. Tumor incidence highlights differences in tumor latency between groups and may be plotted as either the percentage of mice bearing any papilloma over time or as fractional tumor-free survival over time. Overall tumor burden is measured as tumor multiplicity over time. Tumor multiplicity represents average number of papillomas per mouse (calculated as the total number of papillomas detected in a group divided by the total number of mice in the group). Typically, papilloma incidence and multiplicity are calculated by dividing the number of tumor-bearing mice or total number of papillomas, by the number of mice alive at each week, respectively. In case of SCC, SCC incidence and multiplicity are calculated cumulatively, and the total number of SCCs is divided by the number of mice alive at the time. Statistical analysis of differences in

tumor multiplicity should be carried out using nonparametric methods such as the Mann–Whitney U test. The χ^2 test is suitable for comparing tumor incidence between groups [3].

8. To monitor progression of papillomas to SCC, draw a map of tumors and note any changes. Typical papillomas have exophytic growth: extending outward, well-demarcated, symmetrical, pedunculated or dome-shaped, without erosion and ulcers. Visual signs of SCC conversion include growing downward, poorly demarcated, asymmetrical, nonpedunculated or doughnut-shaped papules or nodules with erosion and ulceration [23].
9. During early stages of TPA treatment, typical histological changes are presented with epidermal hyperplasia, while the first signs of papilloma development become apparent by 6–8 weeks of promotion in FVB mice. A small percentage of papillomas begin to convert to SCCs by 20 weeks with signs of invasions of malignant cells deeper into the dermis. Histologically, SCC cells show anaplastic changes and become disorganized losing normal polarity and markers of differentiation. Depending on the degree of cellular architecture disorganization, SCCs are referred to as well-differentiated as opposed to poorly differentiated SCCs. Immunohistochemical methods are commonly used to demonstrate loss in expression of differentiation markers (i.e., keratin 1, keratin 10, loricrin) and aberrant expression of keratin 8 and keratin 13 during papilloma conversion to SCCs (Fig. 2) [3].

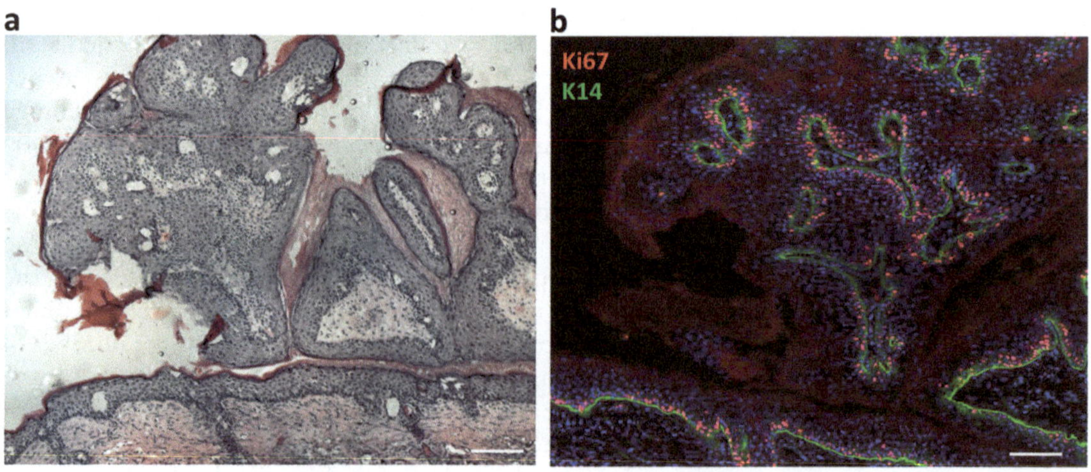

Fig. 2 Morphological assessment of DMBA/TPA induced skin tumors. H&E staining demonstrates epidermal hyperplasia and exophytic growth typical for papilloma (**a**). Visualization of undifferentiated K14-expressing (green) and proliferating Ki67-positive (red) cells using immunofluorescent method (**b**). Scale bar, 200 μm

References

1. Neagu M, Caruntu C, Constantin C, Boda D, Zurac S, Spandidos DA, Tsatsakis AM (2016) Chemically induced skin carcinogenesis: updates in experimental models (Review). Oncol Rep 35(5):2516–2528. https://doi.org/10.3892/or.2016.4683
2. Huang PY, Balmain A (2014) Modeling cutaneous squamous carcinoma development in the mouse. Cold Spring Harb Perspect Med 4 (9):a013623. https://doi.org/10.1101/cshperspect.a013623
3. Abel EL, Angel JM, Kiguchi K, DiGiovanni J (2009) Multi-stage chemical carcinogenesis in mouse skin: fundamentals and applications. Nat Protoc 4(9):1350–1362. https://doi.org/10.1038/nprot.2009.120
4. Ise K, Nakamura K, Nakao K, Shimizu S, Harada H, Ichise T, Miyoshi J, Gondo Y, Ishikawa T, Aiba A, Katsuki M (2000) Targeted deletion of the H-ras gene decreases tumor formation in mouse skin carcinogenesis. Oncogene 19(26):2951–2956. https://doi.org/10.1038/sj.onc.1203600
5. Rehman I, Lowry DT, Adams C, Abdel-Fattah R, Holly A, Yuspa SH, Hennings H (2000) Frequent codon 12 Ki-ras mutations in mouse skin tumors initiated by N-methyl-N′-nitro-N-nitrosoguanidine and promoted by mezerein. Mol Carcinog 27(4):298–307
6. Nasti TH, Rudemiller KJ, Cochran JB, Kim HK, Tsuruta Y, Fineberg NS, Athar M, Elmets CA, Timares L (2015) Immunoprevention of chemical carcinogenesis through early recognition of oncogene mutations. J Immunol 194 (6):2683–2695
7. van de Glind GC, Out-Luiting JJ, Rebel HG, Tensen CP, de Gruijl FR (2016) Lgr5+ stem cells and their progeny in mouse epidermis under regimens of exogenous skin carcinogenesis, and their absence in ensuing skin tumors. Oncotarget 7(32):52085–52094. https://doi.org/10.18632/oncotarget.10475
8. van de Glind GC, Rebel HG, Out-Luiting JJ, Zoutman W, Tensen CP, de Gruijl FR (2016) Lgr6+ stem cells and their progeny in mouse epidermis under regimens of exogenous skin carcinogenesis, and their absence in ensuing skin tumors. Oncotarget 7 (52):86740–86754. https://doi.org/10.18632/oncotarget.13436
9. de Gruijl FR, Tensen CP (2018) Pathogenesis of skin carcinomas and a stem cell as focal origin. Front Med (Lausanne) 5:165. https://doi.org/10.3389/fmed.2018.00165
10. Verma AK, Wheeler DL, Aziz MH, Manoharan H (2006) Protein kinase C epsilon and development of squamous cell carcinoma, the non-melanoma human skin cancer. Mol Carcinog 45(6):381–388. https://doi.org/10.1002/mc.20230
11. Su ZJ, Song JX, Wang ZY, Zhou L, Xia YQ, Yu SB, Sun Q, Liu SS, Zhao L, Li SY, Wei L, Carson DA, Lu DS (2018) Tumor promoter TPA activates Wnt/beta-catenin signaling in a casein kinase 1-dependent manner. Proc Natl Acad Sci U S A 115(32):E7522–E7531. https://doi.org/10.1073/pnas.1802422115
12. DiGiovanni J (1995) Genetic factors controlling responsiveness to skin tumor promotion in mice. Prog Clin Biol Res 391:195–212
13. Aldaz CM, Trono D, Larcher F, Slaga TJ, Conti CJ (1989) Sequential trisomization of chromosome-6 and chromosome-7 in mouse skin premalignant lesions. Cancer Genet Cytogen 41(2):235–235
14. Zenklusen JC, Hodges LC, Conti CJ (1997) Loss of heterozygosity on murine chromosome 6 in two-stage carcinogenesis: evidence for a conserved tumor suppressor gene. Oncogene 14(1):109–114. https://doi.org/10.1038/sj.onc.1200806
15. Ruggeri B, Caamano J, Goodrow T, Dirado M, Bianchi A, Trono D, Conti CJ, Kleinszanto AJP (1991) Alterations of the P53 tumor suppressor gene during mouse skin tumor progression. Cancer Res 51(24):6615–6621
16. Chan KS, Sano S, Kataoka K, Abel E, Carbajal S, Beltran L, Clifford J, Peavey M, Shen J, DiGiovanni J (2008) Forced expression of a constitutively active form of Stat3 in mouse epidermis enhances malignant progression of skin tumors induced by two-stage carcinogenesis. Oncogene 27(8):1087–1094. https://doi.org/10.1038/sj.onc.1210726
17. Navarro P, Gomez M, Pizarro A, Gamallo C, Quintanilla M, Cano A (1991) A role for the E-cadherin cell-cell adhesion molecule during tumor progression of mouse epidermal carcinogenesis. J Cell Biol 115(2):517–533. https://doi.org/10.1083/jcb.115.2.517
18. von Maltzan K, Li Y, Rundhaug JE, Hudson LG, Fischer SM, Kusewitt DF (2016) Role of the slug transcription factor in chemically-induced skin cancer. J Clin Med 5(2). https://doi.org/10.3390/jcm5020021
19. Srivastava J, Rho O, Youssef RM, DiGiovanni J (2016) Twist1 regulates keratinocyte proliferation and skin tumor promotion. Mol Carcinog

55(5):941–952. https://doi.org/10.1002/mc.22335

20. Oft M, Akhurst RJ, Balmain A (2002) Metastasis is driven by sequential elevation of H-ras and Smad2 levels. Nat Cell Biol 4(7):487–494. https://doi.org/10.1038/ncb807
21. Imamoto A, Beltran LM, Fujiki H, Chenicek KJ, DiGiovanni J (1992) Enhanced induction of epidermal ornithine decarboxylase activity in C57BL/6 compared to DBA/2 mice by protein kinase C-activating skin tumor promoters: relevance to genetically mediated differences in promotion susceptibility. Carcinogenesis 13(2):177–182. https://doi.org/10.1093/carcin/13.2.177
22. Schmidt SAJ, Lo S, Hollestein LM (2018) Research techniques made simple: sample size estimation and power calculation. J Invest Dermatol 138(8):1678–1682. https://doi.org/10.1016/j.jid.2018.06.165
23. Filler RB, Roberts SJ, Girardi M (2007) Cutaneous two-stage chemical carcinogenesis. CSH Protoc 2007:pdb.prot4837. https://doi.org/10.1101/pdb.prot4837

Chapter 20

Investigation of Skin Wound Healing Using a Mouse Model

Jianmin Wu and Ning Xu Landén

Abstract

Wound repair is a fundamental physiological process to keep the integrity of the skin, and its dysregulation results in diseases, such as chronic nonhealing wounds or excessive scarring. To study the underlying cellular and molecular mechanisms and identify new therapeutic targets, animal models are often used in the wound healing research. In this chapter, we describe an easy step-by-step protocol to generate skin wounds in a mouse model. Briefly, two full-thickness wounds extending through the panniculus carnosus are made on the dorsum on each side of the midline of a mouse, which is followed by monitoring and quantifying the wound closure. Moreover, the biopsy tissues of skin and wound-edges are collected at different time points for subsequent histology and gene expression analysis.

Key words Skin wound healing, Mouse model, Chronic wound, Histological analysis, Gene expression analysis

1 Introduction

Skin is the outermost layer protecting us from the hostile environment [1]. Once the skin gets injured, the wound-healing cascade initiates right away to recover the barrier. This process has been divided into four stages: hemostasis (0 to several hours after injury), inflammation (1–3 days), proliferation (4–21 days), and remodeling (21 days to 1 year) [2, 3]. Dysregulation of any of these steps results in diseases, such as chronic nonhealing wounds or excessive scarring [1, 4]. Chronic wounds often occur in patients with one or several underlying disorders, such as chronic mechanical pressure, arterial or venous insufficiency, or diabetes mellitus [5]. This presents a significant health and economic burden on our society. Nowadays approximately 1–2% of individuals in developed countries experience a complex hard-to-heal wound [6], which already consumes 2–4% of health care budgets [7–9]. Due to limited understanding of their pathophysiological mechanisms, there is a lack of efficient targeted treatments for chronic wounds [4, 7, 10].

Natalia V. Botchkareva and Gillian E. Westgate (eds.), *Molecular Dermatology: Methods and Protocols*, Methods in Molecular Biology, vol. 2154, https://doi.org/10.1007/978-1-0716-0648-3_20,
© Springer Science+Business Media, LLC, part of Springer Nature 2020

Wound healing is a biological process conserved in all multicellular organisms [11]. To understand the complex cellular and molecular mechanisms of wound healing, animal models are often used to complement findings from in vitro assays [12, 13]. Animal models are also valuable to identify new therapeutics toward the goal of accelerating wound repair [14, 15]. Due to the similarity to human skin, in vivo pig models are often used in preclinical trials of potential treatments. However, limited genetic tools, complicated surgical procedures, and high cost restrict their use [13, 16]. Drosophila and zebrafish models are also used in live imaging or genome-wide screening studies to understand the fundamental tissue repair mechanisms [17, 18]. Nevertheless, to date, the mouse is the most commonly used animal model in wound healing research. Mice can be easily maintained and bred in an animal facility, and it is also feasible to genetically manipulate mice for mechanistic studies [14, 19, 20]. Utilizing genetically modified transgenic or knockout mice models; hundreds of protein-coding genes important for skin wound healing have been identified so far [21, 22]. Recently, non-protein-coding RNAs (ncRNAs) including microRNAs and long noncoding RNAs (lncRNAs) have been shown to be important regulators of skin wound healing [23–27]. For example, microRNA-132 facilitates the transition from the inflammatory to the proliferative phase during skin wound healing [27]. The wound and keratinocyte migration-associated lncRNA 1 (WAKMAR1) regulates wound healing by enhancing keratinocyte migration [25]. Accumulating evidence also reveals increasing numbers of ncRNAs involved in the pathogenesis of chronic nonhealing wounds, suggesting that they may serve as potential therapeutic targets [23–26, 28, 29]. To this end, the mouse model will be valuable for evaluating the in vivo function and therapeutic potential of targeting these novel regulators of wound healing. In this chapter, we describe an easy step-by-step protocol to generate a mouse wound model.

2 Materials

1. C57BL/6N (9–10-week-old) male mice.
2. Isoflurane.
3. Hair removal (depilatory) cream.
4. Ultrapure water.
5. 70% ethanol.
6. Phosphate buffered saline (PBS) (pH 7.4).
7. Chlorhexidine Fresenius Kabi.
8. Formaldehyde: 10% formaldehyde in ultrapure water.

9. Paraffin.
10. TRIzol.
11. Chloroform.
12. Isopropanol.
13. Diethyl pyrocarbonate (DEPC)-treated water.
14. Dry ice.
15. Biosafety hoods.
16. Small animal anesthesia system.
17. Electric clippers.
18. 4 and 8 mm diameter circular biopsy punch.
19. Tissue scissors and forceps (surgical grades).
20. High-resolution digital camera.
21. ImageJ software.
22. Microscope.
23. TissueLyser LT (Qiagen).
24. 1.5 and 2 ml DNase- and RNase-free tubes.

3 Methods

3.1 Generation of Skin Wounds in Mice (See Note 1)

1. C57BL/6N (9–10-week-old) male mice (*see* **Note 2**) are caged individually for 1 week with free access to food and water, handled daily before surgical procedure (*see* **Note 3**).
2. Perform general anesthesia of the mice with 3% isoflurane for at least 5 min using small animal anesthesia system (Fig. 1a).
3. After anesthetizing the mice, place the mouse in the prone position.
4. Remove hair from the dorsal side of the mice with an electric clipper (Fig. 1b).
5. Hair removal (depilatory) cream is smeared using a cotton swab. Within 2 min, gauze swabs must be used to remove all cream and remaining fur (Fig. 1c) (*see* **Note 4**).
6. General anesthesia is continued as described above to ensure that the mice do not wake up during the following surgical procedure.
7. Wipe the skin with 70% ethanol. After drying, rinse the skin with PBS.
8. Make two full-thickness wounds extending through the panniculus carnosus on the dorsum on each side of the midline using a 4-mm biopsy punch. Briefly, hold the skin from the mid-dorsal line, push the sterile biopsy punch into the skin

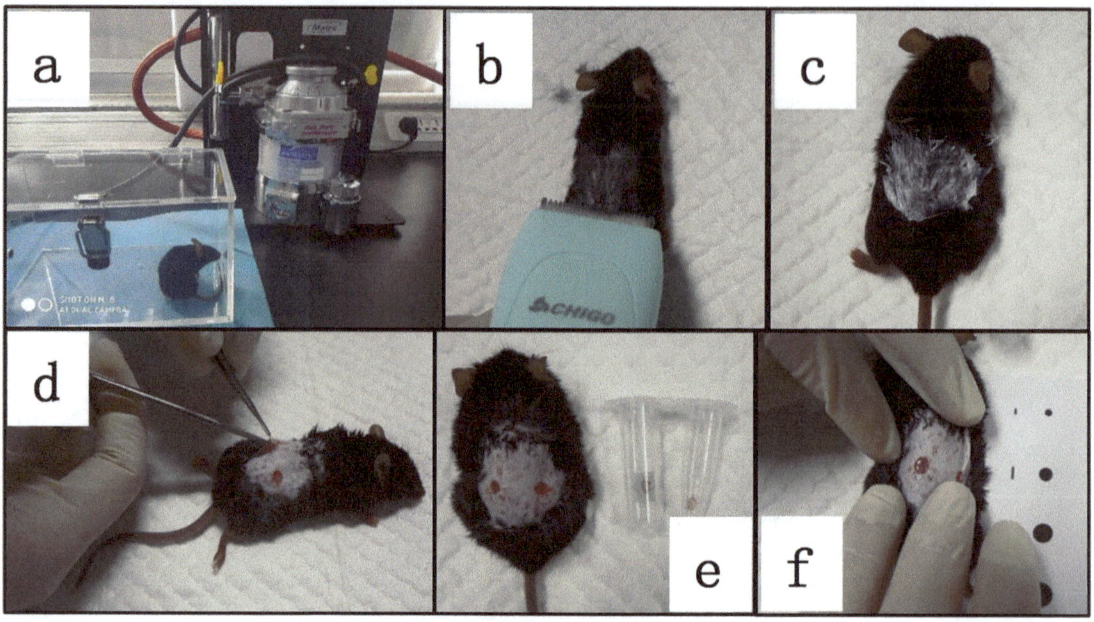

Fig. 1 Surgery procedure of generation of skin wounds in mice. (**a**) General anesthesia is performed in mice. (**b**) Hair from the dorsal side of the mice is shaved with an electric clipper. (**c**) Hair removal (depilatory) cream is applied to remove the remaining fur. (**d**) Wound is generated with a 4 mm-biopsy punch. The skin is lifted with forceps and removed with scissors. (**e**) The excised skin is saved as a baseline control in the subsequent analysis. (**f**) The wound alongside with a circular reference is photographed using a digital camera

with a twisting motion left and right on each side of midline of the mouse, use the surgical forceps by one hand to lift the skin in the middle of punch and the surgical scissors by the other hand to create a full-thickness wound to excise the circle piece of the skin (Fig. 1d). Save the excised skin as a baseline control (Fig. 1e).

9. Repeat the procedure described above to make a wound on the other side of the midline.
10. After the surgery, subcutaneously inject buprenorphine (0.03 mg/kg) for relief of any possible distress caused by the surgical procedure.
11. Cage mice individually.

3.2 Monitoring of the Wound Healing Process (See Note 5)

1. Anesthetize mice as described in **step 3** of Subheading 3.1.
2. Wipe wounds with Chlorhexidine Fresenius Kabi for sterilization, and then carefully remove the scab with sterile forceps just before the photograph (*see* **Note 6**).
3. Prepare a circular reference and place it alongside the wound (*see* **Note 7**).
4. Take photographs of wound using a digital camera every other day until full closure of the wound.

5. Mice are single-caged until the next time of the photograph.
6. Calculate the wound area in pixels with ImageJ software (National Institutes of Health), corrected for the area of the reference circle and expressed as a percentage of the original area.
7. Wound closure is quantified and presented as % wound closure = 100% minus the percentage of the initial wound area size (*see* **Note 8**).

3.3 Histological and Gene Expression Analysis

As an alternative to following the wound healing process macroscopically, one could collect tissue samples before wounds are fully closed (*see* **Note 9**).

1. Euthanize the mice with an overdose of anesthesia at the specified time points after injury and collect wound edges for analysis.
2. Use an 8-mm biopsy punch to excise the wound completely along with surrounding wound margin skin.
3. Put flat the piece of skin (keep the epidermal side up) on a piece of Whatman filter paper and split into two halves using surgical scissors. One half is immediately frozen in dry ice for later RNA isolation, and the other is used for fixation and paraffin embedding for histological analysis (*see* **Note 10**).
4. For gene expression analysis, the wound and skin samples are disrupted and homogenized by TissueLyser LT, then RNA is isolated using TRIzol reagent. Gene expression could be analyzed by qRT-PCR or profiling techniques such as microarray or RNA-sequencing (*see* **Note 11**).
5. For histology analysis, the conventional fixation method using 10% formaldehyde solution for 24–48 h is recommended, and tissues should be fixed in freshly made fixative. Tissue is processed into paraffin wax and sectioned at 6 μm in thickness across the midline of the wound. Paraffin-embedded tissue sections are deparaffinized and rehydrated followed by hematoxylin and eosin (H&E) staining as well as immunohistochemical or immunofluorescence staining with specific antibodies, as appropriate (*see* **Note 12**).
6. After H&E staining, the length of the newly formed epithelial tongue in mice wounds is quantified using the software ImageJ (Fig. 2).

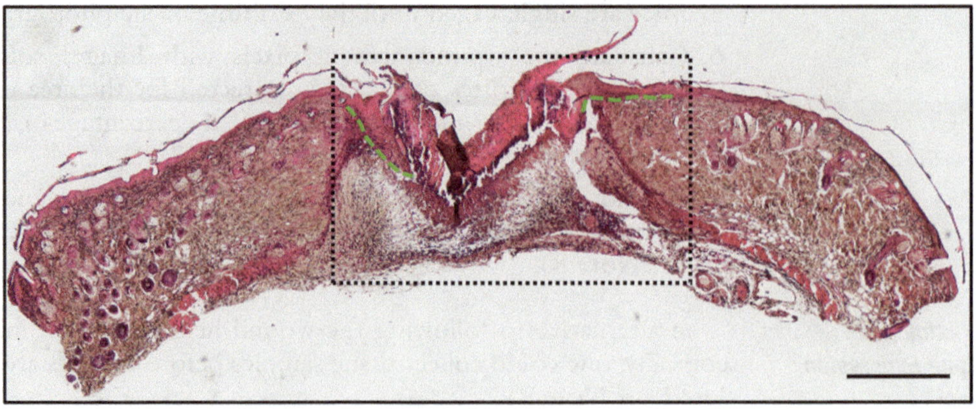

Fig. 2 Histomorphology of mouse wound, 6 days postwounding, H&E (hematoxylin and eosin) staining. Dashed lines mark the newly formed epithelial tongue. Scale bar: 500 μm

4 Notes

1. All procedures involving mice require approval by the Institutional Ethics Committee and an appropriate animal license for the principle investigator.
2. It has been known that the hair cycle can influence the wound healing dynamics. Here we use 9–10-week-old male mice, in which time most hair follicles are in the telogen phase [29, 30].
3. This wound healing model can be performed on wild-type mice or mice with impaired healing ability (e.g., leptin receptor-deficient mice (diabetes model), which imitate some clinically relevant key aspects of chronic wounds). For diabetic mice, one should measure blood sugar and weight before grouping them [29].
4. Depilatory cream dissolves hair at the surface of the skin. Leaving the cream on the skin for a too long time or any residual cream on the skin may cause a chemical burn, affecting the subsequent wound healing.
5. The wound can be treated with the studied drug or compound at the specified time points after injury via appropriate routes, such as intradermal injection or topical application.
6. When analysing wound closure macroscopically, we remove scabs right before every photograph to expose the wound edge. However, if the wound tissue will be used for histological analysis, the scab-removing procedure will be omitted, as it will disturb the newly formed wound tongue, which can be identified using histology under a microscope.
7. A circular reference is placed alongside the wound for quantitative purpose. The wound area calculated with ImageJ must be normalized to the area of the reference circle in the same

photograph. Therefore, there is no need to fix the camera with same distance to the wound during the experimental progress.

8. Alternatively, wound closure may be presented as the percentage of the initial wound area size.
9. One could perform the first round of the experiment to follow the wound healing process macroscopically. In the second round of experiment, tissue samples could be collected at the time points that the group with treatment is different from the control group.
10. If any systemic effect for the local wound treatments will be studied, internal organs of mice (e.g., the liver, lung, kidney, and spleen) can be collected for morphology and histology analysis [29].
11. Gene expression analysis can be performed on wound samples collected at the different time points after wounding, to unravel its dynamic change during the wound healing process.
12. Marker proteins for cell proliferation (e.g., Ki-67 and PCNA), keratinocyte differentiation (e.g., KRT10 and involucrin), angiogenesis (e.g., CD31), and immune cell infiltration (e.g., Gr-1 for neutrophils, CD68 for macrophages) can be detected by immunohistochemistry or immunofluorescence staining. Granulation tissue formation can be evaluated by Masson's trichrome staining.

Acknowledgments

We would like to thank Dongqing Li for critical reading of the manuscript. This work was supported by National Natural Sciences Foundation of China (grant NSFC-81472651 to J.W.).

References

1. Landen NX, Li D, Stahle M (2016) Transition from inflammation to proliferation: a critical step during wound healing. Cell Mol Life Sci 73(20):3861–3885. https://doi.org/10.1007/s00018-016-2268-0
2. Reinke JM, Sorg H (2012) Wound repair and regeneration. Eur Surg Res 49(1):35–43. https://doi.org/10.1159/000339613
3. Gurtner GC, Werner S, Barrandon Y, Longaker MT (2008) Wound repair and regeneration. Nature 453(7193):314–321. https://doi.org/10.1038/nature07039
4. Sen CK, Gordillo GM, Roy S, Kirsner R, Lambert L, Hunt TK, Gottrup F, Gurtner GC, Longaker MT (2009) Human skin wounds: a major and snowballing threat to public health and the economy. Wound Repair Regen 17(6):763–771. https://doi.org/10.1111/j.1524-475X.2009.00543.x
5. Medina A, Scott PG, Ghahary A, Tredget EE (2005) Pathophysiology of chronic nonhealing wounds. J Burn Care Rehabil 26(4):306–319. https://doi.org/10.1097/01.Bcr.0000169887.04973.3a
6. Jarbrink K, Ni G, Sonnergren H, Schmidtchen A, Pang C, Bajpai R, Car J (2016) Prevalence and incidence of chronic wounds and related complications: a protocol

for a systematic review. Syst Rev 5(1):152. https://doi.org/10.1186/s13643-016-0329-y

7. Frykberg RG, Banks J (2015) Challenges in the treatment of chronic wounds. Adv Wound Care 4(9):560–582. https://doi.org/10.1089/wound.2015.0635
8. Gottrup F, Holstein P, Jorgensen B, Lohmann M, Karlsmar T (2001) A new concept of a multidisciplinary wound healing center and a national expert function of wound healing. Arch Surg 136(7):765–772
9. Richmond NA, Maderal AD, Vivas AC (2013) Evidence-based management of common chronic lower extremity ulcers. Dermatol Ther 26(3):187–196. https://doi.org/10.1111/dth.12051
10. Mustoe TA, O'Shaughnessy K, Kloeters O (2006) Chronic wound pathogenesis and current treatment strategies: a unifying hypothesis. Plast Reconstr Surg 117(7 Suppl):35S–41S. https://doi.org/10.1097/01.prs.0000225431.63010.1b
11. Eming SA, Martin P, Tomic-Canic M (2014) Wound repair and regeneration: mechanisms, signaling, and translation. Sci Transl Med 6 (265):265sr266. https://doi.org/10.1126/scitranslmed.3009337
12. Pereira RF, Bartolo PJ (2016) Traditional therapies for skin wound healing. Adv Wound Care 5(5):208–229. https://doi.org/10.1089/wound.2013.0506
13. Grada A, Mervis J, Falanga V (2018) Research techniques made simple: animal models of wound healing. J Invest Dermatol 138 (10):2095–2105.e2091. https://doi.org/10.1016/j.jid.2018.08.005
14. Sami DG, Heiba HH, Abdellatif A (2019) Wound healing models: a systematic review of animal and non-animal models. Wound Med 24(1):8–17. https://doi.org/10.1016/j.wndm.2018.12.001
15. Pastar I, Liang L, Sawaya AP, Wikramanayake TC, Glinos GD, Drakulich S, Chen V, Stojadinovic O, Davis SC, Tomic-Canic M (2018) Preclinical models for wound-healing studies. In: Marques AP, Pirraco RP, Cerqueira MT, Reis RL (eds) Skin tissue models. Academic Press, Boston, pp 223–253. https://doi.org/10.1016/B978-0-12-810545-0.00010-3
16. Lindblad WJ (2008) Considerations for selecting the correct animal model for dermal wound-healing studies. J Biomater Sci Polym Ed 19(8):1087–1096. https://doi.org/10.1163/156856208784909390
17. Razzell W, Wood W, Martin P (2011) Swatting flies: modelling wound healing and inflammation in Drosophila. Dis Model Mech 4 (5):569–574. https://doi.org/10.1242/dmm.006825
18. Henry KM, Loynes CA, Whyte MKB, Renshaw SA (2013) Zebrafish as a model for the study of neutrophil biology. J Leukoc Biol 94 (4):633–642. https://doi.org/10.1189/jlb.1112594
19. Elliot S, Wikramanayake TC, Jozic I, Tomic-Canic M (2018) A modeling conundrum: murine models for cutaneous wound healing. J Invest Dermatol 138(4):736–740. https://doi.org/10.1016/j.jid.2017.12.001
20. Wong VW, Sorkin M, Glotzbach JP, Longaker MT, Gurtner GC (2011) Surgical approaches to create murine models of human wound healing. J Biomed Biotechnol 2011:969618. https://doi.org/10.1155/2011/969618
21. Grose R, Werner S (2004) Wound-healing studies in transgenic and knockout mice. Mol Biotechnol 28(2):147–166. https://doi.org/10.1385/MB:28:2:147
22. Fang RC, Mustoe TA (2008) Animal models of wound healing: utility in transgenic mice. J Biomater Sci Polym Ed 19(8):989–1005. https://doi.org/10.1163/156856208784909327
23. Herter EK, Xu Landen N (2017) Non-coding RNAs: new players in skin wound healing. Adv Wound Care 6(3):93–107. https://doi.org/10.1089/wound.2016.0711
24. Li D, Landen NX (2017) MicroRNAs in skin wound healing. Eur J Dermatol 27(S1):12–14. https://doi.org/10.1684/ejd.2017.3040
25. Li D, Kular L, Vij M, Herter EK, Li X, Wang A, Chu T, Toma MA, Zhang L, Liapi E, Mota A, Blomqvist L, Gallais Serezal I, Rollman O, Wikstrom JD, Bienko M, Berglund D, Stahle M, Sommar P, Jagodic M, Landen NX (2019) Human skin long noncoding RNA WAKMAR1 regulates wound healing by enhancing keratinocyte migration. Proc Natl Acad Sci U S A 116(19):9443–9452. https://doi.org/10.1073/pnas.1814097116
26. Herter EK, Li D, Toma MA, Vij M, Li X, Visscher D, Wang A, Chu T, Sommar P, Blomqvist L, Berglund D, Stahle M, Wikstrom JD, Xu Landen N (2019) WAKMAR2, a long noncoding RNA downregulated in human chronic wounds, modulates keratinocyte motility and production of inflammatory chemokines. J Invest Dermatol 139 (6):1373–1384. https://doi.org/10.1016/j.jid.2018.11.033

27. Li D, Wang A, Liu X, Meisgen F, Grunler J, Botusan IR, Narayanan S, Erikci E, Li X, Blomqvist L, Du L, Pivarcsi A, Sonkoly E, Chowdhury K, Catrina SB, Stahle M, Landen NX (2015) MicroRNA-132 enhances transition from inflammation to proliferation during wound healing. J Clin Invest 125 (8):3008–3026. https://doi.org/10.1172/JCI79052
28. Wu J, Li X, Li D, Ren X, Li Y, Herter EK, Qian M, Toma MA, Wintler AM, Serezal IG, Rollman O, Stahle M, Wikstrom JD, Ye X, Landen NX (2019) MicroRNA-34 family enhances wound inflammation by targeting LGR4. J Invest Dermatol 140:465. https://doi.org/10.1016/j.jid.2019.07.694
29. Li X, Li D, Wang A, Chu T, Lohcharoenkal W, Zheng X, Grunler J, Narayanan S, Eliasson S, Herter EK, Wang Y, Ma Y, Ehrstrom M, Eidsmo L, Kasper M, Pivarcsi A, Sonkoly E, Catrina SB, Stahle M, Xu Landen N (2017) MicroRNA-132 with therapeutic potential in chronic wounds. J Invest Dermatol 137 (12):2630–2638. https://doi.org/10.1016/j.jid.2017.08.003
30. Muller-Rover S, Handjiski B, van der Veen C, Eichmuller S, Foitzik K, McKay IA, Stenn KS, Paus R (2001) A comprehensive guide for the accurate classification of murine hair follicles in distinct hair cycle stages. J Invest Dermatol 117 (1):3–15. https://doi.org/10.1046/j.0022-202x.2001.01377.x

Chapter 21

Human Wound Healing Ex Vivo Model with Focus on Molecular Markers

Jennifer Gherardini, Max van Lessen, Ilaria Piccini, Janin Edelkamp, and Marta Bertolini

Abstract

Wound healing is a complex, multifactorial process that is divided in sequential and overlapping phases in order to restore the skin barrier. For the study of wound healing, different in vivo, in vitro, and ex vivo models have been used in the past. Here we describe in detail the methodology of the human skin punch-in-a-punch ex vivo wound healing model.

Key words Wound healing, Human skin, Ex vivo

1 Introduction

The healing of wounds is a complex, multifactorial process that remains incompletely understood and is divided into four phases: (1) hemostasis, (2) inflammation, (3) tissue formation, and (4) tissue remodeling [1]. While in the ex vivo wound healing model the hemostasis is hardly appreciable, because of lack of blood perfusion, the different other processes can be observed [2–4].

The ex vivo wound healing model described below was adopted and further improved in our laboratories in order to develop an experimental model, which can be used for testing wound healing promoting compounds [2, 4–6].

The recruitment and activation of immune cells, namely, macrophages and mast cells, to the site of the wound can be visualized with CD14/CD68 or c-kit staining, respectively [7].

Moreover, several markers have been established for studying reepithelialization and extracellular matrix remodeling, namely, Ki-67/TUNEL, cytokeratin 6 (K6) [8], collagen I/III ratio, and MMP-2 [9, 10]. These parameters can be assessed over the course

Natalia V. Botchkareva and Gillian E. Westgate (eds.), *Molecular Dermatology: Methods and Protocols*, Methods in Molecular Biology, vol. 2154, https://doi.org/10.1007/978-1-0716-0648-3_21,
© Springer Science+Business Media, LLC, part of Springer Nature 2020

of wound healing as well as reepithelization by assessment at different time points using immunofluorescent microscopy [2].

Although the ex vivo wound healing model yielded many insights into physiological wound healing, there is currently no model to analyze pathological (chronic) wound healing, which is typically observed in type 2 diabetes (T2D), nonhealing diabetic wounds [11, 12], and smokers [13].

Considering the superiority of the human ex vivo wound healing over in vitro models, the common use is often hindered by an insufficient supply of fresh the skin and, depending on the country, rigorous ethical guidelines.

2 Materials

Prepare all solutions under sterile conditions. Store all reagents according to manufacturer's instructions. Diligently follow all waste disposal regulations, especially with regard to human tissue left overs.

2.1 Instruments and Tools

- Sterile gauze pads.
- Scissors and forceps.
- Biopsy punches (2 and 6 mm).
- 100 mm cell culture dish.
- 12-Well plate.
- Pasteur pipette.
- Disposable scalpel.
- Surgical gloves.

2.2 William's E Culture Medium

Prepare 100 ml of WCM in a sterile glass bottle by adding, 2 mM L-glutamine, 10 ng/ml hydrocortisone, 10 μg/ml insulin, and 1% penicillin–streptomycin mix to 100 ml of William's E medium (WEM) at room temperature (RT) under sterile conditions. Store the medium at +4 °C and reequilibrate the medium at RT before using it (*see* **Note 1**).

3 Methods

3.1 Ex Vivo Wounding

The tissue preparation can be done in a nonsterile environment; however, the culture medium used must be prepared under the cell culture hood.

For this assay it is possible to use skin derived from different body regions (scalp skin, abdominal skin, inner thigh skin, and inner arm skin).

1. Prepare the culture media using red William's E medium supplemented with 2 mM L-glutamine, 10 ng/ml hydrocortisone, 10 μg/ml insulin, and 1% penicillin–streptomycin.
2. Take the 12-well plate under the hood and place sterile gauze pads (precut in order to fit into each well). Add 2 ml of medium into each well.
3. Place the skin in a 100 mm petri dish with fresh medium.
4. Transfer the skin onto the lid of a 10 mm petri dish (without medium) with the epidermis facing down. With large curved forceps hold the skin firm and using a scalpel, remove the underlying fat tissue until the dermis (*see* **Note 2**).
5. Flip the skin, with the epidermis facing up, moist the skin by dipping it into the medium.
6. Touch the skin surface with a 2 mm biopsy punch, and gently punch into skin, penetrating the epidermis and the upper part of the dermis by clock-wise rotation of the biopsy punch in order to create a wound (*see* **Note 3**).
7. Use the forceps to pull up the 2 mm skin punch and excise at the dermis level using scissors in order to create a central wound (*see* **Note 4**).
8. Incise the skin around the 2 mm wound with a 6 mm biopsy punch (the 2 mm wound should be in the center of the 6 mm punch) to create a punch-in-a-punch.
9. Using forceps, carefully place the punch-in-a-punch (avoid touching the inner wound punch) on top of the gauze pad (the well has previously been filled with medium) (Fig. 1).
10. Ensure that the epidermis of the punch-in-a-punch is kept at the air–liquid interface. Control that no air pockets are present between the skin punch and the gauze.
11. Incubate for 24 h at 37 °C and 95% air–5% CO_2.
12. Replace the medium and topically apply a candidate test substance (*see* **Note 5**).
13. Replace medium and treatment every 48 h up to 7 days culture (*see* **Note 6**) (Table 1).

The experiment can be terminated at multiple time points (*see* **Note 6**) for the analysis of immune cell activation (*see* **Note 7**), reepithelialization (*see* **Note 8**), and extracellular matrix remodeling (*see* **Note 9**).

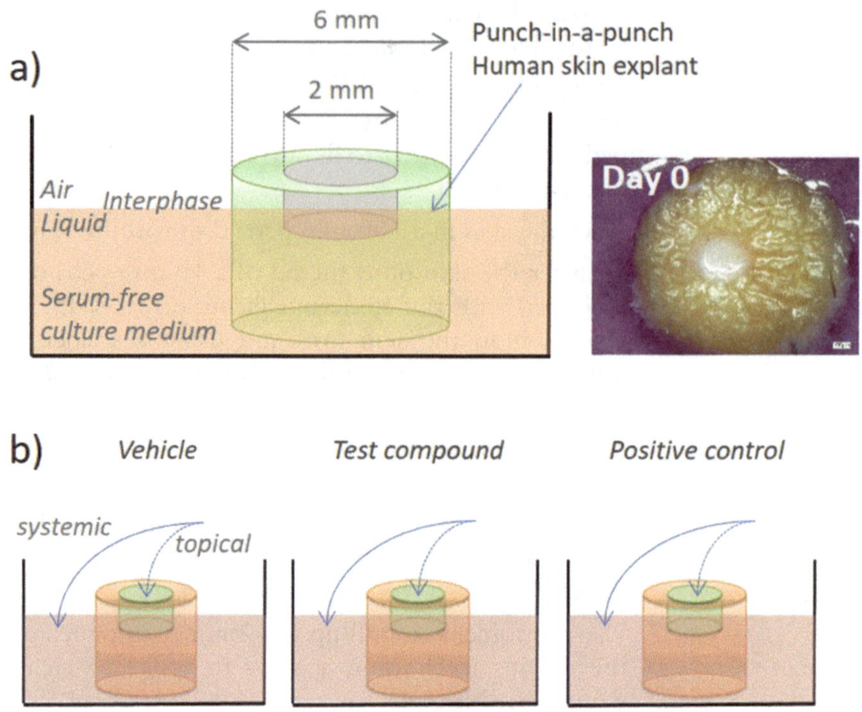

Fig. 1 Human skin punch-in-a-punch ex vivo wound healing model—substance application. (**a**) Schematic illustration of the punch-in-a-punch ex vivo wound healing model (left). Firstly, a 2 mm inner biopsy punch is generated by gently punching into skin, penetrating the epidermis and the upper part of the dermis by clockwise rotation of the biopsy punch in order to create a wound. Forceps are used to pull up the skin punch and excise at the dermis level using scissors to create a central wound. Subsequently, the outer punch is prepared, using a 6 mm biopsy punch around the wound. (**b**) Different options of substance application. Test compounds can be applied either systemically, within the culture medium, or topically within the wound, or on the surrounding epidermal surface. In order to study wound healing promoting effects, it should be considered to include a positive control in the experimental setup, along with a vehicle control of the test compound. (Scheme modified from [15])

4 Notes

1. William's E culture medium can be reused if stored at +4 °C. It is important to carefully check the color of the medium. It can be reused as long as the color remains red.
2. During this step keep the skin moist by adding some drops of medium with a Pasteur pipette.
3. When performing the 2 mm punch incision, do not insert the biopsy punch too deep in the dermis. It is important to incise only the upper part of the dermis. Moreover, take care that wounds of different punch samples are of the same depth.
4. Pull on the skin punch gently. Moreover, while you incise the skin punch using scissors, keep the scissor blades parallel to the epidermis.

Table 1
Experimental procedure of the human skin punch-in-a-punch ex vivo wound healing model

Day 0	Day 1	Day 2	Day 3	Day 4	Day 5	Day 6/Day 7
Prepare a 2 mm inner punch surrounded by a 6 mm outer biopsy punch. Place the skin samples at air–liquid interface (on a gauze pad) within 2 ml/well of WCM in a 12-well plate. Incubate at 37 °C and 95% air–5% CO_2	Assign the skin punches to the treatment groups. Change the culture medium. Apply the vehicle or test substance to the corresponding experimental group. Incubate at 37 °C and 95% air–5% CO_2	Rest	Change the culture medium. Apply the vehicle or test substance to the corresponding experimental group. Incubate at 37 °C and 95% air–5% CO_2	Rest	Change the culture medium. Apply the vehicle or test substance to the corresponding experimental group. Incubate at 37 °C and 95% air–5% CO_2	Terminate the culture

5. In order to study, for example, wound healing–promoting effects ex vivo, it is advisable to include a positive control. Among others, thyrotropin-releasing hormone (TRH), T4, and estrogens have been shown to promote wound healing ex vivo [4, 8, 14].
6. The skin can be kept in culture up to 10 days; however, we recommend stopping the experiment after 1, 3, and 6 or 7 days, in order to able to see the partial wound closure but to avoid off target effects done by the degradation of the skin. At the end of the culture, the skin punches should be processed for analysis. The skin can be immediately snap-frozen in liquid nitrogen, stored in RNA later or lysis buffer and snap frozen in liquid nitrogen, placed into OCT and snap frozen in liquid nitrogen, or fixed in 4% paraformaldehyde.
7. For the analysis of immune cell activation and their phenotype in situ [2], the skin should be placed into OCT and snap-frozen in liquid nitrogen, and frozen sections of 7 μm should be cut. Sections should be collected once the middle of the punch is reached (wound region). The following histochemical- and immune-staining should be performed: c-Kit and toluidine blue to identify the total number or degranulation–maturation–activation of mast cells, CD14/CD68 markers for monocytes or macrophages, TGF, FGF, PDGF, TNF alpha, and IL1b inflammatory cytokines.

8. For the analysis of epidermal reepithelialization, the following histochemical staining and immunostaining should be performed: Periodic Acid Schiff (PAS) and H&E for skin morphology, Ki-67/TUNEL or Ki-67/caspase 3 for proliferation and apoptosis of epidermal keratinocytes, keratins K6-K16-K17 as markers for keratinocyte migration, K5 and K14 for identifying newly formed basal layer keratinocytes, and K15/Ki-67 for epidermal stem cell proliferation.
9. The extracellular matrix remodeling can be investigated by the analysis of the following markers: collagen I/III ratio, MMP-2 and MMP-9 expression/activity, fibronectin, and TGF beta.

References

1. Lindley LE, Stojadinovic O, Pastar I et al (2016) Biology and biomarkers for wound healing. Plast Reconstr Surg 138:18S–28S
2. Nasir NAM, Paus R, Ansell DM (2019) Fluorescent cell tracer dye permits real-time assessment of re-epithelialization in a serum-free ex vivo human skin wound assay. Wound Repair Regen 27:126–133
3. Ud-Din S, Bayat A (2017) Non-animal models of wound healing in cutaneous repair: in silico, in vitro, ex vivo, and in vivo models of wounds and scars in human skin. Wound Repair Regen 25:164–176
4. Zhang GY, Langan EA, Meier NT et al (2019) Thyroxine (T4) may promote re-epithelialisation and angiogenesis in wounded human skin ex vivo. PLoS One 14: e0212659
5. Stone R, Wall JT, Natesan S (2018) PEG-plasma hydrogels increase epithelialization using a human ex vivo skin model. Int J Mol Sci 19:3156
6. Liao T, Lehmann J, Sternstein S et al (2019) Nestin+ progenitor cells isolated from adult human sweat gland stroma promote reepithelialisation and may stimulate angiogenesis in wounded human skin ex vivo. Arch Dermatol Res 311:325–330
7. Krzyszczyk P, Schloss R, Palmer A et al (2018) The role of macrophages in acute and chronic wound healing and interventions to promote pro-wound healing phenotypes. Front Physiol 9:419. https://doi.org/10.3389/fphys.2018.00419
8. Meier NT, Haslam IS, Pattwell DM (2013) Thyrotropin-releasing hormone (TRH) promotes wound re-epithelialisation in frog and human skin. PLoS One 8(9):e73596
9. Krishnaswamy VR, Mintz D, Sagi I (2017) Matrix metalloproteinases: the sculptors of chronic cutaneous wounds. Biochim Biophys Acta 1864:2220–2227
10. Robins SP, Milne G, Duncan A et al (2003) Increased skin collagen extractability and proportions of collagen type III are not normalized after 6 months healing of human excisional wounds. J Invest Dermatol 121:67–272
11. Baltzis D, Eleftheriadou I, Veves A (2014) Pathogenesis and treatment of impaired wound healing in diabetes mellitus: new insights. Adv Ther 31:817–836
12. Boniakowski AE, Kimball AS, Jacobs BN et al (2017) Macrophage-mediated inflammation in normal and diabetic wound healing. J Immunol 199:17–24
13. Lassig AAD, Bechtold JE, Lindgren BR et al (2018) Tobacco exposure and wound healing in head and neck surgical wounds. Laryngoscope 128:618–625
14. Campbell L, Emmerson E, Davies F et al (2010) Estrogen promotes cutaneous wound healing via estrogen receptor β independent of its antiinflammatory activities. J Exp Med 207:1825–1833
15. Chéret J, Lebonvallet N, Buhé V et al (2014) Influence of sensory neuropeptides on human cutaneous wound healing process. J Dermatol Sci 74:193–203

Chapter 22

Method for Investigation of Photobiological Effects of Light on Human Skin Cells Mediated by Low Doses of Light

Charles Mignon and Natallia E. Uzunbajakava

Abstract

Driven by evolution, human skin cells have developed an extraordinary ability both to sense and to respond to the photons of sunlight through a plethora of photobiological interactions, activating intracellular signalling cascades and regulating skin cells homeostasis. It has recently been reported that some of these photobiological responses triggered by low levels of light (or the so-called photobiomodulation) could initiate beneficial therapeutic effects. Identification of these effective light-based therapeutic solutions requires in-depth understanding of the parameter space. The physical, biological, and chemical conditions that need to be fulfilled to facilitate such positive photobiological effects are to be carefully deciphered. Here, we provide the protocols that were specifically developed to investigate multidimensional parameter space driving photobiological interactions triggered by light (photobiomodulation) in the skin cells. The approach is based on the so-called design of experiment (DoE), a statistical method, which allows for the investigation of multidimensional parameters landscapes. This goes hand in hand with sharing practical tips for the design of light-based devices inducing these effects. To exemplify practical applications of the developed methods and light-based devices, we disclose experimental data sets and emphasize robustness and reproducibility of the results.

Key words Photobiomodulation, Photobiology of human cells, Human skin cells, Light-based devices, Photobiology, Photodermatology

1 Introduction

Skin conditions and disorders were ranked to be the fourth leading cause of nonfatal burden worldwide [1]. As effective and efficient treatment is still often lacking, they exert a significant impact on the daily lives of patients [2–4]. Application of the low levels of light in the order of 1–100 mW/cm^2 emitted throughout visible to near-infrared spectrum is emerging as a novel, noninvasive, and inherently safer therapy when recalling potential side effects of topical and systemic drugs [5]. The foundation of this new low-level light-based therapeutic modality is rooted in the fact that human skin cells sense and respond to the photons of light throughout a

Natalia V. Botchkareva and Gillian E. Westgate (eds.), *Molecular Dermatology: Methods and Protocols*, Methods in Molecular Biology, vol. 2154, https://doi.org/10.1007/978-1-0716-0648-3_22,
© Springer Science+Business Media, LLC, part of Springer Nature 2020

plethora of photobiological interactions, where energy of light directly activates intracellular signalling cascades, and as a result triggers therapeutic effects. This has been suggested by a growing number of clinical and basic research studies in the field of Photobiomodulation (PBM) [4, 6, 7].

Sadly, despite of steadily progressing translation of PBM to commercial applications, aiming to address needs of patients, its clinical efficacy remains far from satisfactory [4, 6]. The major reason is a lack of understanding of the basic mechanisms governing PBM effects, which in our opinion, is caused by lack of a rationale for the selection of treatment parameters and means to measure them. As a result, an ill-designed and/or characterized light-based device will easily lead to a lack of reproducibility and overall confusion about experimental outcomes. Also, well-defined protocols and overarching systematic approach to the studies are lacking [8].

As a result, existing scientific literature on in vitro and in vivo studies using PBM is flooded with numerous inconsistencies between varying experimental outcomes linked to one predefined treatment parameter [4]. A very recent review from 2019 still reports a lack of rigorousness in optical treatment parameters in PBM studies [9].

Thus, major efforts are still required to develop rigorous and robust yet pragmatic and practical methods to systematically evaluate the impact of low doses of visible and NIR light on biological samples. To assure quality of experimental results such methods should provide a tight control of the parameters of a light-emitting device and of a target biological sample. This includes temperature, culture conditions, cell lineage and more. In this chapter, we will provide practical tips for design and manufacturing of light-based prototypes for treatment of skin cell culture in vitro, essential protocols and more importantly, we disclose the logic behind the selected approach for the design of experiments and data analysis. We will exemplify our guidelines by giving the detailed description of two LED-based devices, which were specifically developed by us and used in our experimental work for treatment of human skin cells in either 24-well plates or individual 35-mm dishes. Images of both devices are shown in Fig. 1.

2 Materials

2.1 Biomaterials

1. Skin cell culture (e.g., primary human epidermal keratinocytes or dermal fibroblasts).
2. Culture medium required for the selected primary cell line.
3. Phosphate-Buffered Saline (PBS).

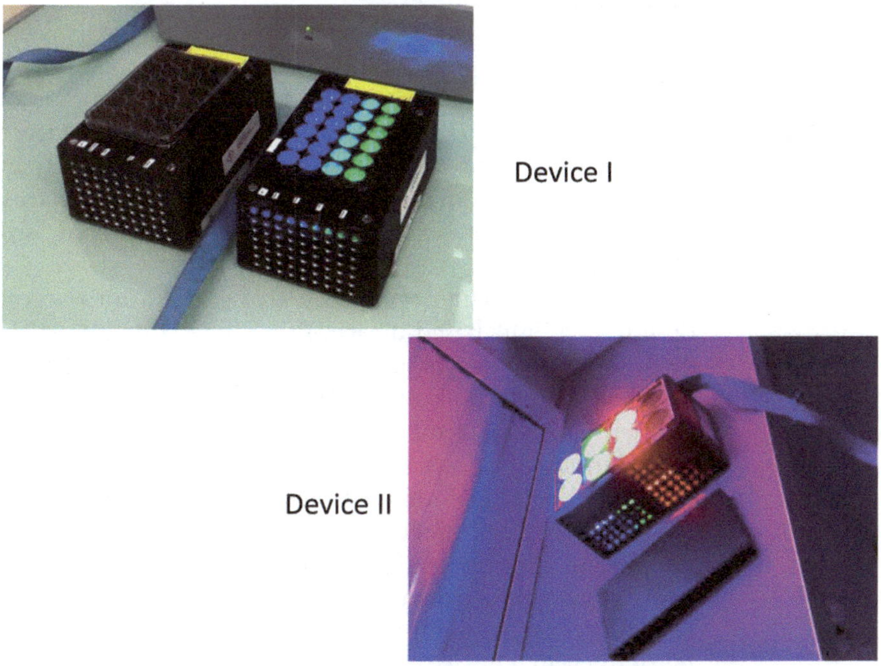

Fig. 1 Photographs of two LED-based device examples: Device I and Device II

4. Alamar Blue Assay (or equivalent assay to test metabolic activity or any other generic assay of interest easily implemented in a large scale experiment involving many, 6 or more well plates with cell culture).
5. Incubator with fixed CO_2 level and adjustable oxygen concentration (the latter condition is optional).

2.2 Mechanical and Optical Components

1. Commercially available light-based devices suitable for work with cell culture.

   [or]

2. Equivalent lab prototypes, which are built from components such as light-emitting diodes (LED) for light generation, mechanical fans and heat sink for cooling of LEDs, electrical current supply to power LEDs and cooling devices, and LED-shaping optics to produce a light beam of desired area and angular directivity.

2.3 Metrology Equipment

1. Thermal camera (e.g., type FLIR camera) to remotely assess temperature of a sample.
2. Optical power meter to measure power of a light beam.
3. Camera (any type from a phone camera to a professional camera) to measure beam homogeneity.

3 Methods

3.1 Design of Light-Based Devices

In many in vitro and in vivo studies on Photobiomodulation, researchers have used customized light-based devices to treat biological samples. These devices include lasers, LEDs and lamps, with or without complementary optical components to shape the optical beam, for example, to create a desired size of an illumination area.

Several key parameters need to be taken into account when designing a light-based device. Primary factors (i.e., those directly defining properties of a light beam) of such a device include: wavelength (nm), power (W), beam area (cm^2), irradiance (W/cm^2), spatial–temporal properties of light (continuous- or pulsed irradiation; coherent or incoherent, collimated or divergent beam, and beam homogeneity across the target area).

Secondary or indirect factors are: temperature (heating) of the device and cooling configuration, compatibility with cell culture dish format(s), automation and software control, and sterility requirements (possibility to use inside the incubator or/and possibility to clean a device) (Fig. 1).

3.1.1 Optical Design

Key design factors are coherent properties of light source (LED vs. laser), pulsed or continuous-wave illumination, wavelength range (nm), beam area (cm^2), irradiance (W/cm^2), and spatial homogeneity across illumination area.

Coherence is the key property that differentiates between a laser and an LED, with laser light being coherent. To our knowledge there is no scientifically plausible theory favoring using coherent light on monolayer of cells [10]. Additionally, from a practical perspective, LEDs are in general cheaper and easier to use, available at various wavelengths over the visible and NIR spectrum at sufficient optical power and easily combined into arrays to illuminate sufficiently large areas. Hence, our preferred choice is to work with LED-based devices.

There are few commercial devices (mostly based on LED panels) generating a light beam homogeneous across its area and stable in time, which is in principle suitable to irradiate a well-plate with a cell culture. Examples include LED panels from major optical components manufacturers such as Edmund Optics and Thorlabs. However, such devices in general do not allow for a sufficiently wide range of optical parameters, which is required to find a "sweet spot" for treatment, and to reach high irradiance levels [4]. Thus, customized solutions are generally more appropriate.

When selecting LEDs or LED arrays, one need to consider the required wavelength range (nm), beam area to illuminate the sample (cm^2), the irradiance (W/cm^2), and then the optical power (W) that such a light source should emit.

As a part of such design choice, angular divergence and spatial homogeneity need to be considered. As previously stated, compared to lasers, LEDs have a large angular beam divergence, while the angular distribution of light intensity is not homogeneous over space (2π steradian). In particular, the central cone counting several tens of degrees around the light emission axis has the highest intensity and is the most homogeneous. When using LED, it is therefore recommended to mechanically restrict the angular emission angle so that the resulting irradiation pattern has sufficient power and is homogeneous over the targeted surface. The angular emission pattern is generally directly available in the LED manufacturer datasheet.

Illustration: Device-I and Device II have both been built using LEDs as shown on panels A and B of Fig. 2. LEDs with emission wavelength over the visible and NIR spectrum have been selected

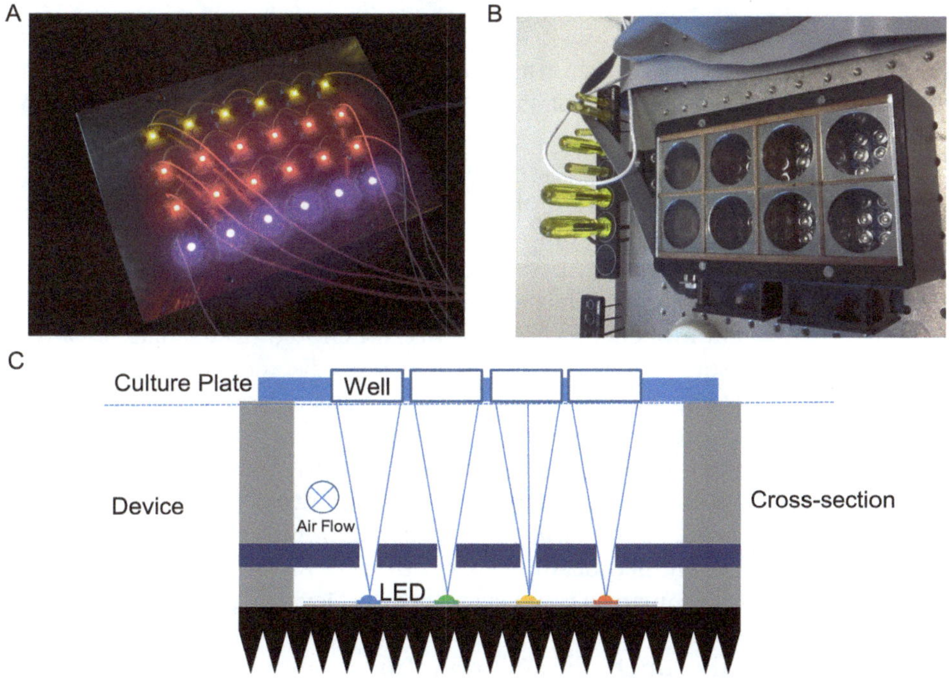

Fig. 2 Device principle: Illumination from the bottom of the well/dish; (**a**) photograph of the LED structure of Device I Version "Red" where the yellow, red and infrared (appears in purple) LEDs are visible; there are 24 LEDs in total, one under each well of a 24-well plate; (**b**) photograph of the LED structure of Device II where nine LEDs are positioned under each well; and (**c**) cross-sectional diagram applicable to both devices

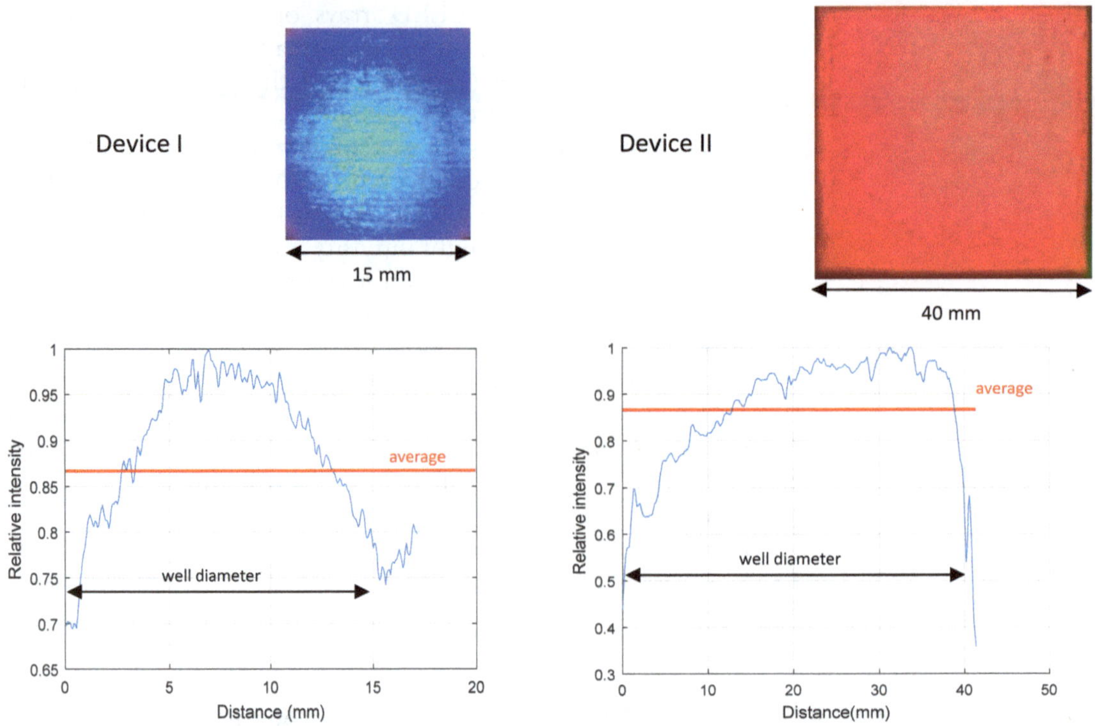

Fig. 3 Typical profiles of one single light beam in the axial *x–y* direction (at 450 nm) in Device I and in Device II measured using a standard camera (shown for 450 nm in Device I and 655 nm in Device II, all wavelengths behave similarly)

to probe the whole range of wavelength reported in most studies in the Photobiomodulation literature [4]. The LEDs selected for devices described here (Luxeon Rebel LED) have an angular emission profile that can be restricted to 30° off axis to create a quasi-homogeneous beam, powerful enough to irradiate areas of several square centimeters at 13 mm distance from the light source (Fig. 3). This is enough to irradiate one single well of a 24-well plate using one single LED (design of Device I). When the area to illuminate is larger, for 35-mm dish for example, it is possible to combine the beams of several LEDs to create a larger homogeneous area. If the target is placed far enough away from the source, the combined beam will provide a highly homogeneous and powerful illumination on a larger area. In Device II, nine LEDs are combined in a 3 × 3 array under each treatment window, placed 50 mm above the LED chips, allowing for a homogeneous irradiation of the 35-mm dish with higher irradiance range. Every LED (except Infrared LEDs) was covered by a small reflector, allowing for better collimation of the light (Reflector OPC1-2-COL from Dialight "Wide" Model) (Fig. 2).

3.1.2 Mechanical Design

Key design factors are light beam directivity (top or bottom illumination), angular light distribution, cooling, and hermetic properties.

Mechanical construction is then needed to contain the light source, to ensure the desired directivity (top or bottom illumination) and angular distribution of illumination on the target by blocking undesired exposure and to support the required cooling of a light source itself and of a biological target.

When making a choice between the illumination from the bottom or from the top of the target (as seen in Fig. 2c and as used in our experiments) we preferred the first option and it has proven itself to be successful. This choice was inspired by the work of Barolet et al. [11], where researchers illuminated human dermal fibroblasts seeded in 24-well plate from the bottom. It is a pragmatic solution, where all the energy emitted by the source is directed to the target and is neither reflected nor absorbed by the support holding a well plate (as this could result in unpredictable overexposure and heating of a sample).

The internal design of the mechanical structure supports the angular distribution of the beam, meaning that light emitted by an LED (over 2π str) is restricted to an angle by a physical aperture and/or wall within the device itself. This selects the central part of the beam with most homogeneous illumination (Fig. 2c).

The cooling of a device is required to maintain LED at an appropriate operating temperature and to prevent from heating the biological target. The cooling is usually implemented exploiting the mechanical structure of the device. As shown in Fig. 2c, the cooling of LED chips can easily be implemented using mechanical fans with directive flow. The generated flow of air directly cools the LED chips' local environment and prevents overheating.

Finally, the entire mechanical enclosure needs to be hermetic so that it can be used inside the incubator and can be cleaned easily. Adding a plastic window at a predefined distance with respect to LEDs is a practical solution that will ensure the correct placement of the well plate with respect to a light source and will physically separate internal components of a light emitting device from a well plate. This physical separation is essential to ensure the sterility requirements involved in the treatment of biological materials such as skin cell culture.

3.1.3 In-House Devices

Illustration: Our in-house devices assured illumination of the samples using several wavelengths of interest (up to six) using a range of irradiances (up to 50 and to 150 mW/cm^2 for Device I and II respectively), where illumination was homogeneous across the sample. It also allowed for sufficient number of technical replicates treated simultaneously, providing data for statistical analysis as seen in Fig. 2c.

In particular, Device I was designed to emit up to six discrete wavelengths: 447, 501, 530, 591, 655, and 850 nm, using two separate submodules. One submodule was emitting 447, 501, and 530 nm covering the "blue" part of the spectrum and the other one emitting 591, 655, and 850 nm covering the "red" part. Each of these submodules was designed to hold a 24-well culture plate positioned on top of them. One LED was illuminating each well of a 24-well culture plate, that is, identical LEDs were placed under 6 wells of the plate (one row) as seen on Fig. 2a. This allowed us to use up to four wavelength/parameter per plate and treat six technical replicates simultaneously. Designed using a similar principle, Device II offered the possibility to treat Petri dishes of larger sizes (up to 35-mm dish) with four different wavelengths: 447, 530, 655, and 850 nm. Device II had eight plastic windows corresponding to eight optical outputs, two of each wavelength. As the area of 35-mm Petri dish is larger compared to a well of a 24-well culture plate, each optical output was formed by an array of nine LEDs (Fig. 2b). The distance between the LED chip and the surface was fixed to 50 mm. In both devices, cooling was performed using mechanical fans with directive flows. The physical separation between the internal part of the device and the Petri dish was obtained using a plastic window to support the sample as seen in Fig. 2c (placed at a calibrated distances from the LEDs).

3.2 Quality Checks Before Executing an Experiment

3.2.1 Check the Irradiance at the Position of the Cell Culture

Prior to irradiating cell culture with light, the irradiance (mW/cm^2) must be measured and logged at the location where the well plate or a Petri dish would be placed. Irradiance is calculated based on measured optical power (mW) and a known area of the sample being irradiated.

This has to be performed for each of the device settings (combinations of wavelength and irradiance). It is also important to log these irradiance levels on a short term, for example, prior to and after execution of an experiment and over a longer period of time. The former one is important to spot any short time fluctuations in light source parameters. The latter one is important as calibrated values might decrease due to ageing or breakdown of LEDs. Although simple, this step is critical to ensure good repeatability and reproducibility of the data.

3.2.2 Check the Homogeneity of the Irradiation Pattern

To check the homogeneity of the irradiation pattern, a camera (such as a phone camera) and a diffuser are sufficient. A diffuser can be as simple as a sheet of paper or semitransparent plastic. In order to check the homogeneity of the beam profile, one should:

1. Place the diffuser on the light-based prototype at the location where the target culture dish would be placed.
2. Use a camera and focus on the diffuser plane.

3. Turn on the light of a prototype at a setting low enough to not saturate the camera.
4. Acquire an image of the diffuser illuminated by light.
5. Process the image with commercial software such as ImageJ or Matlab and directly compute the beam homogeneity across the image. The Plot Profile function of ImageJ can easily give the beam profile.

Illustration: Beam profile measurements of Device I and Device II were taken using a smartphone camera and a sheet of paper placed on top of the plastic window supporting the culture dish during treatment. They are shown in Fig. 3. The homogeneity is around ±10% of an average value over the dish surface, which was assumed to be sufficient for biological experiments.

3.2.3 Check Light-Induced Thermal Impact on the Target

An important consideration in any study in photobiology is the characterization of the light-associated thermal increase that can occur during light treatment. As a cell culture plate and a supporting structure may absorb light it is necessary to verify that the thermal equilibrium of the cells' environment is not disturbed too much. Assessment of a thermal impact can be performed in a noncontact mode using a thermal camera (e.g., FLIR).

To check the thermal impact due to light treatment:

1. Place the cell culture dish on the light-based device/prototype at the calibrated distance from the light-emitting source.
2. Fill in the culture dish with cell culture medium at temperature that is typically used during experiment (*see* **Note 1**).
3. Place the thermal camera above the sample and focus on the cell culture dish.
4. Turn on the light of a prototype and start treatment.
5. At the end or during the light treatment, record the thermal profile of the cell culture dish.
6. Read directly on the image of the camera the final temperature of the culture dish. When several wavelength and/or treatment parameters were tested at the same time, the temperature differences at the end of treatment between the different treatment conditions are immediately obtained.
7. Always perform temperature measurement for samples treated using light and for untreated control.

Illustration: Thermal profile image of the cell culture dish was performed during light treatment using Device I and a FLIR infrared camera (Fig. 4). These tests were performed outside the incubator and using a black 24-well culture dish. Using an infrared camera allowed us to assess if there was a temperature increase at the end of a light treatment.

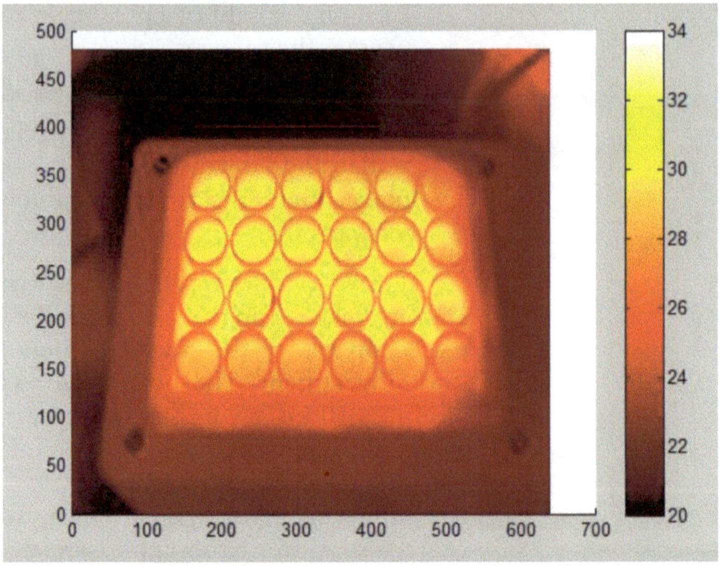

Fig. 4 Example of temperature profile during light treatment. Color axis represents tempearutre of the 24 well-plate. Maximum differences recorded after treatment were 1.5 and 2 °C for green and infrared wavelength, respectively (data not shown). The culture plate was a black 24-well plate and was imaged from above. The focus of the camera was made on the lid of the culture plate

3.2.4 Check a Potential Correlation Between Thermal Impact and Light Treatment

If at the end of the light treatment, there are temperature differences between the control and the treated cell culture wells it is first recommended to ensure that these thermal changes are not correlated with the measured biological impact. Next, it is important to ensure that the temperature of the cell environment never exceeded 37 °C and, ideally, that the wells' temperature was always included in the human skin temperature range [12]. Additionally, it is wise to minimize the impact of small temperature differences between the treated and control wells by limiting the treatment time to short period (<30 min) and replenishing potentially heated culture medium with at the end of treatment with the one at 37 °C.

Illustration: In our experiments [13], light treatment (30 J/cm^2) using Device I with all the available wavelengths (except for the yellow, 590 nm) was associated with a slight increase in temperature (up to 2 °C) compared to control (Fig. 4). However, the biological impacts of the same light dose (30 J/cm^2) at blue, cyan, green, red and infrared wavelengths, measured on the dermal fibroblasts metabolic activity, were very different from each other, and sometimes even triggered an opposite response [13]. Indeed, short visible wavelengths light treatment inhibited the metabolic activity of dermal fibroblasts while long visible and NIR wavelengths treatment resulted in neutral to stimulatory effects on dermal fibroblasts in similar dose range [13]. This suggests that the impact of this small temperature increase was much smaller than the impact of the light treatment itself.

3.2.5 Automated Control of the Light-Based Devices

Automated control is recommended to avoid the propagation of operator errors that will eventually occur over long experimental plans. LabView or similar instrument control software can be used to favor more controlled experiments leading to tighter results.

3.2.6 Check the Impact of the Surrounding Environment

When treating cultured cells with light one needs to pay attention to the optical absorption of the culture media. Absorption of light will lead to heat-induced or photochemically induced effect on the cells and mask the impact of light treatment on its own. Always choose for the medium that is transparent across the selected wavelength range. If optical absorption or transmission data are not available from the supplier, one can use optical spectrometer to acquire the spectrum. As a hint presence of high level, nontransparent culture media having coloration (e.g., red or yellow) should be avoided. For primary fibroblasts, DMEM (Dulbecco's Modified Eagle Medium) without phenol red and supplemented with serum (commonly 10%) is sufficiently transparent across the visible spectrum. As an illustration, the visible absorption spectra of DMEM was measured with and without phenol red (Fig. 5a), as well as the fine UV/visible absorption spectra of DMEM (without phenol red) and serum (Fig. 5b).

In general, when having a closer look at the specific composition of DMEM one can see that it includes inorganic salts, amino acids, other components such as phenol red and some additives such as glucose. Some of them are known to have photochemical properties: amino acids such as tryptophan can have strong UV

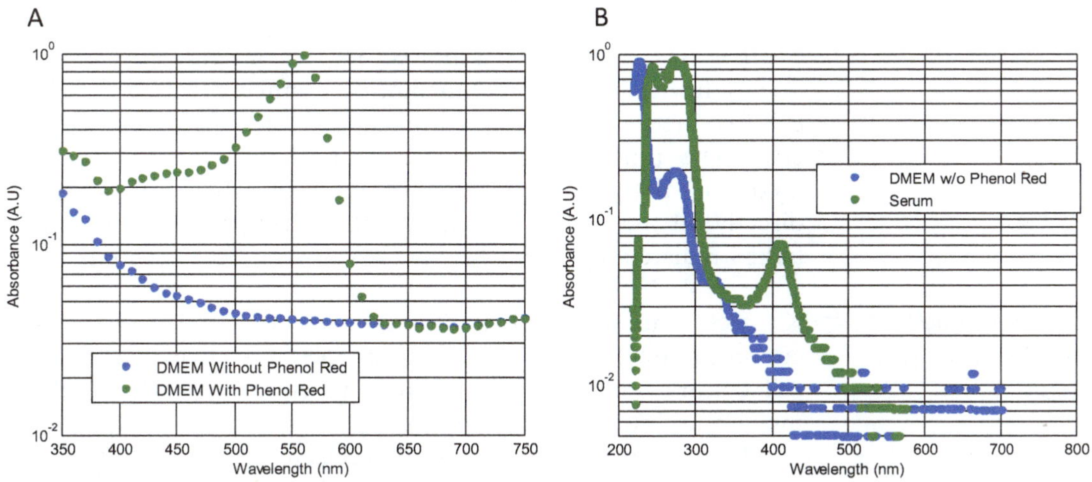

Fig. 5 Absorption spectra of biological media, panel (**a**) DMEM with and w/o phenol red and panel (**b**) DMEM w/o phenol red and serum (pure), measured with NanoDrop 1000 Spectrophotometer

absorbance [14], flavoproteins such as riboflavin and all FAD- or FMN-containing molecules also absorb UV light and have absorption bands extending to visible spectrum till almost 550 nm [15]. Serum also contains vitamin A (at least three forms, retinol, retinal, and retinoic acid) which will also have light-absorption properties depending on which molecule it is coupled to [16]. Thus, light treatment might trigger photochemical reactions in the culture medium and light treatment might affect the cells indirectly via the culture medium.

Interestingly, all three of these light-absorbing molecules could also be present in vivo in human skin [17, 18] and may interact with light. Liebel et al. showed comparable effects of UV and visible light on the induction of reactive oxygen species following irradiation in vivo [19] and Sato et al. [20] demonstrated UV-induced photochemical reaction of riboflavin in vitro.

Knowing that the culture medium contains photoactive proteins, there is a need for protocols to discriminate between the impact of a light treatment occurring (a) directly due to irradiation of the cells' proteins and (b) indirectly due to the irradiated proteins in the culture medium affecting the cells.

In order to distinguish between direct and indirect effects of the light treatment on the cells, several protocols are proposed here.

To prevent any interaction between light and the culture medium, treatment can be performed in PBS. After treatment, the PBS is removed and fresh medium is replenished. Two washing steps are needed during this procedure, which should always be applied to control dishes as well.

In order to take into account interaction of light with culture medium, cells can also be irradiated in culture medium but immediately after light treatment, the culture medium of each treated wells can be replenished. This prevents prolonged contact of irradiated culture medium with the cells.

To assess any potential indirect impact of light on cells solely due to the irradiated culture medium, a useful control is to irradiate cell-free media/solution and immediately bring into contact with cells. The time of cell "exposure" to light-irradiated cell-free medium is then set equal to that of treatment when light is applied directly on the cells. The impacts of both light treated cells and light-irradiated medium treated cells can then be compared.

3.2.7 Have a Tight Control of the Biological Sample

Key factors are cell density, cell lineage, carbon dioxide and oxygen concentration, replacement of culture medium, environmental and cell culture temperature, washing and culture medium replacement, and time outside incubator.

The density of cells, cell type and cell culture conditions have all been shown to affect the response of the cell population to light treatment [21]. Therefore, it is important to set these biological parameters to a predefined biologically meaningful value and keep them controlled.

The cell density is an important parameter, which might strongly differ in vivo depending on a cell type or even lineage. For example, epidermal keratinocytes are tightly packed together while dermal fibroblasts populate the dermis at much lower density. It is therefore suggested to culture cells in vitro at densities matching their in vivo behavior.

While it is obvious that the cell type needs to be preselected prior to the experiment, it is also worthwhile to check the sublineage as well. For example, human dermal fibroblasts exist in at least two different subpopulations: reticular and papillary fibroblasts. We have previously shown that these two human fibroblasts' sublineages respond differently to same light treatment [22].

Many other environmental physical factors will play a role in the response of human skin cells to light treatment. They include the environmental temperature and oxygen and carbon dioxide concentrations for example. Moreover, a study has shown the difference between skin surface and mean internal temperature which suggests a nonhomogeneous temperature depth profile and thus could imply that different skin cell populations shall be cultured at different temperatures depending on their depth location [23]. Similarly, studies have also shown the rapid change in oxygen concentration with the skin depth which could also imply specific oxygen concentration culture condition is required, depending on the targeted skin cell population depth [24].

Last but not least, the experimental procedure will be an inherent part of the biological model including washing, period out of the incubator and medium changes. Thus, it is important to always have the control samples undergoing the exact same experimental procedure as the treated wells, including washing, replenishment and time outside the incubator.

3.3 Design of Experiment

Many factors (including those of physical, chemical, and biological nature) can impact experimental outcome when one is attempting to investigate the impact of light on cells. Thus, delivering quality of the results and justification that the observed cellular effects are solely attributed to a direct impact of light is not straightforward. To resolve this potentially dubious situation one needs to investigate the impact of all controllable factors and noise that potentially influence reactions of cells. An efficient method to do this is a mathematical tool called Design the Experiments (DoE) [25]. DoE uses statistical methodology to determine the importance of specific variables affecting performance of a process (e.g., to find cause-and-effect relationship between a parameter and an experimental outcome). Next to understanding how a particular variable affects the outcome, DoE gives information about interactions between different variables.

3.3.1 Nomenclature

We further use the following nomenclature: *output*, *factor*, and factor *level* and *interval*.

The *output* can be understood in the present context as a biological readout. It represents the value that will be assayed to quantify the response of the cells to the applied treatment.

The *factors* are experimental variables for which one wants to understand their impact on the response of skin cells to light treatment. They can include, for example, the wavelength, the radiant exposure or cell density. Each *factor* may have several *levels*.

The factor *interval* represents the difference that exists between two factor *levels*. If an experiment aims at investigating the impact of the wavelength, the factor *levels* will be the selected discrete wavelengths (e.g., 450 and 655 nm) and the *interval* the numerical difference between those wavelengths (e.g., $655 - 450 = 205$ nm).

3.3.2 Select Factors, Levels and Intervals for the Design of Experiment

The factors related to light treatment (i.e., optical) include wavelength (nm), irradiance (mW/cm^2), radiant exposure or dose (J/cm^2). Biological factors include serum concentration (% FBS), cell density (cells per units of surface), oxygen level (%), and treatment protocols (with or without replenishment of media after treatment, treatment with irradiated cell-free media). Repeatability is defined as a technical repeat during the experiment. Reproducibility is then defined as experiment repetition on a different day with a different sample.

Each factor needs to be varied among predefined levels. The level intervals should be selected large enough so that one expects an impact when the considered factor is varied from a level to the next one. An example of selected factors and levels for optical parameters are given in Fig. 6.

To assure accuracy of results an appropriate number of experimental outcomes are required for validity. This is usually visible in the size of the standard error (e.g., error bars). If the sample size it too small, it will yield large standard deviation and thus large uncertainly of the results. When there is sufficient number of data collected, it is theoretically possible to infer the impact of a specific combination of factors without performing the experiment (fractional factorial design).

3.4 Robust Measurement of the Impact of Visible/NIR Light on Human Skin Cells

We recommend using a general marker reflecting physiological state of cells when assessing the impact of various factors. For example, metabolic activity (e.g., using Alamar Blue® assay, Thermofisher, DAL1100) can be used to understand the impact of optical and biological factors on the light treatment of skin cells. Advantage of the Alamar Blue® assay is its ease of use and applicability to large-scale experiments involving many well plates [24].

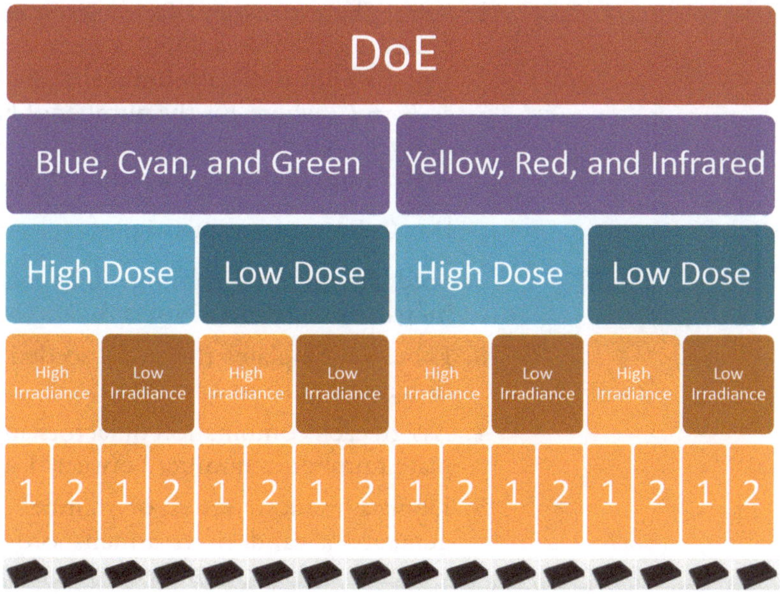

Fig. 6 An example of a typical DoE with three factors (wavelength, dose and irradiance) at two levels (high and low) and two repeats, which in total gives 16 experimental runs. For this specific case this would require sixteen 24-well plates

3.4.1 Measurement Method Analysis

Measurement method analysis (MSA) is needed prior to executing the main experiment. It allows to understand inaccuracies from the measurement method itself and to potentially improve it. Thus, when using the Alamar Blue assay (or any assay that needs incubation time), it is advised to determine the ideal concentration and incubation time first (*see* **Note 2**).

Next, it is recommended to systematically test the full experimental procedure and tighten any variability and systematic errors (bias or offset) that can arise from external factors not related to the light treatment and biological readout procedure (*see* **Note 3**).

3.4.2 Experimental Procedure

The experimental procedure can be divided into three steps: (a) preparation of the skin cells and seeding in the adequate culture dish format, (b) light treatment according to defined systematic protocols, and (c) readout with the selected assay.

Below, we exemplify this experimental procedure developed to identify the experimental factors impacting the response of human dermal fibroblasts to light. The procedure was based on using metabolic activity assay to measure the outcome [20]. The detailed description that follows is based on using 24-well plates for cell culture. The cells are seeded on Day 0. Treatments are applied during three consecutive days (Days 1, 2, and 3). Readout is performed 24 h after each light treatment (Day 4). If needed, this procedure can further be easily adapted to different time sequences.

1. Select the factors and factor levels to be used in the DoE as presented in Subheading 3.3.2. An example of DoE is shown in Fig. 6. The minimal number of experiments is then defined as $N = l^F$, where F is the number of factors and l is the number of levels. When each experiment is to be reproduced twice, this will require doubling the number of experiments.

Day 0: Preparation of sample:

2. Seed the required amount of 24-well cell culture plates with human skin cells (*see* **Notes 3–5**).
3. Label all the plates that were seeded with a unique identification number from 1 to *N*.
4. Assign each combination of factors and factor levels to a unique plate number (a number between 1 and *N*).
5. Draw a random series of number from 1 to *N*. This will generate a random treatment order, that is, the order in which the plates, and factor levels' combination, are treated. This helps to avoid any systematic bias in the results solely due to treatment order.

Days 1, 2 and 3: perform light treatment (with replenishment of culture medium after treatment). Repeat each day of treatment at the same time during the day:

6. Prepare a solution of fresh cell culture medium sufficient to replenish all the wells of the cell culture plates. Bring up to 37 °C in a water bath.
7. Apply light treatment with the selected settings corresponding to the plate label (*see* **Notes 6** and **7**).
8. Replenish the cell culture medium of each well of the treated plate (optional depending on treatment protocol choice, *see* Subheading 3.2.6).
9. Put the treated plate back in incubator and take the next one following the treatment order list (*see* **Notes 8** and **9**).
10. Repeat **steps 6–9** until all plates have been treated.

Day 4: Readout

11. Prepare a sufficient amount of Alamar Blue® solution in fresh medium at the required concentration. Bring the prepared solution to a desired temperature (37 °C) in water bath to warm up (*see* **Note 10**).
12. Add 500 μL of the Alamar Blue solution to each well of the 24-well plates following the treatment order (**step 5**).
13. Incubate the 24-well plate for 3–5 h depending on the confluency of cells (*see* **Note 2**).

Table 1
Example of representation of DoE results

Factor a	Factor b	Variable
Level 1	Level 1	Exp. Pt. 1
Level 1	Level 2	Exp. Pt. 2
Level 2	Level 1	Exp. Pt. 3
Level 2	Level 2	Exp. Pt. 4

14. Read the fluorescence using a plate reader such as FLUOstar Omega II (ex: 544 nm, em: 590 nm) in the same order as the plates were put in contact with Alamar Blue (i.e., the treatment order generated in **step 5**).

3.4.3 Statistical Analysis and Visualization of the Data

The experimental results of a DoE can generally be represented as shown in Table 1. This is an example with only two factors with two levels each (i.e., $l = 2$, $F = 2$, and $N = 4$). Thus, the variable would be measured four times only. When using a 24-well plate for each experiment, means that there will be 24 technical repeats for each condition (each combination of factor and level). The structuring of the table is straightforward: one column for each factor and one column for the outcome; one row for each of the experimental conditions tested. As we consider the full factorial design of experiment, all the possible combinations of the levels of factors are tested.

Illustration: The variable (or the readout) in the experiment presented in Subheading 3.4 would be the relative metabolic activity, defined as a ratio between the average Alamar Blue fluorescence counts of the treated groups and the control group. The factors would be wavelength, irradiance and radiant exposure as seen in Fig. 6.

Once the table with factors and experimental outcomes is made (e.g., using Microsoft Excel or Open Office software or Minitab), statistical data analysis can be performed using functions integrated in such a data analysis software. Visualization of the statistical data analysis can be easily implemented using Matlab, R package DoE. base and/or Minitab for example (*see* **Note 11**).

When visualizing the outcome of the DoE it is useful to look at (a) main effects plot and (b) interaction plot. They are shown in Fig. 7. While the first shows the mean effects of each individual factor on the outcome, the second shows how the interaction between the factors impacts the outcome. Each effect is associated with a statistical test to see if the impact of a specific factor is or is not significant. The statistical analysis is usually performed using ANOVA.

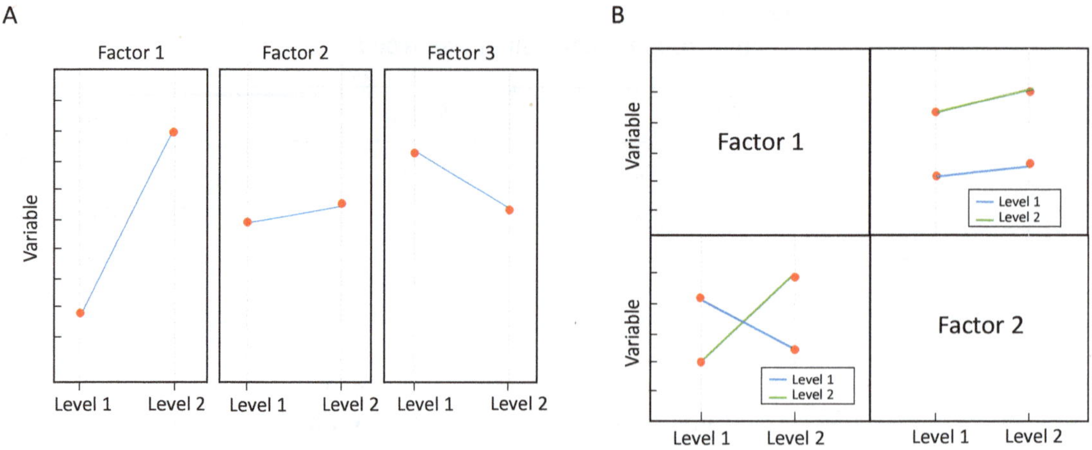

Fig. 7 Schematic representation of the information conveyed by main effects (**a**) and interaction (**b**) plots resulting from a full-factorial DoE experiment

4 Notes

1. The presence of cells in the well plate or a Petri dish is not required to assess the thermal increase due to the light treatment. This is because cells have a relatively low absorption coefficient compared to that of culture medium ingredients and thus will not significantly contribute to the temperature increase.
2. Prior to the light treatment of cells, it is advised to determine incubation time for the Alamar assay, sufficient to result in a detectable color change upon a reaction with cells. This time depends on the hardware (plate reader), cell type and cell density. Ideally, the fluorescence counts read by the reader should be in the middle of the accessible range. Too low values may be strongly impacted by noise and too high values may fall into a range where response of a system is no longer linear.
3. Prior to experiments, it is also worth to perform measurement system analysis (MSA) by at least carrying out a Gage Reproducibility and Repeatability (Gage R&R) test. This allows to assess the impact of operator, measuring instrument and method (plate reader, etc.), microplate type (black/white), illumination system, and technical repeatability on experimental outcome. The results of such MSA can be analyzed using methods similar to those applied for the analysis of the impact of light. The results will allow to determine the best conditions for optimal repeatability and reproducibility of the measurement system [26].

4. Immediately after the cells are introduced in the cell culture dishes and before the plates are placed in the incubator, it is recommended to place the dish on a horizontal planar area (e.g., table top) and gently shake the dish along two horizontal orthogonal directions (north/south and east/west) to homogenize the distribution of cells in the wells. Then place the dishes in the incubators. This will significantly improve the homogeneity of the spatial seeding distribution.
5. Always record the passages of the cells. For primary cell cultures keep the passage limited to 6–8, as most primary cell cultures are known to lose/change their specific properties after multiple passages in culture [27].
6. Apply each procedure steps to the control dish as well. In our experiment, the same 24-well plate contained both control and treated groups, as one row of the 24-well plate was always kept as control, that is, not irradiated. This helps to monitor if any unexpected effect occurred during the treatment of a specific plate. Both control and treated groups should always be kept under the same ambient conditions.
7. Light treatment can be conducted outside the culture incubator but shall never extend beyond a maximum of ~45 min at room temperature.
8. It is worth to consider having a sheet of paper or an electronic table sheet with a table showing in columns the plate number and in rows the treatment number. When a plate has been treated, the operator should mark the corresponding box off the list. This is important to avoid double treatment of plates, which can easily occur in large experiments with a repetitive procedure.
9. The incubator temperature and gas concentration is usually significantly disturbed when opened, even if it is to take only one single plate. Thus, for large-scale experiments, it is advised not to open the incubator at each single treatment, as this will disturb the incubator equilibrium over a long period. To avoid such effect, it is recommended to take 4–8 plates out, perform the treatment of these plates without putting them back in incubator, and then simultaneously take the next 4–8 plates and put the treated 4–8 treated plates back in the incubator.
10. It is advised to consider storing the culture medium in a freezer for further analysis, collagen production in dermal fibroblasts' supernatants for example.
11. It is recommended to check the normality of the data distribution points before performing ANOVA analysis of the data.

Acknowledgments

The authors acknowledge Rene Kragt (Philips Innovation Services) and Frank van Abeelen (Philips Research) for assistance with the design and manufacturing of the light-based devices described here.

References

1. Hay RJ, Johns NE, Williams HC et al (2014) The global burden of skin disease in 2010: an analysis of the prevalence and impact of skin conditions. J Invest Dermatol 134:1527–1534
2. Posnett J, Franks PJ (2008) The burden of chronic wounds in the UK. Nurs Times 104:44–45
3. Nouri K (2012) Lasers in dermatology and medicine. Springer, New York
4. Mignon C, Botchkareva NV, Uzunbajakava NE et al (2016) Photobiomodulation devices for hair regrowth and wound healing: a therapy full of promise but a literature full of confusion. Exp Dermatol 25:745
5. Wolverton SE (1992) Monitoring for adverse effects from systemic drugs used in dermatology. J Am Acad Dermatol 26:661–679
6. Chung H, Dai T, Sharma SK et al (2012) The nuts and bolts of low-level laser (light) therapy. Ann Biomed Eng 40:516–533
7. Rinaldi F (2008) Laser: a review. Clin Dermatol 26:590–601
8. Hadis MA, Zainal SA, Holder MJ et al (2016) The dark art of light measurement: accurate radiometry for low-level light therapy. Lasers Med Sci 31:789–809
9. Serrage H, Heiskanen V, Palin WM et al (2019) Under the spotlight: mechanisms of photobiomodulation concentrating on blue and green light. Photochem Photobiol Sci 18:1877
10. Karu TI (2003) Low-power laser therapy. In: Biomedical photonics handbook. CRC Press, Boca Raton, FL, pp 1–26
11. Barolet D, Duplay P, Jacomy HH et al (2010) Importance of pulsing illumination parameters in low-level-light therapy. J Biomed Opt 15:048005
12. Olesen BW (1982) Thermal comfort. Tech Rev 2:3–37
13. Mignon C, Uzunbajakava NE, Raafs B et al (2016) Photobiomodulation of distinct lineages of human dermal fibroblasts: a rational approach towards the selection of effective light parameters for skin rejuvenation and wound healing. Proc SPIE 9695:969508–969516
14. Edelhoch H (1967) Spectroscopic determination of tryptophan and tyrosine in proteins. Biochemistry 6:1948–1954
15. Massey V, Ganther H (1965) On the interpretation of the absorption spectra of flavoproteins with special reference to D-amino acid oxidase. Biochemistry 4:1161–1173
16. Dartnall HJA, Bowmaker JK, Mollon JD (1983) Human visual pigments: microspectrophotometric results from the eyes of seven persons. Proc R Soc Lond B Biol Sci 220 (1218):115–130
17. Lee TH, Lerner AB, Halberg RJ (1953) Water soluble vitamins in normal human skin. J Invest Dermatol 20:19–26
18. Vahlquist A, Lee J, Michaelson G et al (1982) Vitamin a in human skin: II Concentrations of carotene, retinol and dehydroretinol in various components of normal skin. J Invest Dermatol 79:94–97
19. Liebel F, Kaur S, Ruvolo E et al (2012) Irradiation of skin with visible light induces reactive oxygen species and matrix-degrading enzymes. J Invest Dermatol 132:1901–1907
20. Sato K, Taguchi H, Maeda T et al (1995) The primary cytotoxicity in ultraviolet-a-irradiated riboflavin solution is derived from hydrogen peroxide. J Invest Dermatol 105:608–612
21. Mignon C, Uzunbajakava NE, Raafs B et al (2017) Photobiomodulation of human dermal fibroblasts in vitro: decisive role of cell culture conditions and treatment protocols on experimental outcome. Sci Rep 7:2797
22. Mignon C, Uzunbajakava NE, Castellano-Pellicena I et al (2018) Differential response of human dermal fibroblast subpopulations to visible and near-infrared light: potential of photobiomodulation for addressing cutaneous conditions. Lasers Surg Med 50:859

23. Liu Y, Wang L, Liu J et al (2013) A study of human skin and surface temperatures in stable and unstable thermal environments. J Therm Biol 38:440–448
24. van Tonder A, Joubert AM, Cromarty AD (2015) Limitations of the 3-(4,5-dimethylthiazol-2-yl)-2,5-diphenyl-2H-tetrazolium bromide (MTT) assay when compared to three commonly used cell enumeration assays. BMC Res Notes 8:47
25. Wagner JR, Mount EM, Giles HF et al (2014) Design of experiments. In: Extrusion. Elsevier Inc., Amsterdam, pp 291–308
26. [Editorial] (2016) Reality check on reproducibility. Nature 533:437–437
27. Janson D, Saintigny G, Mahé C et al (2013) Papillary fibroblasts differentiate into reticular fibroblasts after prolonged in vitro culture. Exp Dermatol 22:48–53

INDEX

Natalia V. Botchkareva and Gillian E. Westgate (eds.), *Molecular Dermatology: Methods and Protocols*, Methods in Molecular Biology, vol. 2154, https://doi.org/10.1007/978-1-0716-0648-3,
© Springer Science+Business Media, LLC, part of Springer Nature 2020

FSC
www.fsc.org
MIX
Papier | Fördert gute Waldnutzung
FSC® C083411

Kolibri 360 GmbH
Ferdinand-Jühlke-Straße 7
99095 Erfurt, Deutschland
produktsicherheit@kolibri360.de